Lecture Notes in Electrical Engineering

Volume 764

The book series *Lecture Notes in Electrical Engineering* (LNEE) publishes the latest developments in Electrical Engineering - quickly, informally and in high quality. While original research reported in proceedings and monographs has traditionally formed the core of LNEE, we also encourage authors to submit books devoted to supporting student education and professional training in the various fields and applications areas of electrical engineering. The series cover classical and emerging topics concerning:

- Communication Engineering, Information Theory and Networks
- Electronics Engineering and Microelectronics
- Signal, Image and Speech Processing
- Wireless and Mobile Communication
- Circuits and Systems
- Energy Systems, Power Electronics and Electrical Machines
- Electro-optical Engineering
- Instrumentation Engineering
- Avionics Engineering
- Control Systems
- Internet-of-Things and Cybersecurity
- Biomedical Devices, MEMS and NEMS

For general information about this book series, comments or suggestions, please contact leontina.dicecco@springer.com.

To submit a proposal or request further information, please contact the Publishing Editor in your country:

China

Jasmine Dou, Editor (jasmine.dou@springer.com)

India, Japan, Rest of Asia

Swati Meherishi, Editorial Director (Swati.Meherishi@springer.com)

Southeast Asia, Australia, New Zealand

Ramesh Nath Premnath, Editor (ramesh.premnath@springernature.com)

USA, Canada:

Michael Luby, Senior Editor (michael.luby@springer.com)

All other Countries:

Leontina Di Cecco, Senior Editor (leontina.dicecco@springer.com)

**** This series is indexed by EI Compendex and Scopus databases. ****

More information about this series at http://www.springer.com/series/7818

Reji Kumar Pillai · Atul Dixit · Suhas Dhapre
Editors

ISUW 2019

Proceedings of the 5th International Conference and Exhibition on Smart Grids and Smart Cities

Springer

Editors
Reji Kumar Pillai
India Smart Grid Forum
New Delhi, India

Atul Dixit
New Delhi, India

Suhas Dhapre
Mumbai, India

ISSN 1876-1100 ISSN 1876-1119 (electronic)
Lecture Notes in Electrical Engineering
ISBN 978-981-16-1298-5 ISBN 978-981-16-1299-2 (eBook)
https://doi.org/10.1007/978-981-16-1299-2

This Springer imprint is published by the registered company Springer Nature Singapore Pte Ltd.
The registered company address is: 152 Beach Road, #21-01/04 Gateway East, Singapore 189721, Singapore

Contents

Forecasting Short Term Peak Loads of Distribution Transformer (DT) Using Machine Learning and Computational Statistics—Various Methodologies and Their Pros and Cons

Arghya Roy

Abstract Machine Learning (ML) is a technique that employs computational statistics to learn the pattern of data and from there it tends to predict. Over last few years, abundance of data have been made available in standard format and therefore it is compelling to use ML techniques to achieve better prediction. In several instances we have witnessed 98–99% accuracy in prediction levels using ML techniques. In Utility Industry, one of the key parameter every DISCOM would like to forecast is Peak Load. Since HT (>11 kVA) and EHT (Extra High Tension > 66 kVA) customers in the designated DISCOM region demand continuous power supply, reducing interruption hours and feeder outages are two most important factors draw attentions of Utility Service Providers. It has been found that accurate prediction of Peak Load Demand for a day ahead can remarkably reduce the above two parameters. In fact, knowing the peak load in advance can help in optimizing the load distribution at substation levels. While predicting peak load, we should consider weather (Temperature), time of the day (expressed in HH:MM format), Day type (Weekdays or Weekends) and finally kWH of a DT.

Keywords Smart meters · Neural net · Tree based methods · Real time data gathering · ETL · Data modelling · Forecasting · Smart grid

1 Introduction

The Indian power sectors at the cusp of a transformation process from being manually operated to digitalization. The Distribution Utilities face new challenges every day and have to improve profits by reducing operating and maintenance costs, while providing customers with a reliable power supply and a broad range of services. India is currently battling the highest *transmission and distribution losses* (T & D Loss or AT & C Loss) in the world which is close to around 25%. It has been proved that the crux of reducing AT & C Loss lies in managing the demand properly.

A. Roy (✉)
India Power Corporation Limited, Kolkata 700091, India
e-mail: arghya.roy@indiapower.com

R. K. Pillai et al. (eds.), *ISUW 2019*, Lecture Notes in Electrical Engineering 764,
https://doi.org/10.1007/978-981-16-1299-2_1

Essentially, demand management or *demand response* (DR) is a function of *load forecasting* and *Peak Load Management* (PLM). Load forecasting is concerned with different horizon scales: hourly, daily, weekly, monthly, and annual values for both overall demand and peak demand. Importantly, short-term forecasting has attracted a great deal of attention in the literature due to its applicability in power system management, including resource planning and taking control actions towards load balancing.

In this paper, we have examined *Machine Learning* (ML) driven some important load forecasting techniques. We have used ML model to forecast the peak load of each Distribution Transformer (DT) for next day.

The remainder of the paper is structured as follows: a literature review on related problems is provided in Sect. 2. Next we discuss about some Machine Learning Techniques and then focus on Neural Net based ML models (Sect. 3). We then outline the *data pipeline* process i.e. fetching data from DT, cleaning and performing necessary *Exploratory Data Analysis* (EDA) and finally fit the data to ANN and LSTM RNN model (Sect. 4). In Sect. 5 we discuss about performance metrics of the above two model and chalk out the deployment plan for further use. In Sect. 6 we discuss on the business benefits of the load prediction.

2 Literature Review

A considerable amount of research has been dedicated to electricity price forecasting. As investigated by Weron [8], there were more than 800 publications related to electricity price forecasting in the years 1989–2013, as indexed in the Web of Science and Scopus databases. Energy forecasting models can be systematized in several ways, categorized as static or dynamic, univariate or multivariate, or as involving various techniques ranging from simple naive methods through a wide range of times series methods to complex hybrid and artificial intelligence models. For the purpose of the review, the approaches are categorized under the following headings: (a) Statistical and Time Series Models; (b) Computational Intelligence models; and (c) hybrid models.

a. **Statistical and Time Series Models**

Time series models are the most frequently used methods to represent the future values based on previous observations. The models based on time series have many forms adequate for forecasting electricity consumption volume and peak demand load in the electrical grid. These models are often applied to demand forecasting at the regional/national level. *Regression models* are built to capture a relationship between a dependent variable, which is the energy usage, and one or more independent variables (the predictors). *Econometric models* are built to investigate dependencies between the energy demand and other macro-economic variables such as energy price, gross national product (GNP), technology, investments, and population size.

Some authors like Sengupta et al., have demonstrated that econometric models are effective for forecasting energy patterns in developing countries like India.

b. **Computational Intelligence Models**

Within the category of computational intelligence methods, we use many Machine Learning Models like *Support Vector Machines* (SVM), *Tree Based methods with Gradient Boosting*, and ANN and its variants to model complex and dynamic systems. Machine Learning based techniques as mentioned here are quite promising because of (i) High Accuracy and Precession, (ii) Fast computation, (iii) Easy to deploy and (iv) Can manage non-linearity amongst Predictor and Dependent variables.

c. **Hybrid Models**

Based on the available literature, there is a clear and increasingly recognizable research interest that looks at the application of various hybrid models, combining techniques from two or more of the groups listed above. This stream is aimed at increasing the load forecasting accuracy by benefiting from the best features associated with different approaches and their synergy. For instance, it has been found that the combination of Econometric forecasting model like ARIMA (seasonality adjusted), and the weighted SVM can effectively account for the nonlinearity and seasonality and produce quite accurate forecasting results.

3 Important Machine Learning Techniques for Load Forecasting

There are many ML models which are quite popular and useful. And many of them are being extensively used in Retails, Healthcare and other domains for predictions. ML models which are used for predictions are termed as *Supervised Learning Models*. Few very popular amongst them are Tree Based methods, Random Forests, Support Vector Machines (SVM) and Neural net or Perceptron Model (ANN). For quite a large dataset, we found SVM and ANN are quite robust and accurate.

a. **Support Vector Machines (SVM)**

Support vector machines (SVMs) are a set of supervised learning methods used for classification, regression and outliers detection. In SVM, a *Maximal Margin Hyperplane* (also known as the optimal separating Hyperplane), which is the separating *Hyperplane* that is farthest from the training observations is introduced and we compute the (perpendicular) distance from each training observation to a given separating Hyperplane; the smallest such distance is the minimal distance from the observations to the Hyperplane, and is known as the *margin*.

To construct a Maximal Margin Hyperplane, margin value to be maximized.

Maximum M
$\beta_0, \beta_1, \ldots, \beta_p$
subject to

$$\sum_{j=1}^{p} \beta_j^2 = 1$$

$$y_i\left(\beta_0 + \beta_1 \cdot x_{i1} + \beta_2 \cdot x_{i2} + \cdots + \beta_p \cdot x_{ip}\right) \geq M(1 - \varepsilon_i)\ \forall i = 1, 2, 3, \ldots, n$$

$$\varepsilon_i \geq 0, \sum_{i=1}^{n} \varepsilon_i \leq C$$

M represents the margin of our Hyperplane, and the optimization problem chooses $\beta_0, \beta_1, \ldots, \beta_p$ to maximize M. Typically generating the beta values are complex and beyond the scope of this article. 'ε' is the slack variable and C is the penalty parameter. We use *Kernel* trick to derive those betas. In the above equation, 'y' is the dependent or predictor variable.

The advantages of support vector machines are:

i. Effective in high dimensional spaces.
ii. Uses a subset of training points in the decision function (called support vectors), so it is also memory efficient.
iii. Versatile: different Kernel functions can be specified for the decision function. Common kernels are provided, but it is also possible to specify custom kernels.

b. **Artificial Neural Network (ANN)**

An Artificial Neural Network (ANN) is a computational model that is inspired by the way biological neural networks in the human brain process information. Typically an ANN consisted of three elements viz. Input Layer (or Input Node), Hidden Layer and Output Layer. A neural network is a *two-stage regression or classification model*, typically represented by a network diagram as shown below. This network applies both to regression or classification (Fig. 1).

For K-class classification, there are K units at the top, with the kth unit modelling the probability of class k. There are K target measurements. Y_k, $k = 1, \ldots, K$, each being coded as a $0 - 1$ variable for the kth class. Derived features $a_1, a_2, \ldots, a_k$ are created from linear combinations of inputs (X) then target Y_k is modelled as linear combinations of a_m

$$a_m = \varphi \cdot \left(\alpha_{0m} + \alpha_m^T \cdot X\right), m = 1, 2, \ldots, M$$

$$t_k = \beta_{0k} + \beta_k^T \cdot a, k = 1, 2, \ldots, K$$

$$Y_k = f(X) = g_k(t), k = 1, 2, \ldots, K$$

Fig. 1

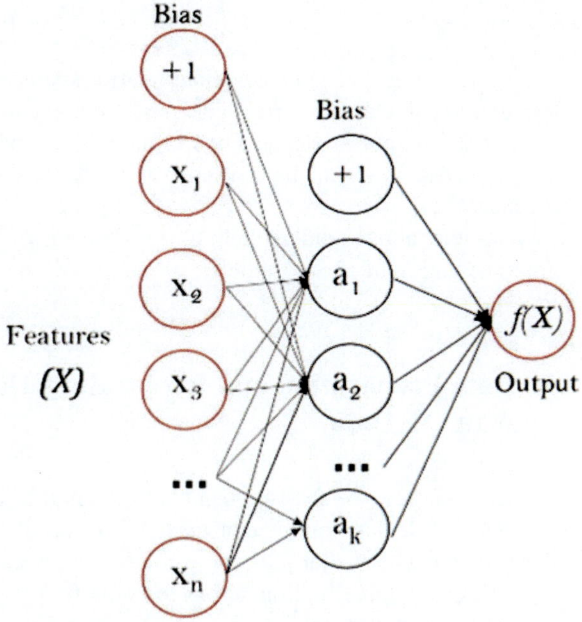

where a = (a_1, a_2, a_3, . . . , a_k) and t = (t_1, t_2, . . . , t_k).

'φ' is called the activation function. Beta values are derived by a technique named back-propagation i.e. comparing the value of Y_k with the actual Y and then again retrain the whole hidden layers by adjusting 'α'.

RNNs are called recurrent because they perform the same task for every element of a sequence, with the output being depended on the previous computations. Here is what a typical RNN looks like (Fig. 2).

Instead of neurons, LSTM networks have memory blocks that are connected through layers.

x_t is the input at time step t. s_t is the hidden state at time step t. It's the "memory" of the network. s_t is calculated based on the previous hidden state and the input at the current step:

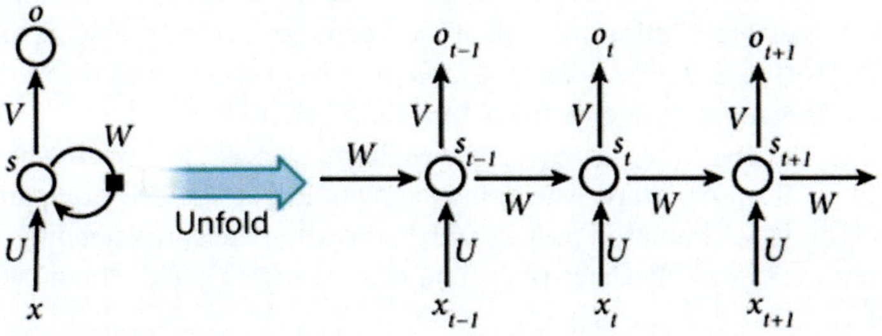

Fig. 2

$$s_t = f(Ux_t + Ws_{t-1}).$$

The function f usually is a nonlinearity such as tanh or ReLU. s_{t-1} which is required to calculate the first hidden state, is typically initialized to all zeroes.

o_t is the output at step t. For example, if we wanted to predict the next word in a sentence it would be a vector of probabilities across our vocabulary. $o_t = \text{softmax}(Vs_t)$.

With the above understanding we now move forward data preparation and implementation of above models.

4 Data Preparation and Exploratory Data Analysis (EDA) with DT Data

As stated earlier, we shall be using data related to a single DT. Data are captured through IIoT devices mounted on the DT. Data is collected at *every 15 min Frequency*. Based on the data availability, some graphs have been plotted. We wanted to examine the kVA or load distribution across the time frame for each DT. We also wanted to see the Harmonic distortion of each DT on a given time scale. Then we explored the statistical correlation between Harmonic Distortion (Voltage Level) and Load for Each DT. We have also plotted the DT Oil temperature against kVA (load). We have identified a kWH value as peak load when it is more than 99 percentile of previous 30 days value. We have marked this in Red (Fig. 5). We scaled this down to daily basis and identified the peak loads.

Based on the data as collected above and the graph so plotted, we forecast the next level of kWH and it's corresponding time in the day it may occur. Specifically we are trying to *predict two variables* viz. *kWH (directly) and Hour (retrieved from database based on predicted kWH)* given the data set as shown in Table 1.

Collecting data from IIoT is dependent on the signal quality of the communication channel. In case of Comm. Network congestion, data may not be retrieved properly. In order to tackle this, we have used a '*conditional imputation of missing values*' technique. In this technique, we impute the value of kWH and kVA by its last *3 observation's rolling mean* if the missing data is present only at a *single occurrence i.e. only at one instance*. But if we observe that for consecutive 3 times, the kWH and kVA values are coming zero, then we keep them as it is. We are not going to reduce or resort to any *dimensionality reduction techniques* as all the X variables are important. This input has come from Technical Team.

Based on the above X variables, we would be forecasting kWH values for every 15 min interval on next day. The values we would get shall be compared with the peak kWH and kVA values in past 10 and 30 days and the corresponding 'hour' shall also be retrieved from database (based on nearest match). This 'hour' would be the *peak hour.*

In the dataset, we have 98193 Row counts. So the dimension of X is (98193, 9)

Table 1

X variables	Data type	Frequency	Unit	Atomic/derived
Avg voltage level harmonic distortion	Float64	30 min interval daily	%	Derived
Avg. current level harmonic distortion	Float64	30 min interval daily	%	Derived
Day of the week	Integer	Daily		Derived
Hour of the day	Integer	Daily		Derived
kVA	Float64	30 min interval daily		Atomic
Current	Integer	30 min interval daily	A	Atomic
Voltage	Integer	30 min interval daily	Vt	Atomic
Oil temperature	Integer	30 min interval daily	°C	Atomic
Oil level	Boolean	30 min interval daily		Derived

As mentioned above, we have used four Machine Learning Models to predict the kWH. Specifically we have used Linear Regression, Support Vector Machines and Neural Net Model. Within Neural Net we have experimented with a special case of ANN named LSTM RNN.

5 Model Performance and Deployment for Business Use

Table 2 summarizes the performance metrics. Since we are predicting kWH which is a numerical value (data type: Float64) two performance metrics have been used viz. multiple R^2 and Root Mean Squared Error (RMSE)

As it is clearly visible from Table 2, ANN and LSTM RNN are the two winners with highest accuracy level. Even SVM also scores well in accuracy.

Table 2

Model name	Multiple R-squared (%)	Root mean squared error (RMSE)
Multi-layer perceptron or artificial neural net (ANN)	98.19	573.330
Support vector machines	92.55	590.458
LSTM RNN	97.22	579.573
Linear regression	77.25	899.223

The only drawbacks of these ML models are training time. In a situation like this, it took around 9–10 h for each model to get trained (LSTM RNN was quite fast though, took only 3 h) (Figs. 3, 4, 5, 6, 7 and 8).

Deployment Strategy

Any ML model which is finally selected must be deployed for further prediction on new data set. For deployment, we have chosen Flask RESTFul API service through which we perform the following steps:

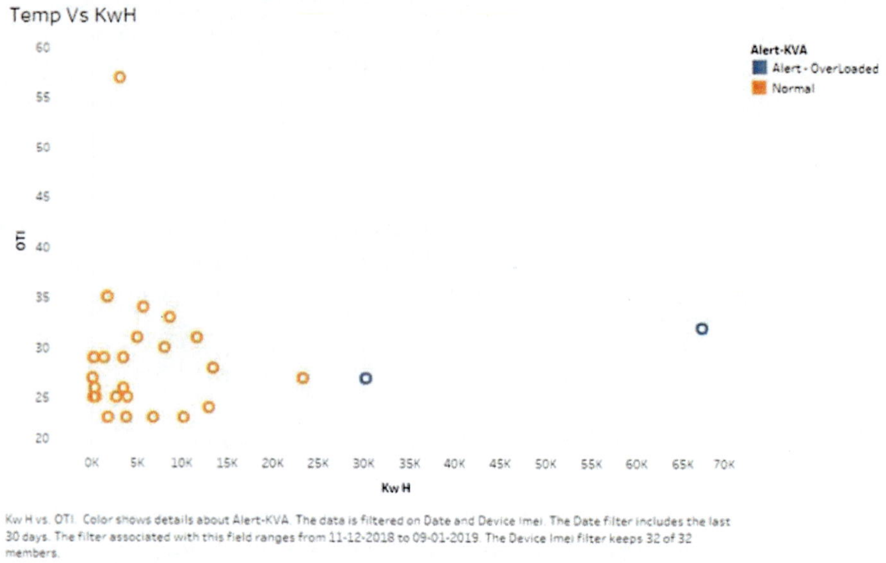

Fig. 3

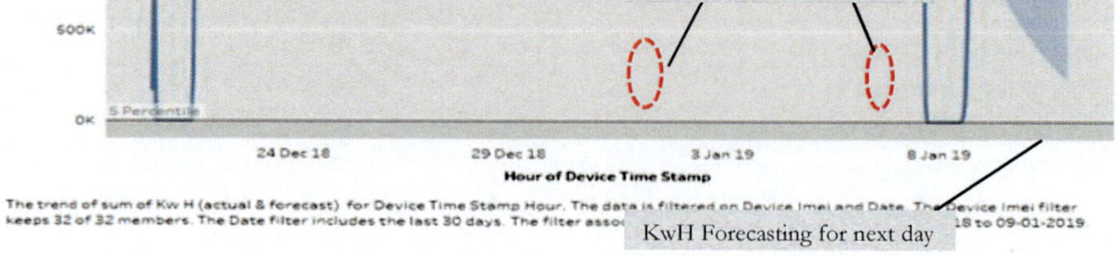

Fig. 4

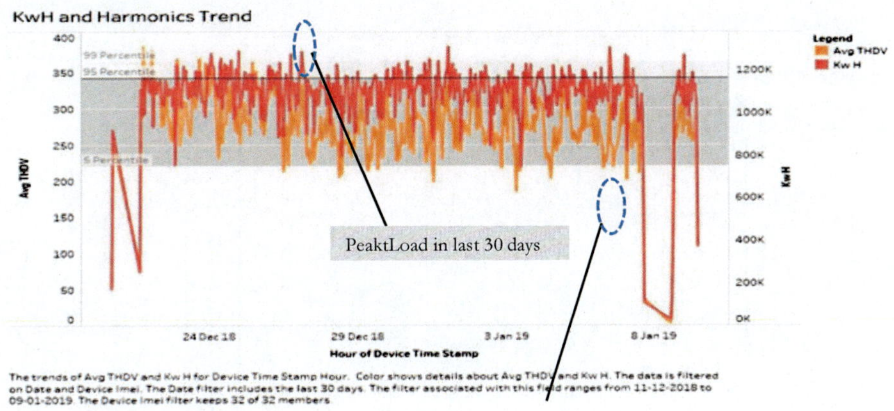

Fig. 5

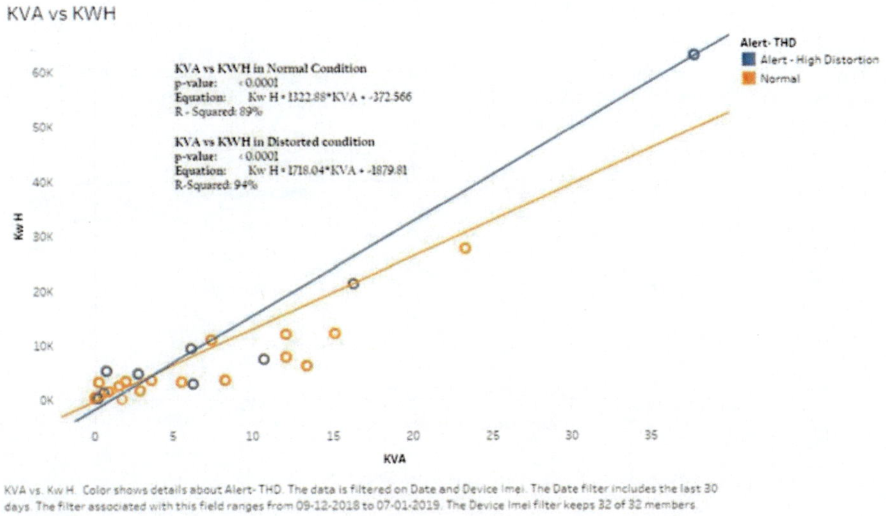

Fig. 6

a. Save the model on the local drive. Specifically we save the model parameters as derived while training time.
b. On a server, call the model by providing full file path.
c. Accept the user input through a GUI in JSON format via AJAX Query.
d. Use prediction function of the model called in (b).
e. Pass on the predicted value, kWH to the end point (i.e. Front End) via AJAX Method.

Users can see the predicted value (next day's kWH value) in the analytics dashboard as a grey area with solid line (please refer Fig. 5). User can also see the '*hour*'

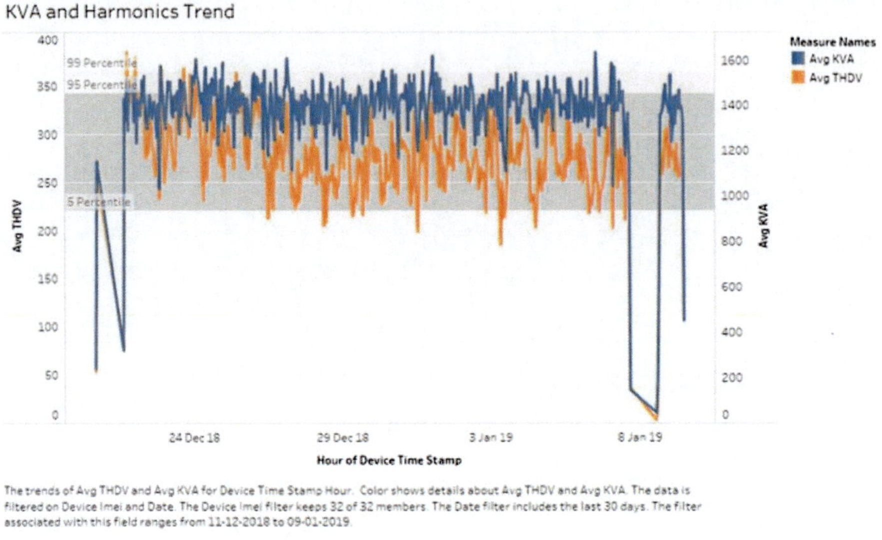

Fig. 7

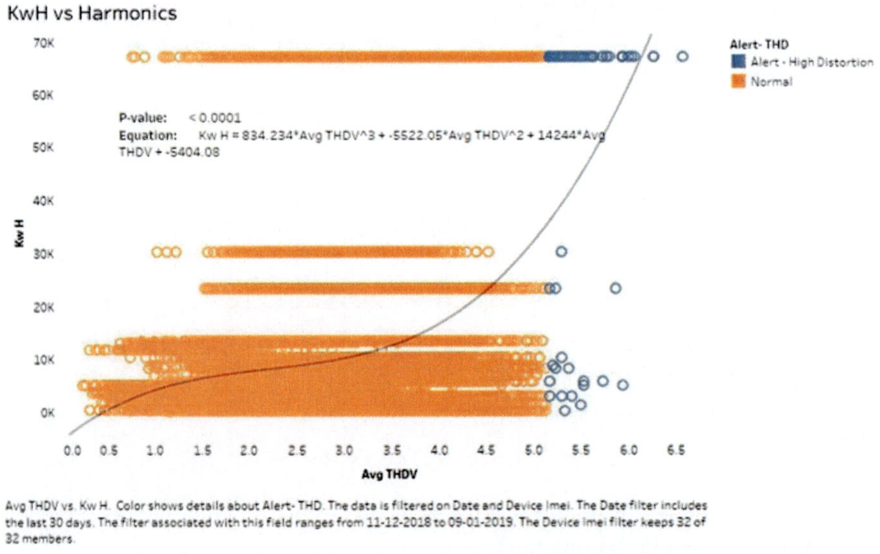

Fig. 8

corresponding to this kWH value. So, using the ML and Data Wrangling techniques, we are able to predict (i) kWH and (ii) Peak Hour. In case that kWH is *peak load* (nearest match to 99 percentile based on previous day and 30 days), we are also able to detect the corresponding 'hour' as *peak hour*.

6 Benefits

The predicted kWH values can be aggregated for all DTs to arrive at total load demand on next day. We can derive the '*hour range*' based on the predicted value. Accurate forecasting of kWH and retrieving the corresponding 'hour' of the day can help in

a. Scheduling the Load balancing activity well in advance (in this case 24 h).
b. Preventing unwanted Interruption.
c. Deriving insight on Transformer Load.
d. Preventing high Harmonic Distortion in Transformer thereby reducing maintenance cost.
e. Paving way the Demand Response and Management efficiently.

7 Conclusion

Load forecasting can indeed benefit the Utility business and therefore it is very important to have a robust and accurate load forecasting techniques in place. Adding to this is deployment strategy which is equally crucial. Since the DT data is of near real time, it is very important that data is well received, processed efficiently as explained in EDA step. Post this the prediction value would be derived based on the model selected above. And all these steps must be accomplished within 10 min of receiving data. Because of the requirement of such fast computing, Utility industry should adopt Hadoop based Big Data framework. Overall, we see a promising future for ML driven techniques in various aspects of Power Distribution, Generation and Transmission Process.

References

1. Hastie T, Tibshirani R, Friedman J The elements of statistical learning. Springer, Singapore
2. Gajowniczek K, Zabkowski T Two-stage electricity demand modelling using machine learning algorithms. In: MDPI energies paper series
3. Grant J, Eltoukhy M, Asfour S Short-term electrical peak demand forecasting in a large government building using artificial neural networks. In: MDPI energies paper series
4. Smart grid handbook for regulators and policy makers, November 2017
5. IDC Technologies Industrial automation—pocket guide
6. India Country Report—Research, development, demonstration and deployment of smart grids in India, Ministry of Science and Technology, Govt. of India
7. Power distribution handbook—ABB
8. Weron R (2014) Electricity price forecasting: a review of the state-of-the-art with a look into the future. Int J Forecast 30:1030–1081

Forecasting Short Term Peak Loads of Distribution Transformer (DT) Using Machine Learning and Computational Statistics—Various Methodologies and Their Pros and Cons

Arghya Roy

Abstract Machine Learning (ML) is a technique that employs computational statistics to learn the pattern of data and from there it tends to predict. Over last few years, abundance of data have been made available in standard format and therefore it is compelling to use ML techniques to achieve better prediction. In several instances we have witnessed 98–99% accuracy in prediction levels using ML techniques. In Utility Industry, one of the key parameter every DISCOM would like to forecast is Peak Load. Since HT (>11 kVA) and EHT (Extra High Tension > 66 kVA) customers in the designated DISCOM region demand continuous power supply, reducing interruption hours and feeder outages are two most important factors draw attentions of Utility Service Providers. It has been found that accurate prediction of Peak Load Demand for a day ahead can remarkably reduce the above two parameters. In fact, knowing the peak load in advance can help in optimizing the load distribution at substation levels. While predicting peak load, we should consider weather (Temperature), time of the day (expressed in HH:MM format), Day type (Weekdays or Weekends) and finally kWH of a DT.

Keywords Smart meters · Neural net · Tree based methods · Real time data gathering · ETL · Data modelling · Forecasting · Smart grid

1 Introduction

The Indian power sectors at the cusp of a transformation process from being manually operated to digitalization. The Distribution Utilities face new challenges every day and have to improve profits by reducing operating and maintenance costs, while providing customers with a reliable power supply and a broad range of services. India is currently battling the highest *transmission and distribution losses* (T & D Loss or AT & C Loss) in the world which is close to around 25%. It has been proved that the crux of reducing AT & C Loss lies in managing the demand properly.

A. Roy (✉)
India Power Corporation Limited, Kolkata 700091, India
e-mail: arghya.roy@indiapower.com

R. K. Pillai et al. (eds.), *ISUW 2019*, Lecture Notes in Electrical Engineering 764,
https://doi.org/10.1007/978-981-16-1299-2_1

Essentially, demand management or *demand response* (DR) is a function of *load forecasting* and *Peak Load Management* (PLM). Load forecasting is concerned with different horizon scales: hourly, daily, weekly, monthly, and annual values for both overall demand and peak demand. Importantly, short-term forecasting has attracted a great deal of attention in the literature due to its applicability in power system management, including resource planning and taking control actions towards load balancing.

In this paper, we have examined *Machine Learning* (ML) driven some important load forecasting techniques. We have used ML model to forecast the peak load of each Distribution Transformer (DT) for next day.

The remainder of the paper is structured as follows: a literature review on related problems is provided in Sect. 2. Next we discuss about some Machine Learning Techniques and then focus on Neural Net based ML models (Sect. 3). We then outline the *data pipeline* process i.e. fetching data from DT, cleaning and performing necessary *Exploratory Data Analysis* (EDA) and finally fit the data to ANN and LSTM RNN model (Sect. 4). In Sect. 5 we discuss about performance metrics of the above two model and chalk out the deployment plan for further use. In Sect. 6 we discuss on the business benefits of the load prediction.

2 Literature Review

A considerable amount of research has been dedicated to electricity price forecasting. As investigated by Weron [8], there were more than 800 publications related to electricity price forecasting in the years 1989–2013, as indexed in the Web of Science and Scopus databases. Energy forecasting models can be systematized in several ways, categorized as static or dynamic, univariate or multivariate, or as involving various techniques ranging from simple naive methods through a wide range of times series methods to complex hybrid and artificial intelligence models. For the purpose of the review, the approaches are categorized under the following headings: (a) Statistical and Time Series Models; (b) Computational Intelligence models; and (c) hybrid models.

a. Statistical and Time Series Models

Time series models are the most frequently used methods to represent the future values based on previous observations. The models based on time series have many forms adequate for forecasting electricity consumption volume and peak demand load in the electrical grid. These models are often applied to demand forecasting at the regional/national level. *Regression models* are built to capture a relationship between a dependent variable, which is the energy usage, and one or more independent variables (the predictors). *Econometric models* are built to investigate dependencies between the energy demand and other macro-economic variables such as energy price, gross national product (GNP), technology, investments, and population size.

Some authors like Sengupta et al., have demonstrated that econometric models are effective for forecasting energy patterns in developing countries like India.

b. **Computational Intelligence Models**

Within the category of computational intelligence methods, we use many Machine Learning Models like *Support Vector Machines* (SVM), *Tree Based methods with Gradient Boosting*, and ANN and its variants to model complex and dynamic systems. Machine Learning based techniques as mentioned here are quite promising because of (i) High Accuracy and Precession, (ii) Fast computation, (iii) Easy to deploy and (iv) Can manage non-linearity amongst Predictor and Dependent variables.

c. **Hybrid Models**

Based on the available literature, there is a clear and increasingly recognizable research interest that looks at the application of various hybrid models, combining techniques from two or more of the groups listed above. This stream is aimed at increasing the load forecasting accuracy by benefiting from the best features associated with different approaches and their synergy. For instance, it has been found that the combination of Econometric forecasting model like ARIMA (seasonality adjusted), and the weighted SVM can effectively account for the nonlinearity and seasonality and produce quite accurate forecasting results.

3 Important Machine Learning Techniques for Load Forecasting

There are many ML models which are quite popular and useful. And many of them are being extensively used in Retails, Healthcare and other domains for predictions. ML models which are used for predictions are termed as *Supervised Learning Models.* Few very popular amongst them are Tree Based methods, Random Forests, Support Vector Machines (SVM) and Neural net or Perceptron Model (ANN). For quite a large dataset, we found SVM and ANN are quite robust and accurate.

a. **Support Vector Machines (SVM)**

Support vector machines (SVMs) are a set of supervised learning methods used for classification, regression and outliers detection. In SVM, a *Maximal Margin Hyperplane* (also known as the optimal separating Hyperplane), which is the separating *Hyperplane* that is farthest from the training observations is introduced and we compute the (perpendicular) distance from each training observation to a given separating Hyperplane; the smallest such distance is the minimal distance from the observations to the Hyperplane, and is known as the *margin.*

To construct a Maximal Margin Hyperplane, margin value to be maximized.

Maximum M
$\beta_0, \beta_1, \ldots, \beta_p$
subject to

$$\sum_{j=1}^{p} \beta_j^2 = 1$$

$$y_i\left(\beta_0 + \beta_1 \cdot x_{i1} + \beta_2 \cdot x_{i2} + \cdots + \beta_p \cdot x_{ip}\right) \geq M(1 - \varepsilon_i)\ \forall i = 1, 2, 3, \ldots, n$$

$$\varepsilon_i \geq 0, \sum_{i=1}^{n} \varepsilon_i \leq C$$

M represents the margin of our Hyperplane, and the optimization problem chooses $\beta_0, \beta_1, \ldots, \beta_p$ to maximize M. Typically generating the beta values are complex and beyond the scope of this article. 'ε' is the slack variable and C is the penalty parameter. We use *Kernel* trick to derive those betas. In the above equation, 'y' is the dependent or predictor variable.

The advantages of support vector machines are:

i. Effective in high dimensional spaces.
ii. Uses a subset of training points in the decision function (called support vectors), so it is also memory efficient.
iii. Versatile: different Kernel functions can be specified for the decision function. Common kernels are provided, but it is also possible to specify custom kernels.

b. **Artificial Neural Network (ANN)**

An Artificial Neural Network (ANN) is a computational model that is inspired by the way biological neural networks in the human brain process information. Typically an ANN consisted of three elements viz. Input Layer (or Input Node), Hidden Layer and Output Layer. A neural network is a *two-stage regression or classification model*, typically represented by a network diagram as shown below. This network applies both to regression or classification (Fig. 1).

For K-class classification, there are K units at the top, with the kth unit modelling the probability of class k. There are K target measurements. Y_k, k = 1, ..., K, each being coded as a 0 − 1 variable for the kth class. Derived features $a_1, a_2, \ldots, a_k$ are created from linear combinations of inputs (X) then target Y_k is modelled as linear combinations of a_m

$$a_m = \varphi \cdot \left(\alpha_{0m} + \alpha_m^T \cdot X\right), m = 1, 2, \ldots, M$$

$$t_k = \beta_{0k} + \beta_k^T \cdot a, k = 1, 2, \ldots, K$$

$$Y_k = f(X) = g_k(t), k = 1, 2, \ldots, K$$

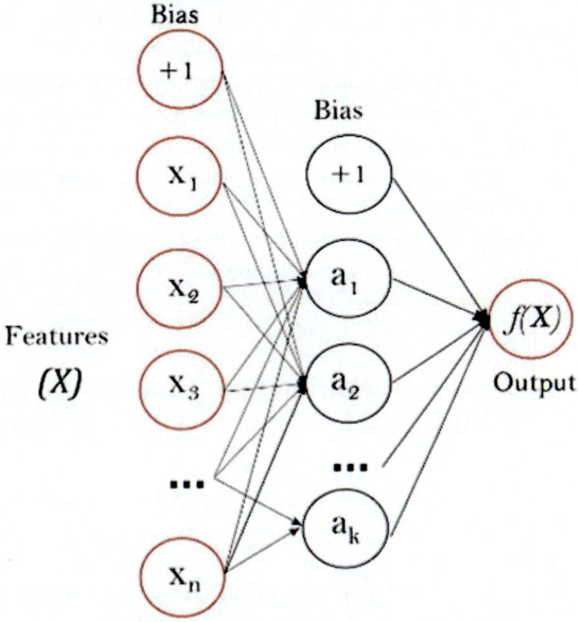

Fig. 1

where a = (a_1, a_2, a_3, …, a_k) and t = (t_1, t_2, …, t_k).

'φ' is called the activation function. Beta values are derived by a technique named back-propagation i.e. comparing the value of Y_k with the actual Y and then again retrain the whole hidden layers by adjusting 'α'.

RNNs are called recurrent because they perform the same task for every element of a sequence, with the output being depended on the previous computations. Here is what a typical RNN looks like (Fig. 2).

Instead of neurons, LSTM networks have memory blocks that are connected through layers.

x_t is the input at time step t. s_t is the hidden state at time step t. It's the "memory" of the network. s_t is calculated based on the previous hidden state and the input at the current step:

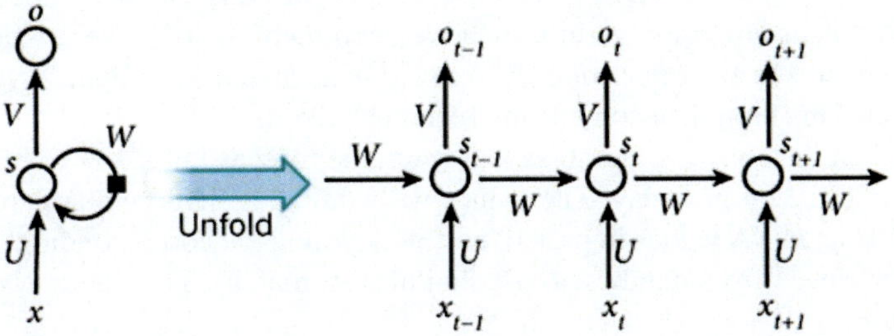

Fig. 2

$$s_t = f(Ux_t + Ws_{t-1}).$$

The function f usually is a nonlinearity such as tanh or ReLU. s_{t-1} which is required to calculate the first hidden state, is typically initialized to all zeroes.

o_t is the output at step t. For example, if we wanted to predict the next word in a sentence it would be a vector of probabilities across our vocabulary. $o_t = \text{softmax}(Vs_t)$.

With the above understanding we now move forward data preparation and implementation of above models.

4 Data Preparation and Exploratory Data Analysis (EDA) with DT Data

As stated earlier, we shall be using data related to a single DT. Data are captured through IIoT devices mounted on the DT. Data is collected at *every 15 min Frequency.* Based on the data availability, some graphs have been plotted. We wanted to examine the kVA or load distribution across the time frame for each DT. We also wanted to see the Harmonic distortion of each DT on a given time scale. Then we explored the statistical correlation between Harmonic Distortion (Voltage Level) and Load for Each DT. We have also plotted the DT Oil temperature against kVA (load). We have identified a kWH value as peak load when it is more than 99 percentile of previous 30 days value. We have marked this in Red (Fig. 5). We scaled this down to daily basis and identified the peak loads.

Based on the data as collected above and the graph so plotted, we forecast the next level of kWH and it's corresponding time in the day it may occur. Specifically we are trying to *predict two variables* viz. *kWH (directly) and Hour (retrieved from database based on predicted kWH)* given the data set as shown in Table 1.

Collecting data from IIoT is dependent on the signal quality of the communication channel. In case of Comm. Network congestion, data may not be retrieved properly. In order to tackle this, we have used a '*conditional imputation of missing values*' technique. In this technique, we impute the value of kWH and kVA by its last *3 observation's rolling mean* if the missing data is present only at a *single occurrence i.e. only at one instance.* But if we observe that for consecutive 3 times, the kWH and kVA values are coming zero, then we keep them as it is. We are not going to reduce or resort to any *dimensionality reduction techniques* as all the X variables are important. This input has come from Technical Team.

Based on the above X variables, we would be forecasting kWH values for every 15 min interval on next day. The values we would get shall be compared with the peak kWH and kVA values in past 10 and 30 days and the corresponding 'hour' shall also be retrieved from database (based on nearest match). This 'hour' would be the *peak hour.*

In the dataset, we have 98193 Row counts. So the dimension of X is (98193, 9)

Table 1

X variables	Data type	Frequency	Unit	Atomic/derived
Avg voltage level harmonic distortion	Float64	30 min interval daily	%	Derived
Avg. current level harmonic distortion	Float64	30 min interval daily	%	Derived
Day of the week	Integer	Daily		Derived
Hour of the day	Integer	Daily		Derived
kVA	Float64	30 min interval daily		Atomic
Current	Integer	30 min interval daily	A	Atomic
Voltage	Integer	30 min interval daily	Vt	Atomic
Oil temperature	Integer	30 min interval daily	°C	Atomic
Oil level	Boolean	30 min interval daily		Derived

As mentioned above, we have used four Machine Learning Models to predict the kWH. Specifically we have used Linear Regression, Support Vector Machines and Neural Net Model. Within Neural Net we have experimented with a special case of ANN named LSTM RNN.

5 Model Performance and Deployment for Business Use

Table 2 summarizes the performance metrics. Since we are predicting kWH which is a numerical value (data type: Float64) two performance metrics have been used viz. multiple R^2 and Root Mean Squared Error (RMSE)

As it is clearly visible from Table 2, ANN and LSTM RNN are the two winners with highest accuracy level. Even SVM also scores well in accuracy.

Table 2

Model name	Multiple R-squared (%)	Root mean squared error (RMSE)
Multi-layer perceptron or artificial neural net (ANN)	98.19	573.330
Support vector machines	92.55	590.458
LSTM RNN	97.22	579.573
Linear regression	77.25	899.223

The only drawbacks of these ML models are training time. In a situation like this, it took around 9–10 h for each model to get trained (LSTM RNN was quite fast though, took only 3 h) (Figs. 3, 4, 5, 6, 7 and 8).

Deployment Strategy

Any ML model which is finally selected must be deployed for further prediction on new data set. For deployment, we have chosen Flask RESTFul API service through which we perform the following steps:

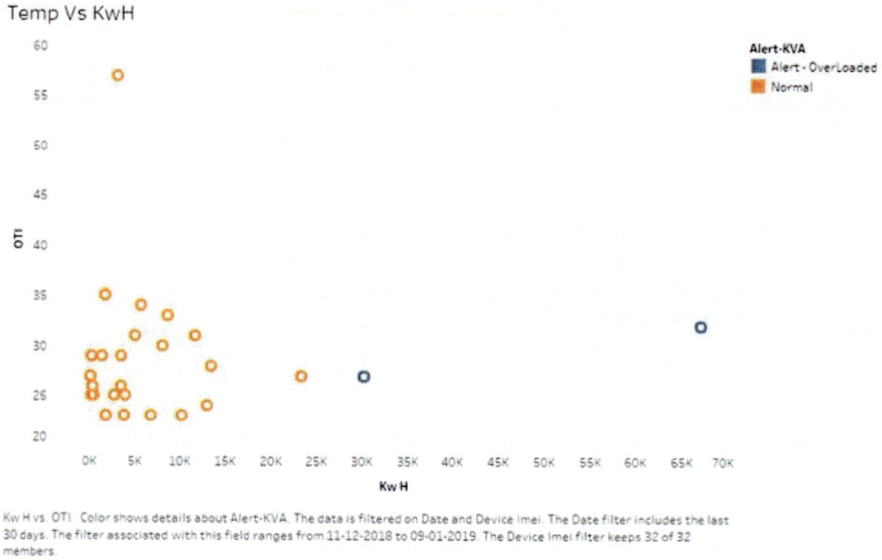

Fig. 3

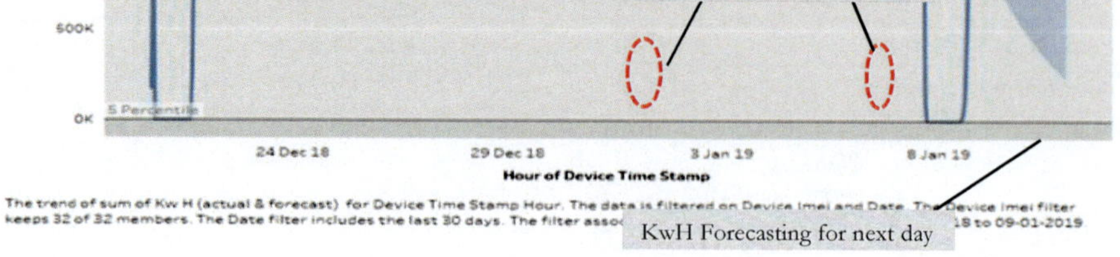

Fig. 4

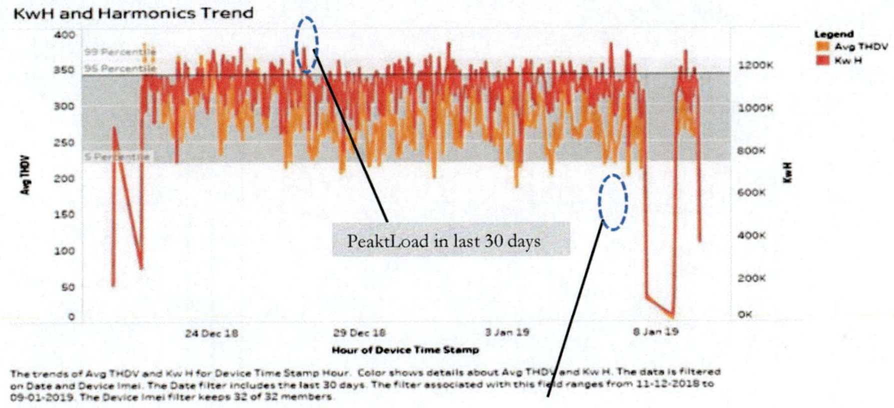

Fig. 5

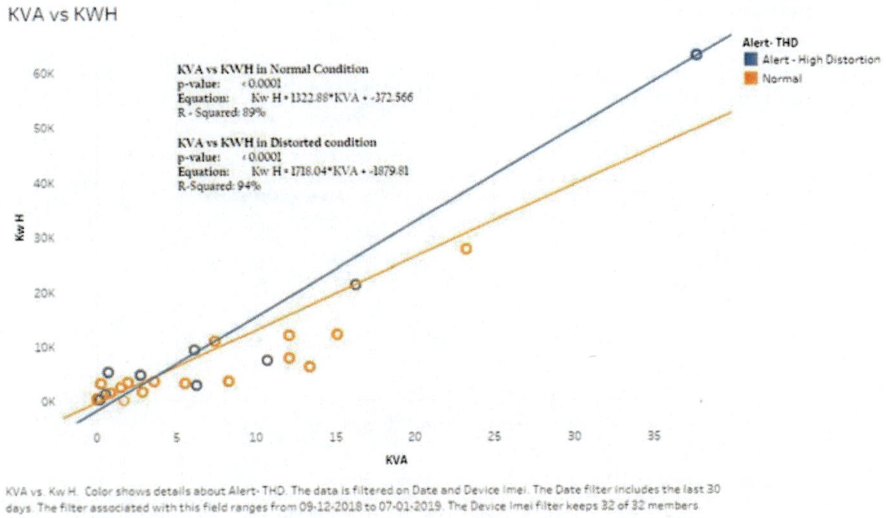

Fig. 6

a. Save the model on the local drive. Specifically we save the model parameters as derived while training time.
b. On a server, call the model by providing full file path.
c. Accept the user input through a GUI in JSON format via AJAX Query.
d. Use prediction function of the model called in (b).
e. Pass on the predicted value, kWH to the end point (i.e. Front End) via AJAX Method.

Users can see the predicted value (next day's kWH value) in the analytics dashboard as a grey area with solid line (please refer Fig. 5). User can also see the '*hour*'

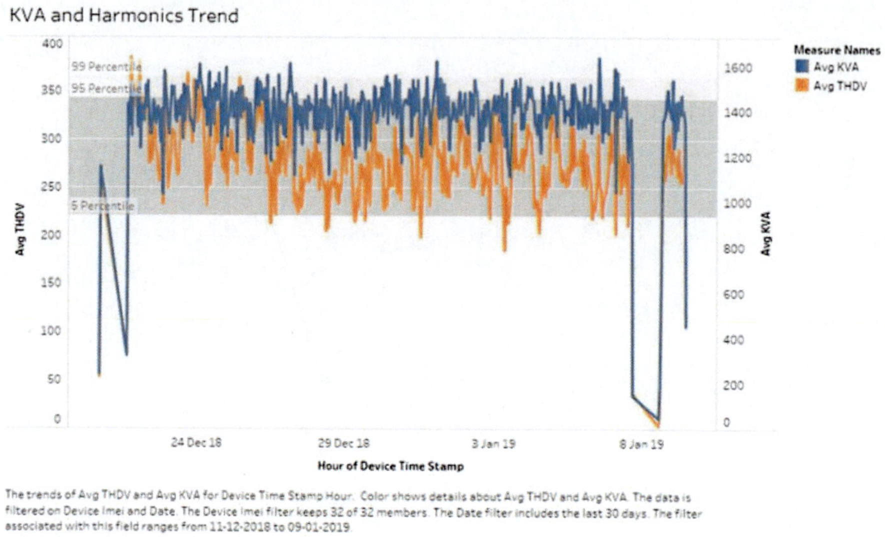

Fig. 7

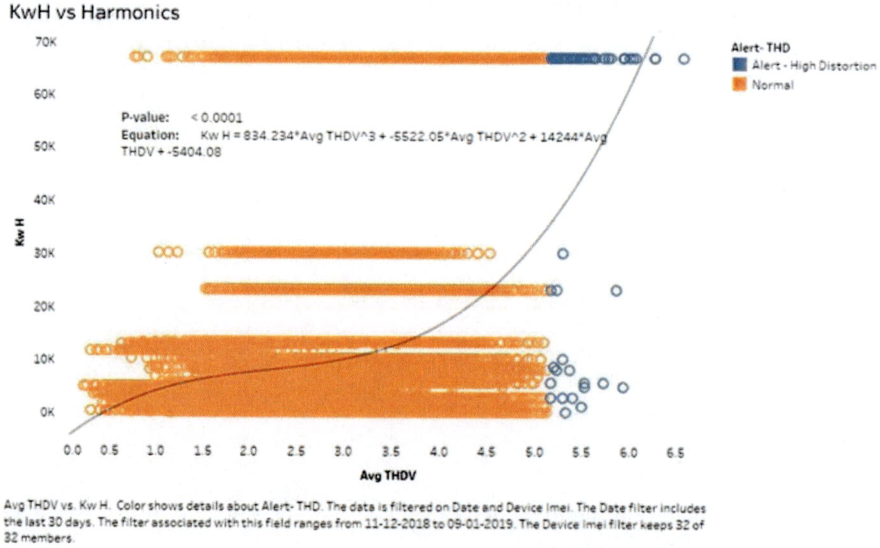

Fig. 8

corresponding to this kWH value. So, using the ML and Data Wrangling techniques, we are able to predict (i) kWH and (ii) Peak Hour. In case that kWH is *peak load* (nearest match to 99 percentile based on previous day and 30 days), we are also able to detect the corresponding 'hour' as *peak hour*.

6 Benefits

The predicted kWH values can be aggregated for all DTs to arrive at total load demand on next day. We can derive the '*hour range*' based on the predicted value. Accurate forecasting of kWH and retrieving the corresponding 'hour' of the day can help in

a. Scheduling the Load balancing activity well in advance (in this case 24 h).
b. Preventing unwanted Interruption.
c. Deriving insight on Transformer Load.
d. Preventing high Harmonic Distortion in Transformer thereby reducing maintenance cost.
e. Paving way the Demand Response and Management efficiently.

7 Conclusion

Load forecasting can indeed benefit the Utility business and therefore it is very important to have a robust and accurate load forecasting techniques in place. Adding to this is deployment strategy which is equally crucial. Since the DT data is of near real time, it is very important that data is well received, processed efficiently as explained in EDA step. Post this the prediction value would be derived based on the model selected above. And all these steps must be accomplished within 10 min of receiving data. Because of the requirement of such fast computing, Utility industry should adopt Hadoop based Big Data framework. Overall, we see a promising future for ML driven techniques in various aspects of Power Distribution, Generation and Transmission Process.

References

1. Hastie T, Tibshirani R, Friedman J The elements of statistical learning. Springer, Singapore
2. Gajowniczek K, Zabkowski T Two-stage electricity demand modelling using machine learning algorithms. In: MDPI energies paper series
3. Grant J, Eltoukhy M, Asfour S Short-term electrical peak demand forecasting in a large government building using artificial neural networks. In: MDPI energies paper series
4. Smart grid handbook for regulators and policy makers, November 2017
5. IDC Technologies Industrial automation—pocket guide
6. India Country Report—Research, development, demonstration and deployment of smart grids in India, Ministry of Science and Technology, Govt. of India
7. Power distribution handbook—ABB
8. Weron R (2014) Electricity price forecasting: a review of the state-of-the-art with a look into the future. Int J Forecast 30:1030–1081

Modular Microgrid Controller on OpenFMB Standard

Mohan Pavan Kumar Bailapudi

Abstract The smart grid future is forecasted to be mix of traditional central generation with a large network of distributed energy resources such as rooftop and community solar, microgrids and responsive loads. By utilizing micro-Distributed Energy Resources (DER), such as solar PV, energy storage, renewable energy sources fossil-fueled generators and combined heat and power plants, microgrids can supply local electrical and heat loads in local areas in an economic and environment friendly way. Significant technical challenges arise in the operation, planning and control of microgrids, due to the intermittent behaviour of renewable power generation and the minimum SOC limit of energy storage devices. The two-way communication and coordination between DER nodes capability of the microgrid provides a feasible solution to can address these challenges. The MicroGrid Central Controller (MGCC) provides autonomous coordination of the DER to serve the critical and non-critical loads economically in islanded and grid-connected modes. The proposed platform can be deployed locally or in a Virtual Private Cloud. The platform has a default optimizer (economic dispatch engine) where the operator can set thresholds and triggers to dispatch based on user-defined rules. The system is designed so that the optimization engine can be readily replaced as more efficient algorithms are developed. Device-to-device communication employs Open Field Message Bus (OpenFMB) measurement and control profiles. OpenFMB is a distribution network communication standard based on IEC CIM and IEC 61850. Use of OpenFMB eases interoperability with system operators and third-party service providers (aggregators).

Keywords MicroGrid · Central controller · Distributed energy resources · OpenFMB · Virtual private cloud

---

M. P. K. Bailapudi (✉)
IoT Team, Amzur Infotech, Visakhapatnam, India
e-mail: mohan.bailapudi@amzurtech.com

R. K. Pillai et al. (eds.), *ISUW 2019*, Lecture Notes in Electrical Engineering 764,
https://doi.org/10.1007/978-981-16-1299-2_2

1 Introduction

The MicroGrid (MG) concept has been proposed for efficient and flexible utilization of Distributed Energy Resources [1]. According to the US Department of Energy (DOE), as well as Electric Power Research Institute (EPRI), a MG is a group of interconnected loads and distributed energy resources within clearly defined electrical boundaries that acts as a single controllable entity with respect to the grid. It may be grid connected or "island" (disconnect) from the grid based on the loss of grid power, power quality challenges, or to contribute to service restoration. In short, it provides a more flexible, reliable and resilient energy system. The capability of integrating different kinds of distributed energy resources and generators (DGs), such as renewable energy, storage system and micro turbines, enhances the sustainability, efficiency and cost-effectiveness of the overall system.

The use of a hierarchical control scheme has been proposed for MGs in order to manage objectives in different time scales, technical fields and significances [2, 3], as it was proposed for manufacturing [4], power systems [5, 6], process systems [7], and in general terms, for large complex systems [8]. Usually a three-level hierarchy is considered comprising primary, secondary and tertiary control levels. Primary control is implemented in local DG controllers (DGCs) including inner voltage/current control loops and power sharing control loops. It ensures the stable operation of the DGs and distributed power sharing among them. Moreover, in order to enhance the system power quality [9–11] and achieve accurate power sharing [12, 13], secondary control approaches can be developed which act over primary control loops by sending adjustment and compensation references. In the tertiary level, optimization and decision making functions can be applied which give optimal set-points to lower level controllers achieving intelligent and more efficient operation of the whole system [11, 14–17]. In addition, the synchronization and reconnection with external grids is based on the cooperation between secondary and tertiary levels. From primary, secondary levels to tertiary control level, the control bandwidths are decreased to achieve the decoupled behavior between layers, which also simplifies the implementation of higher-level controllers as well as the system stability analysis shown in Fig. 1.

Thanks to the advances in information and communication technologies, the real-world implementation of the above mentioned control levels can be deployed centrally, on-site or in a hybrid fashion. Consider the benchmark MG shown in Fig. 2, in which DGs are connected to the system in a distributed way supplying consumers, i.e. university. A point of common coupling (PCC) connects the MG to the Electric Power System. DG Controllers (DGCs) perform primary (autonomous) control functions. The MG central controller (MGCC) coordinates the DGs based on user-set dispatch rules that may optimize for cost, power quality, and environmental and economic considerations, e.g. maximize the on-site consumption of solar rather than export solar generation to the grid.

Having implemented state-of-the-art technologies and advancements in the micro-grid control as mentioned above this proposal has gone a step ahead and include

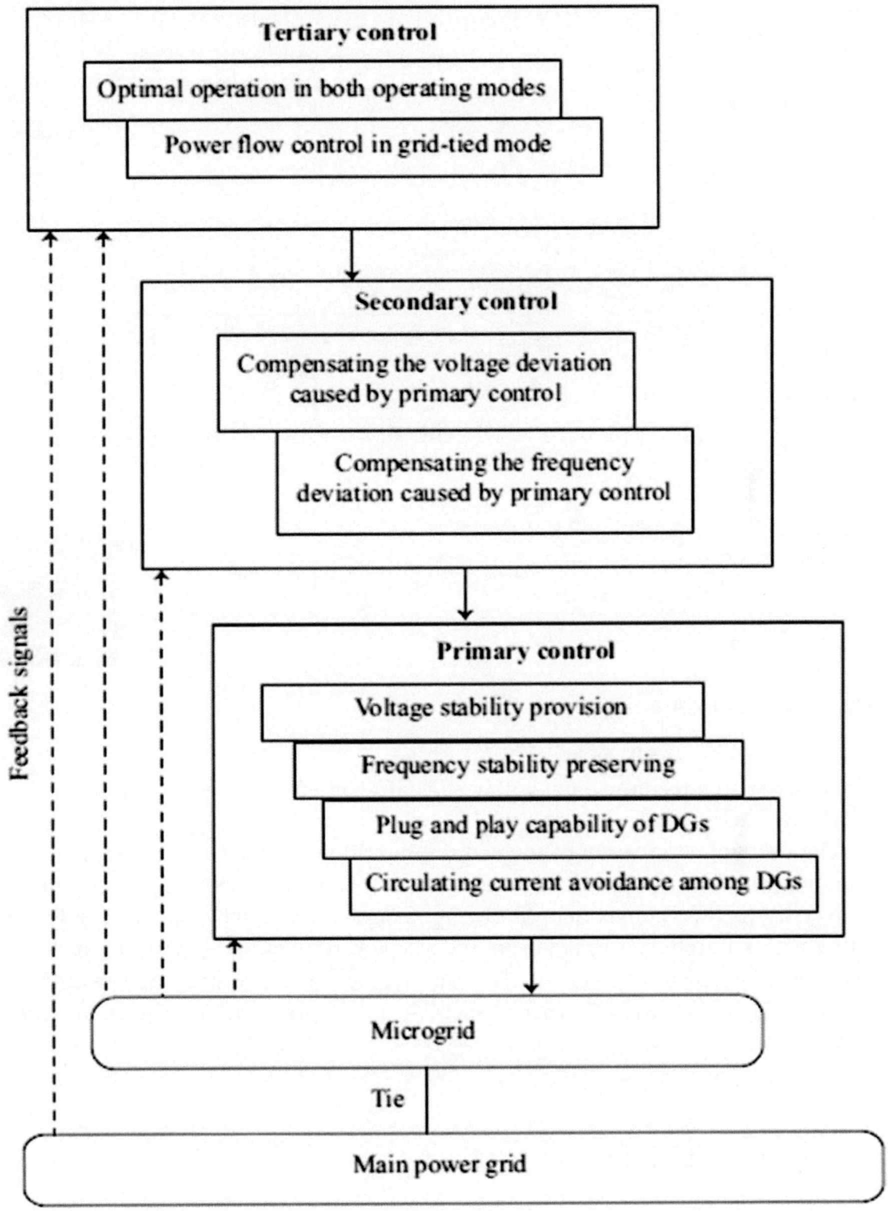

Fig. 1 Hierarchical control levels of microgrid

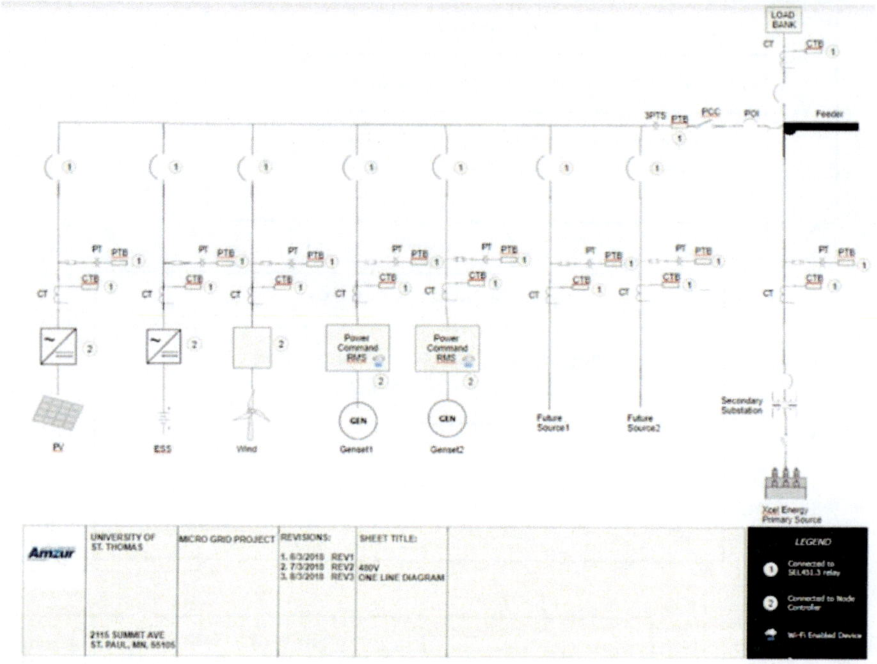

Fig. 2 Microgrid system

cloud in the architecture which will host MGCC and GUI for accessing MGCC and designing MG.

The test and verification of the control algorithms as well as measurements and communication infrastructure require laboratory simulations and testing. Our partner, University of St. Thomas School of Engineering, is a Renewable Energy Facility lab which is introduced in detail in Sect. 2. A study case is established in Sect. 3 and implemented in the system. Section 4 presents the experimental results with discussions over the system performance. Section 5 gives the conclusion and future plan.

2 The USTREF Lab

2.1 *Lab Overview*

The University of St. Thomas Renewable Energy Facility (USTREF) is a multipurpose microgrid research lab and testing facility that includes a 50 kW solar PV array, 10 kW of wind emulation capacity, a 50 kW diesel genset, a 100 kWh energy storage system, SEL 3555 and multiple 451 controllers, and meters and circuit breakers at each node. The facility serves as a platform for advanced microgrid EMS, DERMS, and inverter research, development and deployment. The combination of hierarchical control structure with proper design of the hardware structure formulates a generalized and expandable experimental platform in iMG lab. The details of the lab are introduced in this section.

The SEL Real-time Automation Controller (RTAC) is designed for substation automation and control. It provides power protection, data concentration, event collection, and network security. The 3555 communicates with an SEL 451 at each DER node to provide high speed fault protection, and metering as shown in Fig. 3. Primary protection functions are autonomous and secondary controls (online/offline or device settings) can be executed locally or remotely (in the cloud).

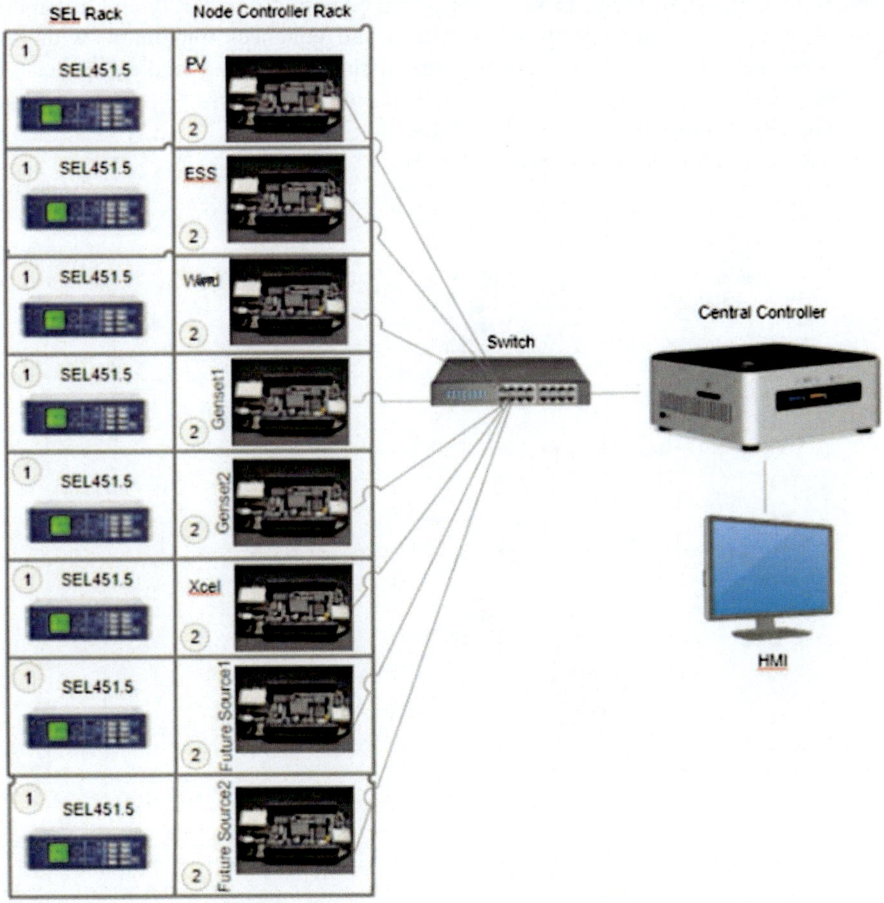

Fig. 3 Communication interface

The power converters used in the testbed are SunSpec compliant Advanced Function inverters hat support islanding, volt/VAR optimization, watt/frequency control, and other advanced features.

2.2 *Communication Architecture and Its Security*

Traditional SCADA systems employ a Master-Slave architecture. OpenFMB, the communications framework employed at the UST microgrid, is a publish-subscribe design that supports peer-to-peer communications and is characterized by low-latency. This framework and reference architecture based on existing data and information models (CIM and 61850) enables grid edge interoperability and distributed intelligence, augments operational systems, and enhances integration with field devices—see Fig. 4. The MGChas a solar, energy storage and load forecaster. The forecasters can be configured to suit specific field requirements.

The cloud-based implementation of the MGCC is secured with AWS (Amazon Web Services) VPC (Virtual Private Cloud).

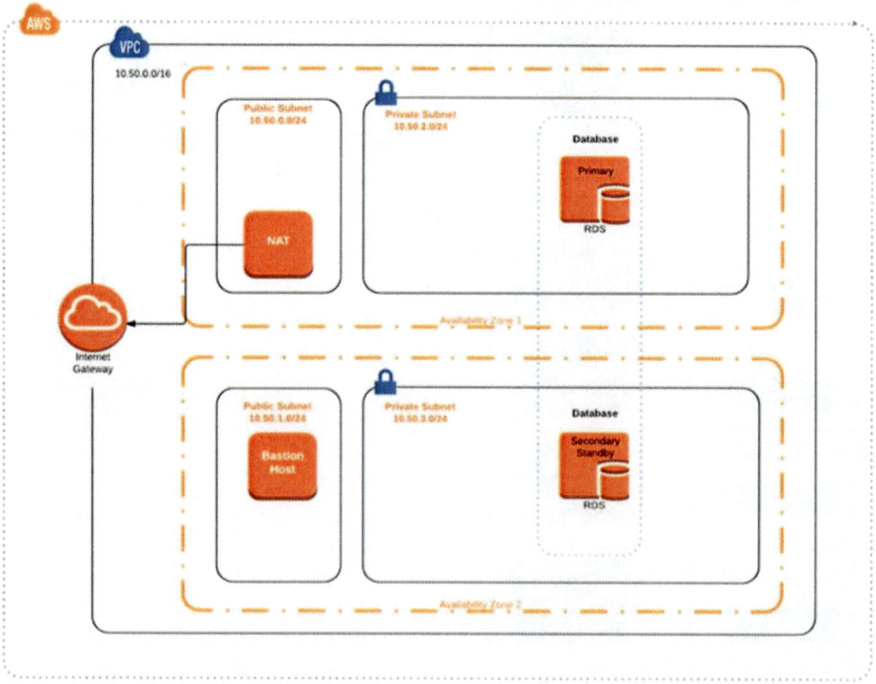

Fig. 4 AWS VPC architecture

2.3 AWS Virtual Private Cloud

Amazon Virtual Private Cloud (Amazon VPC) logically isolates the MG instance from the public internet.

One has complete control over his/her virtual networking environment, including selection of your own IP address range, creation of subnets, and configuration of routing tables and network gateways. For example, one can create a subnet for web servers hosting Energy Dashboards that are publicly accessible and securely isolate backend systems with no public access. You can leverage multiple layers of security, including security groups and network access control lists, to help control access to Amazon EC2 instances in each subnet.

Reason for going with AWS VPC is that it is secure, scalable and reliable.

2.4 OpenFMB Standard

The electric power grid has a wide range of systems and devices to transmit and distribute power from generation to load. These systems and devices use a variety of communication and protocol standards and many systems are effectively siloed from other applications that could leverage streaming data to provide better situational awareness. Very few devices are capable of peer-to-peer communications, let alone exchanging data and information for distributed intelligence and control.

OpenFMB is an open standard for field area communications and interoperability. The standard defines a reference architecture platform comprising internet protocol (IP) networking, Internet of Things (IoT) messaging protocols, standardized common semantic models, messages, and services to enable the secure, reliable, and scalable communications and peer-to-peer information exchange between devices on the electric grid.

With the implementation of the OpenFMB standard, devices from different vendors and utilities will, in fact, be able to interoperate directly or through gateway technologies that act as a bridge, translator, or adapter for legacy protocols. This specification defines

the building blocks that consist of device types, utility legacy protocols (DNP3, Modbus, IEC 61850 MMS/GOOSE, C12, etc.), IoT protocols, and use-case application logic that may be involved in this reference architecture (Figs. 5 and 6).

3 Use Case

Steps using MGCC cloud application to design Microgrid system and setup the online optimizers and its options

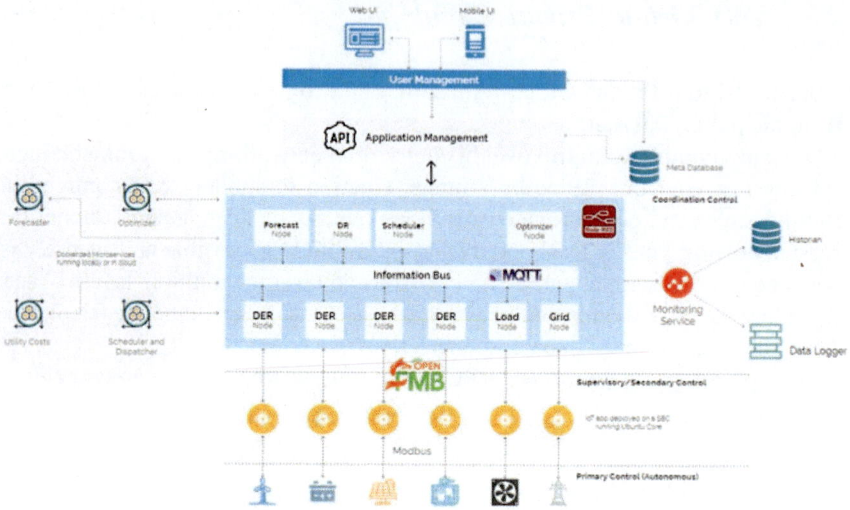

Fig. 5 MGCC communication architecture

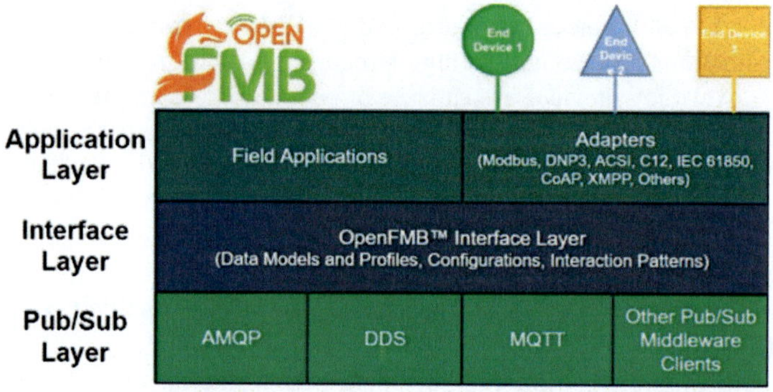

Fig. 6 OpenFMB technical architecture

Step 1: The microgrid system in Fig. 2 is designed in the cloud application with all the necessary information for all the DER's Energy Storage(ES), Generator(GEN), PhotoVoltaic(PV), LOAD and PCC(Point of Common Coupling) parameters as shown in Fig. 7.

Step 2: Select the optimizer and set the optimizer options
The optimizers used in this particular use case is MILP(Mixed Integer Linear Programming) and set optimization options(system cost = 0.7, system voltage = 0.2 and power factor = 0.1).
The cost function is formulated with above mentioned weights for system cost, system voltage and power factor.

Step 3: Select the Location for PV Forecast

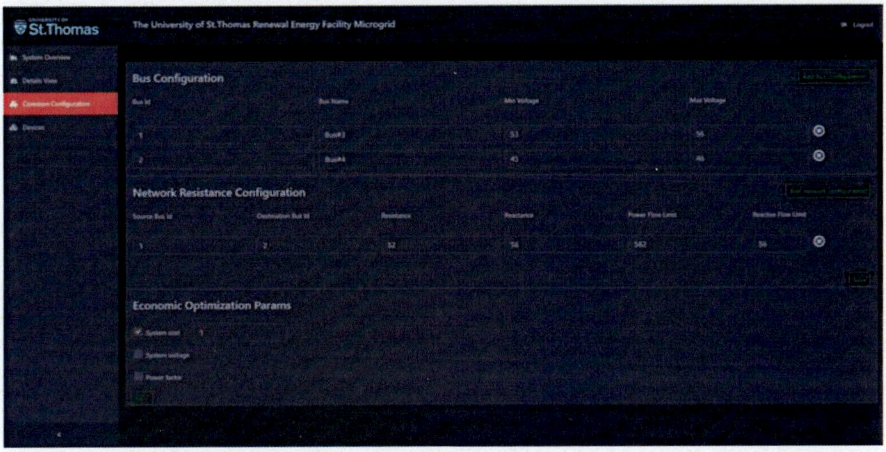

Fig. 7 Application for configuring the microgrid

Latitude and Longitude of site location is given and the other the PV parameters is given for PV forecast date.
The online forecaster used is Solcast API which gives PV forecast for seven days from the time of request.

With all the above three steps completed, launch the MGCC application which runs continuously for every 5 min interval calculating the set points to DER's based in the load forecast.

4 Results

The experimental results showing the performance of primary control at the microgrid bus (PCC) is shown in Figs. 9 and 10 (Fig. 8).

The power converters operate in v/f mode in the primary control during the black start of microgrid and adjusted the frequency to 60 Hz and voltage to 480 V.

The experimental results showing the performance of secondary control is shown in Fig. 10.

The power converters operate in PQ mode, for the load change at 5 s, the set points to DER's are changed to match the load. Then the frequency is adjusted to 60 Hz by the secondary controller.

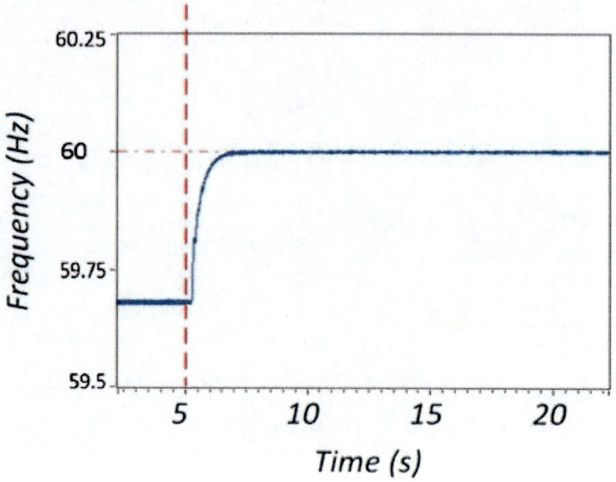

Fig. 8 Frequency at PCC

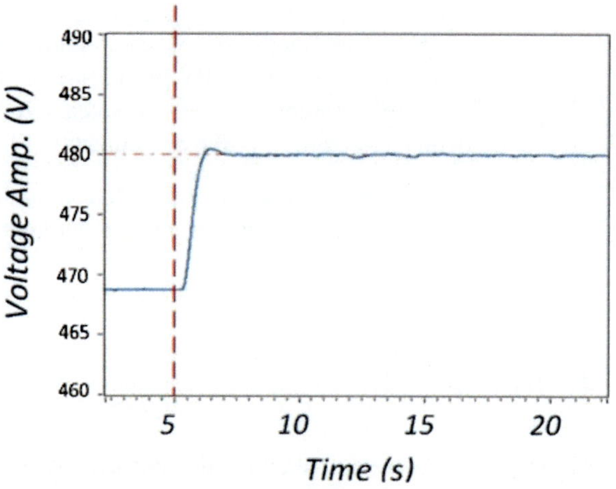

Fig. 9 Voltage at PCC

5 Conclusion

This paper implemented a MicroGrid at University of St. Thomas where the MGCC is running on the cloud with 5 min interval of PV and load forecast with OpenFMB standard in communication with in the field devices. The security of MGCC is fully ensured with AWS VPC. The proposed application has wiped out all the myths on microgrid operations being hosted on cloud. This application had disintegrated the steps in setting up of MicroGrid Central Controller. The application is scalable to any microgrid size with custom or predefined optimizers and forecasters techniques.

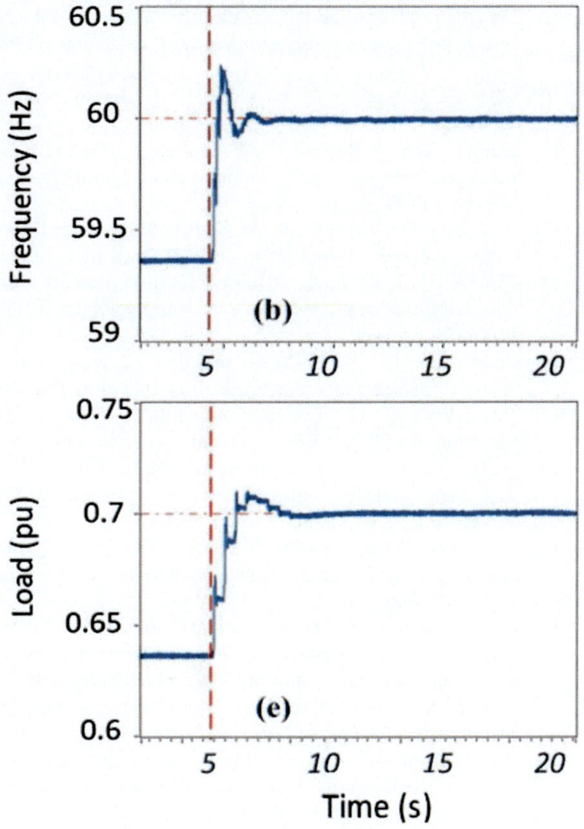

Fig. 10 Load to frequency change at PCC

References

1. IEEE guide for design, operation, and integration of distributed resource island systems with electric power systems, pp 1–54 (2011)
2. Bidram A, Davoudi A (2012) Hierarchical structure of microgrids control system. IEEE Trans Smart Grid 3(4):1963–1976
3. Guerrero JM, Vasquez JC, Matas J, de Vicuna LG, Castilla M (2011) Hierarchical control of droop-controlled AC and DC microgrids—a general approach toward standardization. IEEE Trans Ind Electron 58(1):158–172
4. Hax AC, Meal HC (1973) Hierarchical integration of production planning and scheduling. Logisti 1:53–69 (In Production)
5. Schweppe FC, Mitler S (1972) Hierarchical system theory and electric power systems. In: Handschin (ed) Real-time control electron power system. Elsevier, Amsterdam, pp 259–277
6. Smith NJ, Sage AP (1973) An introduction to hierarchical systems theory. Comput Electr Eng 1(1):55–71
7. Darby ML, Nikolaou M, Nicholson D, Jones J (2011) RTO: an overview and assessment of current practice. J Process Control 21:874–884
8. Mesarović MD, Macko D, Takahara Y (1972) Theory of hierarchical, multilevel systems. Academic Press, New York

9. Savaghebi M, Jalilian A, Vasquez JC, Guerrero JM (2012) Secondary control scheme for voltage unbalance compensation in an islanded droop-controlled microgrid. IEEE Trans Smart Grid 3(2):797–807
10. Mishra S, Mallesham G, Jha AN (2012) Design of controller and communication for frequency regulation of a smart microgrid. IET Renew Power Gener 6(4):248
11. Meng L, Tang F, Savaghebi M, Vasquez JC, Guerrero JM (2014) Tertiary control of voltage unbalance compensation for optimal power quality in islanded microgrids. IEEE Trans Energy Convers PP(99):1–14
12. He J, Li Y, Blaabjerg F (2014) An islanding microgrid reactive power, imbalance power, and harmonic power sharing scheme. IEEE Trans Power Electron PP(99):3389–3401
13. Chamorro HR, Ramos G (2011) Microgrid central fuzzy controller for active and reactive power flow using instantaneous power measurements. In: 2011 IEEE power and energy conference at Illinois, pp 1–6
14. Cho C, Jeon J-H, Kim J-Y, Kwon S, Park K, Kim S (2011) Active synchronizing control of a microgrid. IEEE Trans Power Electron 26(12):3707–3719
15. Tsikalakis AG, Hatziargyriou ND (2011) Centralized control for optimizing microgrids operation. In: 2011 IEEE power and energy society general meeting, pp 1–8
16. Meng L, Dragicevic T, Guerrero JM, Vasquez JC (2013) Optimization with system damping restoration for droop controlled DC-DC converters. In: 2013 IEEE energy conversion congress and exposition, pp 65–72
17. Dragicevic T, Guerrero JM, Vasquez JC, Skrlec D (2014) Supervisory control of an adaptive-droop regulated DC microgrid with battery management capability. IEEE Trans Power Electron 29(2):695–706
18. Papathanassiou S (2005) A benchmark low voltage microgrid network. In: Proceedings of the CIGRE symposium: power systems with dispersed generation, pp 1–8
19. Rasheduzzaman M, Bhaskara SN, Chowdhury BH (2012) Implementation of a microgrid central controller in a laboratory microgrid network. In: 2012 North American Power Symposium (NAPS), pp 1–6
20. Pilo F, Pisano G, Soma GG (2007) Neural implementation of microgrid central controllers. In: 2007 5th IEEE international conference on industrial informatics, vol 2, pp 1177–1182
21. Zhangang Y, Yanbo C, Chengshan W (2009) Construction, operation and control of a laboratory-scale microgrid. In: 2009 international conference on sustainable power generation and supply, pp 1–5
22. Mohammed OA, Nayeem MA, Kaviani AK (2009) A laboratory based microgrid and distributed generation infrastructure for studying connectivity issues to operational power systems. In: IEEE PES general meeting, PES 2010
23. Cornforth DJ, Berry A, Moore T (2011) Building a microgrid laboratory. In: 8th international conference on power electronics—ECCE Asia: "Green world with power electronics", ICPE 2011-ECCE Asia, 2011, pp 2035–2042
24. Buso S, Caldognetto T (2014) Rapid prototyping of digital controllers for microgrid inverters. IEEE J Emerg Sel Top Power Electron PP(99):440–450
25. Sučić S, Havelka JG, Dragičević T (2014) A device-level service oriented middleware platform for self-manageable DC microgrid applications utilizing semantic-enabled distributed energy resources. Int J Electr Power Energy Syst 54:576–588

Co-ordination Control of Hybrid AC/DC Micro Grid Using High Frequency Transformer

K. Baskaran and G. Indhumathi

Abstract A Micro grid is a group of interconnected loads and distributed energy resources within a clearly defined electrical boundaries that act as a single controllable entity with respect to the grid. In AC and DC Micro grids multiple reverse conversions are required in an individual AC/DC loads. This may add additional losses to the system operation and make current home and office appliances more complicated. As a result, system co-ordination becomes too difficult in conventional micro grid. To overcome this drawbacks, hybrid micro grid which is an integral part of the smart grid is the most suitable solution to provide for the increasing penetration of DC-compatible energy sources, storage and loads which is recently prevalent in all Electric Power Industries. One of the most important feature of hybrid micro grid is advanced structure which can facilitate the connection of various AC, DC generation systems with optimal asset utilization and operation efficiency. It consists of both AC and DC networks connected together by multi bi-directional converters. AC sources and loads are connected to the AC networks. Similarly DC sources and loads are connected to the DC networks. Energy storage system can be connected to the DC or AC links. In this paper DC/DC converter with High frequency transformer is used to replace a normal conventional bulky transformer for bus voltage matching and galvanic isolation. Among different DHFT topologies, CLLC-TYPE has been suggested for its bidirectional power flow, seamless transition and low switching losses. DHFT open-loop control has been performed to simplify the systematic co-ordination and provide a smooth power transfer between AC/DC links. It has been designed in order to maximize the conversion efficiency and minimize the output voltage variations in different load conditions. Thus the hybrid micro grid has been simulated and analyzed using Simulink in MATLAB.

Keywords Hybrid AC/DC micro grid · DHFT · CLLC converter · Bidirectional power flow · Conversion efficiency

---

K. Baskaran (✉) · G. Indhumathi
Department of EEE, Alagappa Chettiar Government College of Engineering and Technology, Karaikudi, India
e-mail: drbaskaran@gct.ac.in

R. K. Pillai et al. (eds.), *ISUW 2019*, Lecture Notes in Electrical Engineering 764,
https://doi.org/10.1007/978-981-16-1299-2_3

1 Introduction

In modern years 3 Φ and 1 Φ AC power systems have been in existent for almost 100 years due to their efficient transformation of power at different voltage levels over long distance. In recent times more renewable power conversion systems are connected in low voltage AC distribution systems as distributed generators or AC micro grids due to environmental issues caused by conventional fossil fuel power plants. Similarly, more DC loads such as Electric Vehicles and Light-Emitting Diode (LED) are connected to AC power systems to save more energy and reduce CO emission. Long distance High voltage transmission is not necessary when power is supplied by local renewable resources [1]. To assist the connection of renewable power sources to conventional AC systems AC micro grids are used [2–5]. However, DC power from Photo Voltaic (PV) panels or fuel cells has to be converted into AC using DC/DC boosters and DC/AC inverters in order to connect to an AC grid.

In an AC grid, embedded AC/DC and DC/DC converters are essential for various home and office facilities to supply different DC voltages. DC Micro grid has become standard due to the increasing penetration of DC-compatible loads and Renewable Energy Sources. Recently, DC grids are resurging due to the development of renewable DC power sources and their inherent advantage for DC loads in commercial, industrial and residential applications. However, AC sources have to be converted into DC before connected to a DC grid and DC/AC inverters are required for conventional AC loads. Multiple reverse conversions required in individual AC or DC grids may add additional loss to the system operation and will make the current home and office appliances more complicated. The smart grid concept is currently prevailing in the Electric Power Industry. The objective of constructing a smart grid is to provide reliable, high quality electric power to digital societies in an sustainable way and environmentally friendly. One of most important features of a smart grid is the advanced structure which can facilitate the connections of various AC and DC generation systems, energy storage options, and various AC and DC loads with the optimal asset utilization and operation efficiency. To achieve those goals, power electronics plays a vital role to interface different sources and loads to a smart grid. In [6], describes a DC-micro gird based power generation system with solar cell generation and wind turbine generation also analyzing an autonomous-control method for a DC micro-grid system having distribution power generators. The Control method intended for suppression of circulating current detects only the DC grid voltage. In existing system which consists of a separate AC Grid, DC Grid and a conventional large transformer were multiple power conversion i.e. DC-AC-DC or AC-DC-AC are required. This will leads to power loss, complexity in circuit and less efficiency in the grid system [7]. On comparing with the different types of DHFT topology converter, few converters like the Dual Active Bridge converter (DAB), Resonant DAB Converter has attracted the most interests due to its simple structure, inherent advantage of Zero Voltage Switching (ZVS) and Zero Current Switching (ZCS) and better control flexibility (only unidirectional applications). But due to its inherent high turn off power loss which degrades the system overall efficiency. To make it bidirectional an extra resonant capacitor is needed.

To overcome these drawbacks, DC-DC converter High Frequency Transformer (HFT) with CLLC type converter has been replaced instead of normal large transformer.

2 System Configuration and Description of Block Diagram

The typical layout of hybrid AC/DC micro grid with the implementation of DHFT linked with bidirectional interlinked converter is shown in Fig. 1. DHFT boost the DC bus voltage to the magnitude such that the voltage at DC and AC outputs of BIC can be given with single stage topology. BIC is capable of operating in three modes namely, AC Voltage Regulation Mode, DC Voltage Regulation Mode and Power Dispatch Mode. Capability mode of operation of DHFT with BIC are illustrated in this Layout of Hybrid Micro grid as shown in Fig. 1.

On comparing with the different types of DHFT topology converters [7–10], few converters like the Dual Active Bridge (DAB) converter, Resonant DAB Converter has attracted the most interests due to its simple structure, inherent advantage of Zero Voltage Switching (ZVS) and Zero Current Switching (ZCS) and better control flexibility(only unidirectional applications) [11]. But due to its inherent high turn off power loss which degrades the system overall efficiency. To make it bidirectional an extra resonant capacitor is needed and switching loss is reduced. An extra resonant capacitor is added to the unidirectional LLC resonant converter to recognize bidirectional power flow while ensuring ZVS and ZCS [12–15].

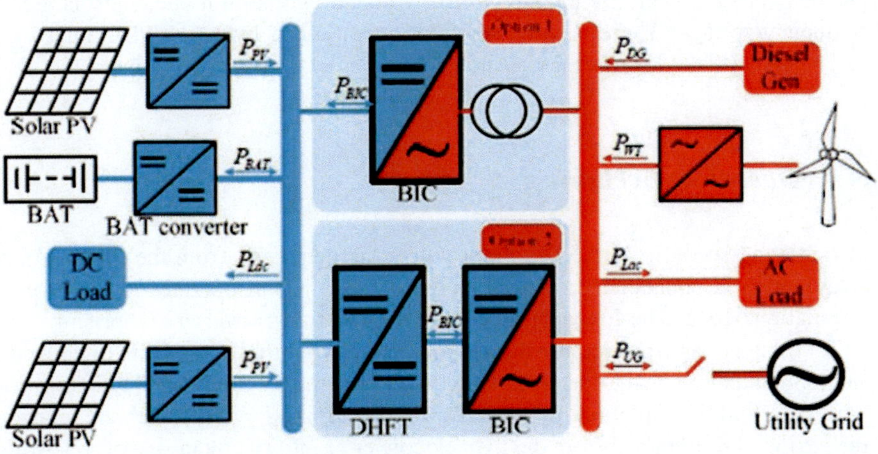

Fig. 1 Layout of hybrid micro grid using HFT

CLLC symmetric resonant topology has also been examined [16]. This topology ensures directional power flow, yet guarantees high-power conversion efficiency in both directions. The transitions between different power flow directions can be realized seamlessly.

The following advantages has been analyzed in the grid consisting of HFT with CLLC Topology:

- The Power flow is Bidirectional.
- Seamless transition.
- Reduce Switching Loss without Snubber—Inherent ZVS, ZCS.

3 Control of Hybrid AC/DC Micro Grid with DHFT

The schematic illustration of hybrid AC/DC micro grid with the enactment of DHFT has been demonstrated. DHFT which is innate with the Flyback converter boosts the DC bus voltage to the magnitude such that the voltage at DC and AC outputs of BIC can be coordinated. The control algorithm of BIC remains unchanged. Based on the availability of AC and DC sources BIC is expert of operating in three modes, namely AC Voltage Regulation Mode (VRM), DC VRM and Power Dispatch Mode (PDM). Capability of multiple-mode operation of DHFT, which is connected in series with BIC. An instant operating mode transition in accordance to BIC operating modes set a challenge for the communication link between the DHFT/BIC and central controller.

Therefore, the system control is simplified and the number of operating scenarios is reduced by preferring the open loop control system of DC-DC converter with HFT. The main control intentions of hybrid micro grid are the voltage regulations of AC and DC grid system. The BIC in DC Voltage Regulation Mode controls the DC bus voltage. Similarly the BIC in AC Voltage regulation mode controls the AC frequency/voltage. Therefore, the working operation of hybrid AC/DC micro grid can be categorized into 3 cases as shown.

3.1 Operating Scenario 1

In operating scenario 1, the DC input voltage is generated from the Photo Voltaic array and it is connected to the flyback converter which operates in DC Voltage Regulation Mode. The Flyback converter is an isolated converter comprising of the HFT which act as Bi-Directional Interlinking Converter (BIC). BIC is organized in such a manner that it will operate in PDM Mode and the LV Side of DC-DC Converter with HFT and the direction of Power flow is determined by BIC. Hence it converts the purified DC obtained from the flyback converter into AC by means of a H-Bridge Inverter. Thus the AC Output voltage is obtained in this operating scenario which is used to run all types of DGs and Wind turbine.

3.2 Operating Scenario 2

In this working operation an AC Input supply voltage is given, were the output is obtained in the previous working operation of the grid system. Now the H-Bridge inverter act as a converter by means of the logical NOT Operator and inverts the pulse. Here the AC voltage/frequency is controlled, while the HV side of DHFT is regulated and the power flow from AC to DC Micro grid through BIC occurs. Thus the DC Output voltage is obtained by converting the AC to DC by means of flyback converter which is used to run all types of DC Loads.

3.3 Operating Scenario 3

In this operating scenario 3, if neither the AC nor DC supply is available in the grid, DC voltage is synchronized by BESS and Bi-directional Interlinked converter and is organized to operate in AC voltage Regulation mode and control the AC micro grid frequency, voltage. It is similar to the operating scenario 1, were the LV side of the DC-DC converter with HFT is synchronized and hence the power flow in bidirectional is obtained in the hybrid micro grid.

The detailed design procedures are demonstrated in Fig. 2.

The schematic diagram of DHFT with CLLC topology is as shown in Fig. 2. The resonant circuit is comprised of the capacitors Cr1, Cr2 and leakage inductance Lr1, Lr2. Power is transferred from port AB to port CD with the magnetizing inductance, *Lm*1. The equivalent circuit diagram of power flow conversions between LV and HV sides is as shown in Fig. 2. *L*1, *C*1, *L*2 and *C*2 denote the equivalent inductance and

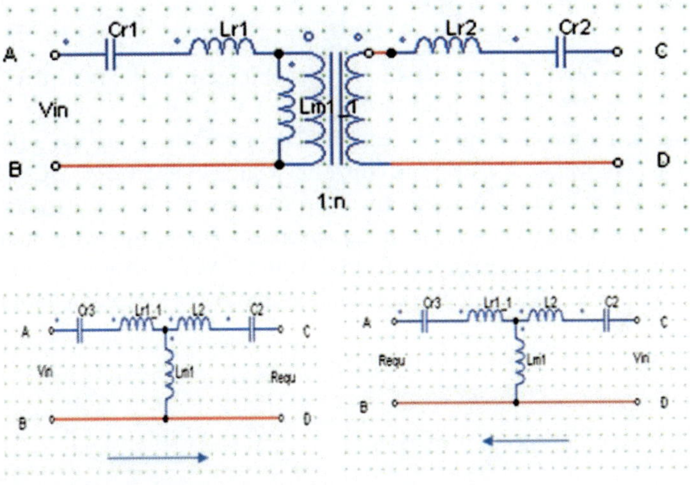

Fig. 2 Schematic diagram of DHFT with CLLC topology

capacitance of the primary and secondary sides, respectively. DHFT operates under the resonant condition to ensure the maximum power transmission and to reduce switching loss.

4 Simulation and Results Discussion

The simulation diagram for co-ordination control of hybrid micro grid is shown in Fig. 3.

5 Operation of Simulation

5.1 *DC Input for AC Load*

An input supply voltage of about 66 V generated from the PV array is given to the Flyback converter which is one type of the DC/DC or AC/DC isolated converter. The DC power from the Flyback converter is then filtered using an LC filter. Then the purified DC obtained is converted into AC by means of an H-bridge inverter were the MOSFET is used as a switch. Hence AC output of about 230 V is obtained which is used to run all types of DGs and wind turbine etc. In this project as an AC load, resistor load is used.

5.2 *AC Input for DC Load*

Thus from the output side of the simulation obtained in the hybrid AC/DC micro grid system, an AC input voltage of about 110 V is given. Now the inverter acts as a converter by means of the logical NOT operator, inverting the pulse. Now the AC obtained is given to the Flyback converter which converts AC/DC. Thus the DC output voltage of about 100 V respectively is obtained which is used to run all types of DC loads. Here battery is used for storage purpose. Therefore by using bidirectional converter with high frequency transformer, Hybrid AC/DC micro grid is simulated. This reduces the number of power conversions and reduce the switch loss and power loss in the system.

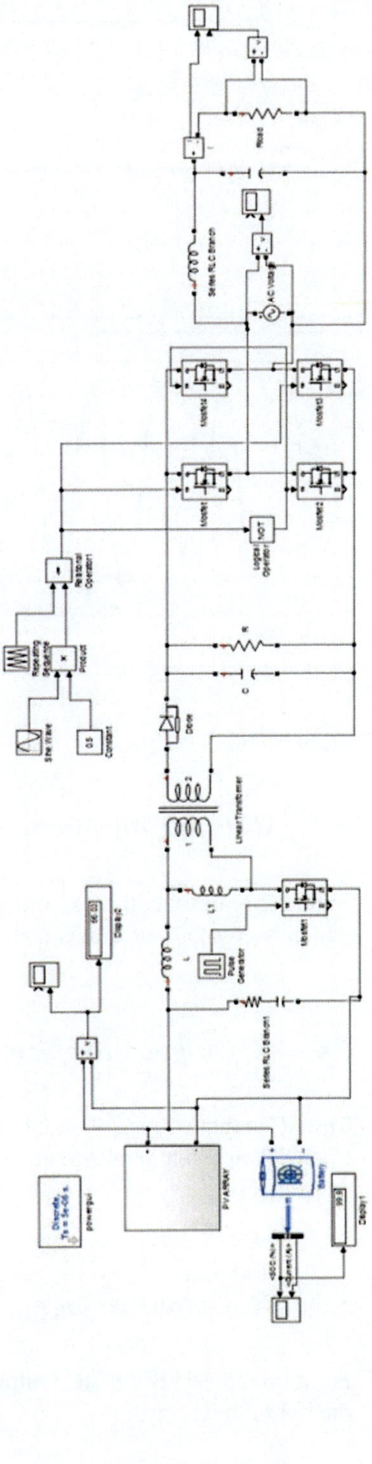

Fig. 3 Simulation of co-ordination control of hybrid micro grid

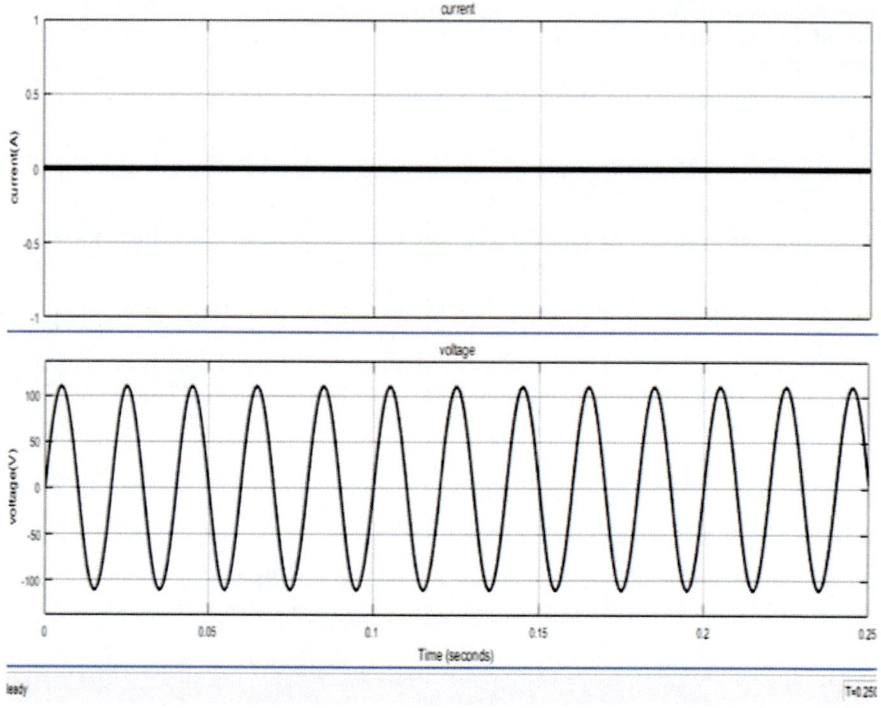

Fig. 4 AC input waveform AC input voltage: 110 V, frequency: 50 Hz

5.3 *AC Input Waveform*

Figure 4 represents the AC input voltage of about 110 V from the AC micro grid which is given along Y-axis and time period along X-axis.

5.4 *AC Output Waveform*

The AC output voltage of about 230 V and output current of about 0.24 A is obtained. Figure 5 represents voltage and current along Y-axis and time period along X-axis respectively.

5.5 *DC Output Voltage*

Figure 6 represents the DC Output Voltage of about 99.9 V which is obtained from the hybrid micro grid.

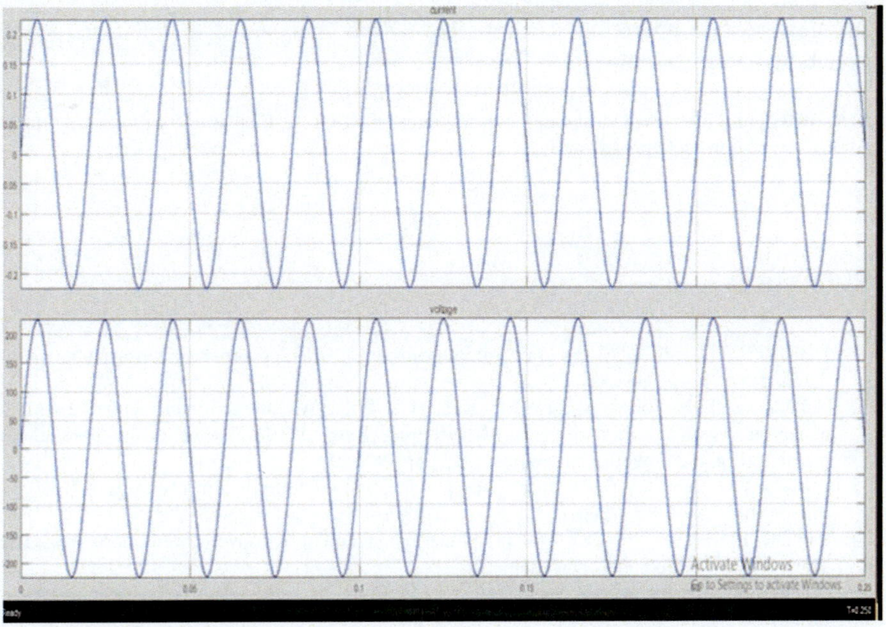

Fig. 5 AC output waveform

Fig. 6 DC output voltage

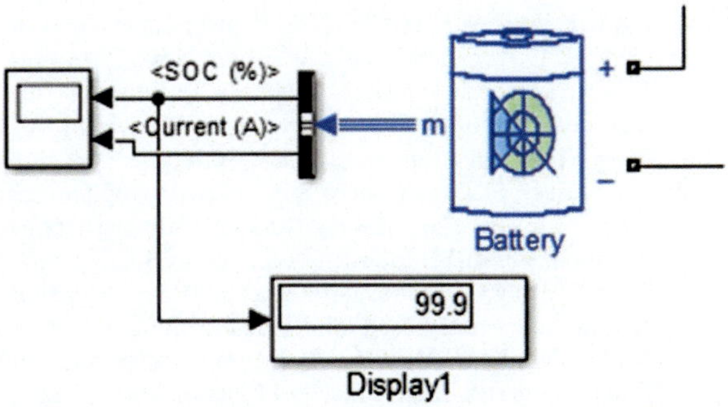

6 Conclusion

In this work, DHFT with CLLC topology has been proposed to replace the conventional large transformer in hybrid AC/DC micro grid for BIC voltage matching and galvanic isolation. Various operating scenarios have been analyzed and DHFT open-loop control has been suggested due to its to simplified system coordination and enhanced reliability. In this, DHFT with CLLC topology has been implemented in hybrid AC/DC micro grid for, higher conversion efficiency, Rated conversion ratio with minimum voltage variations. Therefore by using BIC converter bidirectional power flow is achieved in hybrid micro grid with minimum voltage variations.

Different system operating scenarios have been analysed with Enhanced system Reliability.

Acknowledgements I would like to thank my Supervisor Dr. K. Baskaran for motivating me and guiding me to complete project activities.

References

1. Lasseter RH (Jan 2002) Micro grids. In: Proceedings of IEEE power engineering society winter meeting, vol 1, pp 305–308
2. Zoka Y, Sasaki H, Yorino N, Kawahara K, Liu CC (2004) An interaction problem of distributed generators installed in a micro grid. In: Proceedings IEEE electric utility deregulation, restructuring and power technologies, vol 2, pp 795–799
3. Lasseter RH, Paigi P (2004) Micro grid: a conceptual solution. In: Proceedings of IEEE 35th PESC, vol 6, pp 4285–4290
4. Sao CK, Lehn PW (2008) Control and power management of converter fed micro grids. IEEE Trans Power Syst 23(3):1088–1098
5. Logenthiran T, Srinivasan D, Wong D (2008) Multi-agent coordination for DER in micro grid. In: Proceedings of IEEE international conference on sustainable energy technologies, pp 77–82
6. Ito Y, Zhongqing Y (2004) DC micro-grid based distribution power generation system. In: The 4th international power electronics and motion control conference
7. Liu X, Wang P, Loh PC (2011) A hybrid AC/DC micro grid and its coordination control. IEEE Trans Smart Grid 2(2):278–286
8. Eghtedarpour N, Farjah E (2014) Power control and management in a hybrid AC/DC micro grid. IEEE Trans Smart Grid 5(3):1494–1505
9. Fan H, Li H (2011) High-frequency transformer isolated bidirectional DC–DC converter modules with high efficiency over wide load range for 20 kVA solid-state transformer. IEEE Trans Power Electron 26(12):3599–3608
10. Du Y, Baek S, Bhattacharya S et al (2010) High-voltage high-frequency transformer design for a 7.2 kV to 120 V/240 V 20 kVA solid state transformer. In: IECON 2010-36th annual conference on IEEE Industrial Electronics Society, pp 493–498
11. Zhao T, Wang G, Bhattacharya S et al (2013) Voltage and power balance control for a cascaded H-bridge converter-based solid-state transformer. IEEE Trans Power Electron 28(4):1523–1532
12. Qin H, Kimball JW (2013) Solid-state transformer architecture using AC–AC dual-active-bridge converter. IEEE Trans Ind Electron 60(9):3720–3730
13. Chang CH, Chang EC, Cheng HL (2013) A high-efficiency solar array simulator implemented by an LLC resonant DC–DC converter. IEEE Trans Power Electron 28(6):3039–3046
14. Jiang T, Zhang J, Wu X et al (2015) A bidirectional LLC resonant converter with automatic forward and backward mode transition. IEEE Trans Power Electron 30(2):757–770
15. Chen W, Rong P, Lu Z (2010) Snubberless bidirectional DC–DC converter with new CLLC resonant tank featuring minimized switching loss. IEEE Trans Ind Electron 57(9):3075–3086
16. Chen W, Lu Z (2008) Investigation on topology for type-4 LLC resonant DC-DC converter. In: Proceedings IEEE transactions on power electronics specialists conference, pp 1421–1425
17. Jung JH, Kim HS, Ryu MH, Baek JW (2013) Design methodology of bidirectional CLLC resonant converter for high-frequency isolation of DC distribution systems. IEEE Trans Power Electron 28(4):1741–1755

Stop "Rooftop Solar Projects with Net-Metering" Switch to "Virtual Rooftop Solar Power Projects"

Ajay Chandak

Abstract Virtual Rooftop Solar Power Projects restores balance of subsidizing consumers and subsidized consumers for a DISCOM to operate. Innovation of Retail Solar Power Exchange provides operative mechanism for trading the credited solar power for the investors and other stakeholders.

Keywords Rooftop netmetering · Virtual rooftop solar projects

1 Introduction

Roof Top Solar (RTS) projects with different incentives have been promoted in developed countries in last 2 decades. India adopted the same since 2015 with net-metering policy. As solar power was very expensive till 2014 different incentives were needed. Higher power purchase tariffs were provided to bigger plants and on smaller scale RTS projects were promoted with different financial and tax incentives to fetch retail investment [1]. Situation has changed since 2014 with sudden drop in cost of solar panels, but the policies haven't changed in India. At present utility scale solar power projects provide power cheaper than coal [2] and hence continuation of incentivizing RTS net-metering scheme needs a serious review.

2 Issues with Rooftop Solar Net-Metering

In India, power is treated as essential commodity. People who afford are charged higher tariff with cross subsidy surcharge (CSS), while consumers below poverty level, farmers, charitable organisations etc. are provided subsidized power. The consumers are generally grouped in two categories, 'Subsidizing Consumers' and 'Subsidized Consumers'. Introduction of RTS net-metering scheme kills this very basic financing mechanism ending up subsidizing rich 'Subsidizing Consumers'.

A. Chandak (✉)
Prince, Suman Foundation, Dhule 424005, India

R. K. Pillai et al. (eds.), *ISUW 2019*, Lecture Notes in Electrical Engineering 764,
https://doi.org/10.1007/978-981-16-1299-2_4

Experience in Maharashtra, Gujarat and other states show that the subsidizing consumers with high tariff category are aggressively adopting RTS net-metering scheme for purely financial reasons and in most cases they end up with payback of less than 3 years. As these 'Subsidizing Consumers' practically do not purchase power, the DISCOMs are losing their revenue. The DISCOMs has to maintain grid supply to such consumers and incur all losses and fixed expenses. 'Subsidizing consumer' eventually becomes 'subsidized consumer'. DISCOMs in Maharashtra realized this problem and have approached electricity regulatory commissions to stop RTS net-metering and switch to 'Gross Metering'. If RTS projects continue at current speed then very soon the DISCOMs will start making huge losses on this account and will fail to subsidize the deserving consumers. Hence implementing RTS net-metering in current form will be disaster for the DISCOMs.

Government of India has set target of 40 GW [3] through RTS projects and the progress is not satisfactory. On Dec. 18, 2017 MNRE has come up with a draft to incentivise DISCOMs so as to complete the RTS targets. Financial impact analysis shows that if target of 40 GW is achieved then it will result in losses of 24,000 crore rupees per year for DISCOMs and accumulated losses over life of the project will be whopping 600,000 crore rupees. Calculations are shown in Table 1.

All DISCOMs are financially in a bad shape and GoI already launched schemes like 'UDAY' for their revival [4]. New schemes are also due to revive stressed power assets. This indirect additional subsidy burden because of RTS will be finally passed on to the consumers and will result in tariff hike of 25–50 paise per kWh. Loss of CSS by DISCOMs may result in withdrawal of subsidy by DISCOMs to subsidized consumers like farmers and BPL families and will result in large scale social unrest.

Other problems associated RTS projects are:

Table 1 Loss calculations of GoI and DISCOMs due to RTS

Target of RTS	40,000	MW
Subsidy by GoI	18,000	Per kW
Target residential with CFA	5000	MW
CFA by GoI for residential	9000	Crore Rs
CFA by GoI to DISCOMs	14,450	Crore Rs
Commercial and industrial target	20,000	MW
Tax rebate	4800	Crores Rs
Total revenue loss of GoI CFA + tax rebate	28,250	Crores Rs
Recurring losses for DISCOM		
Power generation through RTS	60,000,000	MWh/year
Average power purchase cost	4	Rs/kWh
Average selling cost to subsidizing consumers	8	Per kWh
Anticipated losses by DISCOM on RTS	4	Rs/kWh
Annual losses by DISCOM	24,000	Crore Rs/year
Cumulative losses by DISCOM over 25 years	600,000	Crore Rs

(a) Our grids are designed for one way transportation of energy. From power plants to grid and from grid to consumer through distribution system. RTS generates power at tail end of the grid and practically causes two way transmissions for which the grids are not designed. Distributed generation makes the task of grid management more and more difficult and we really need a smart grid; which is not in sight in near future.
(b) Most of the people advocating for RTS projects put saving of distribution losses as one of the major advantage. It's not so. Approximately 5% saving is feasible in distribution losses by rooftop project, however such projects produce 10–25% less power than centralized projects and all the advantage of saving in losses is eaten up.
(c) Small systems are executed by small companies and one faces all issues of technical competency, quality and reliability. Many consumers are just not competent to maintain such systems.
(d) Insurance companies do not insure performance of RTS systems as they do in centralized bigger projects.
(e) Capital investment in RTS projects is 20–40% higher than that in centralized MW scale projects. Studies in USA indicate that levelised cost of power through RTS is double that of centralized MW scale projects.

For all these reasons, it makes sense to stop RTS power projects completely. The biggest advantage in RTS projects was to get retail investment in capacity building of solar power. The same can still be achieved by adopting author's innovation of "Virtual Rooftop Solar Power Projects" where retail investment is fetched to setup a grid interactive utility scale projects and the investors get their share of power through the same grid. The business model benefits all stakeholders, especially DISCOMs so that it creates a Win–Win Situation. This innovation can complete 40 GW target in no time.

3 Innovation: "Virtual Rooftop Solar Power Projects"

Author has proposed this innovative business model; "Virtual Rooftop Solar Power Projects". This model fetches huge retail investment from retail investors who get solar power credits as return. As the innovation provides returns to all stakeholders in the form of share in generated power. It creates a Win–Win situation. Biggest beneficiary is DISCOM. It gets 30–50% share in power generated which it can use for subsidizing the deserving categories of consumers. Also 100% Renewable Energy Certificates (RECs) are owned by DISCOM to fulfill their Renewable Power Obligations (RPO).

Mechanism of "Virtual Rooftop Solar Power Projects" is explained below.

Capital for the project is raised from retail investors while other stakeholders pitch in for the services they provide. Other stakeholders include land owner, O & M company, innovator, project developer and DISCOM. All these investors and stakeholders are issued solar power cards against their investments or services as shown in Fig. 1.

As all investors can participate in the project who does not own a physical rooftop but get all benefits as if the owner of conventional solar rooftop projects and hence the name of the innovation is "Virtual Rooftop Solar Power Projects".

Each solar power card has unique consumer ID corresponding to stake in the project. All capital raised from the retail investors is invested in utility scale solar power plant.

Once the plant is ready and start generating power, the generated power is recharged in the solar power cards in proportion of the stake of the investor or other stakeholder on monthly basis. This is shown in Fig. 2.

Solar power card holder can set off the available recharge power in the solar power card with his/her own power bills on the same lines as that of netmetering polity followed for RTS projects. The difference in this innovation is that the DISCOM also gets substantial share in the generated power in the form of CSS (Cross Subsidy Surcharge). For e.g. If a residential consumer sets off his power bill with credit of say 100 units from his solar power card, then say 40% of the credit goes to DISCOM and

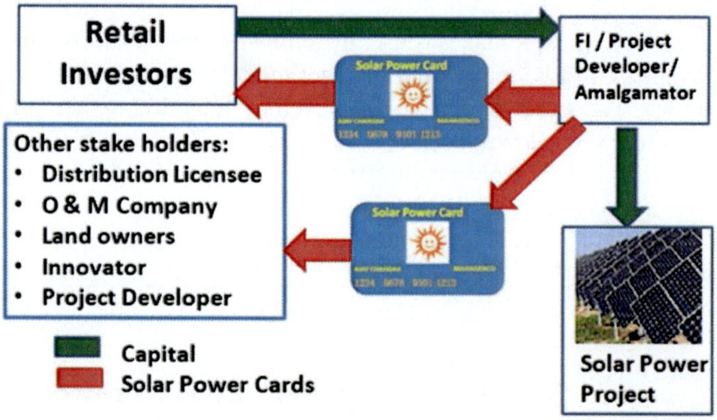

Fig. 1 Capital flow and issue of solar power cards

Fig. 2 Recharge of power in solar power cards

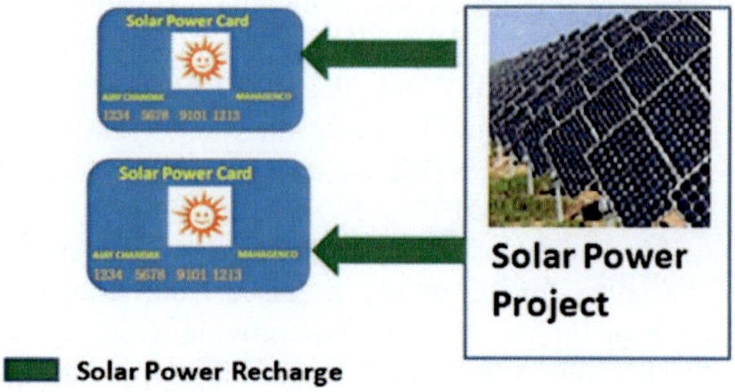

60% to the residential consumer. So DISCOM gets 40% units while the consumer gets set off of 60% units from his bills. The ratio can be something like 30–70% in case of industrial consumer and 50–50% for commercial consumer. DISCOM gets large amount of share in generated solar power at zero cost. This power can be utilized for subsidizing consumers.

Unlike conventional RTS netmetering where the consumer gets 100% credit for the power generated, the DISCOMs lose huge revenue, while in Virtual rooftop solar power projects, DISCOMs get large chunk of power from the solar power generators at no cost. This share of power can be used by DISCOM for subsidizing power to the consumers like BPL category, farmers and charitable organisations. Innovative concept of Virtual rooftop solar power project puts back the balance between subsidizing consumers and subsidized consumers.

Operating mechanism for implementation of the innovation is a solar power card or solar valet. The billing system will operate in the same manner as that of existing RTS net-metering system and no challenges are envisaged at this level.

4 Retail Solar Power Exchange

Further innovation of developing "Retail Solar Power Exchange" will open up solar market completely with 100% transparency and flexibility. The mechanism of retail solar power exchange is shown in Fig. 3.

Under this concept of a 'retail solar power exchange' the investors and stakeholders get flexibility to trade the solar power credits in a market and the rates are decided on more transparent manner. Amount of CSS can be fixed by electricity regulatory commissions for different category of consumers. For e.g. the CSS can be 30% for industrial consumer, 40% for residential and 50% for commercial consumers.

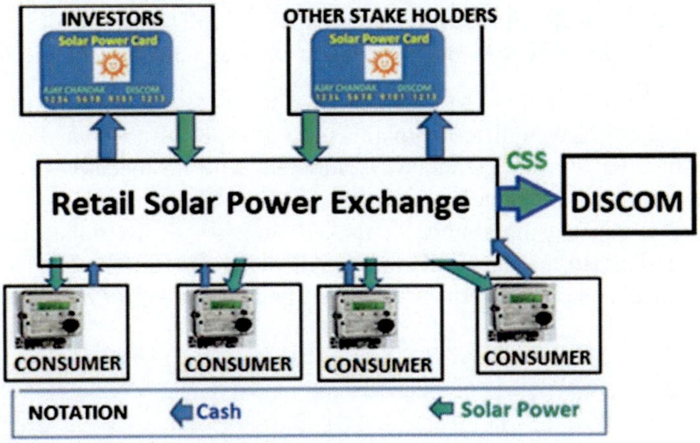

Fig. 3 Concept of solar power exchange

Solar power card holder can trade the solar power credits in the retail solar power exchange and whenever any such transaction occurs power corresponding to the CSS will be credited to the account of DISCOM while balance power will be credited to the consumer who purchases the power from the solar card holders. Payment and delivery mechanism of solar power credits will be exactly on the same lines of share market; solar power credits will be traded in DMAT (Dematerialised) form and payments are made to the seller.

5 Conclusions

The innovative business model has following features which are in line with the Indian Electricity act and current policies of Government of India.

(i) Virtual Rooftop solar power projects is the only business model that provides access to solar power to 100% consumers irrespective of the fact whether they possess a suitable rooftop or not. It also provides opportunity for retail investors for capacity building in solar power segment.
(ii) New proposal reduces the financial losses of DISCOMs to great extent and give them justified returns of minimum 30% plus 100% RECs.
(iii) New proposal provides compensation for DISCOMs for wheeling and CSS. Existing rooftop model does not provide the same.
(iv) New proposal is consumer friendly as the tariff hike of 20–50 paise can be averted or at least reduced to large extent.
(v) 100% flexibility, liquidity and transparency for the investors.
(vi) Virtual Rooftop Solar proposal will separate "Grid" and "Business". Energy minister has shown his intent to do so.
(vii) There is no better model to comply solar power obligation of 10% under clause 6.3 of "Smart Cities".
(viii) The innovation provides better capacity utilization by the solar power companies who can do business at will and need not wait for winning contracts.

Target of 100 GW will look smaller if "Virtual Rooftop Solar Power Projects with Retail Solar Power Exchange" is adopted, with no incentives required from Government of India or state governments.

This innovative business model is ranked amongst top 10 global innovations in 'Renewable Transformation Challenge' by 'Elsevier' and 'International Solar Energy Society' and also has many other credentials and is awaited for implementation.

References

1. Article in Fortune (2016) http://fortune.com/2016/09/12/solar-panel-farms-boom/
2. Study report of Brattle Group USA (2015) http://brattle.com/news-and-knowledge/news/study-by-brattle-economists-quantifies-the-benefits-of-utility-scale-solar-pv
3. CERC order on petition 17/SM/2015 dated 23-3-2016 & petition 3/SM/2016 dated 29-4-2016
4. Power Finance Corporation report for performance of state utilities: 2006–2014

Smart Battery Management System for Enhancing Smart Micro Grid Performance and Energy Management

Rashi Gupta, Bharat Gupta, and Uday Mumbaikar

Abstract Energy storage system (ESS) is an essential component of smart micro grid for compensating intermittent renewable generation and continuous power supply. Batteries are most commonly used in ESS. For optimal energy management of micro grid, the optimization algorithm needs knowledge of battery parameters like state of charge (SOC), voltage, temperature etc. Further for implementing various control and stability strategies, there is need of communication of battery parameters among various components of micro grid. With knowledge of battery parameter, grid operator can make better utilization of available ESS resources and also reduce renewable curtailment. A smart battery management system (BMS) is developed which calculates and communicates battery parameters. Various communication protocols namely Modbus, CAN, Ethernet and Wifi are incorporated in the smart BMS which makes it compatible for many applications. Smart BMS additionally performs active cell balancing using cell to cell balancing topology. The BMS is successfully implemented in a smart micro grid in India and the findings of the implementation are discussed in this paper. They serve as the foundation for further implementation of optimal energy management in the smart micro grid for minimizing operation costs.

Keywords Smart micro grid · Battery management system · Optimization · Energy management system

R. Gupta (✉) · B. Gupta · U. Mumbaikar
Vision Mechatronics Private Limited, Thane, India
e-mail: info@vmechatronics.com

B. Gupta
e-mail: info@vmechatronics.com

U. Mumbaikar
e-mail: info@vmechatronics.com

R. K. Pillai et al. (eds.), *ISUW 2019*, Lecture Notes in Electrical Engineering 764,
https://doi.org/10.1007/978-981-16-1299-2_5

1 Introduction

The growing environmental concerns and depleting reserves of fossil fuels has led to an increase in penetration of renewable generation [1]. It is leading to an increase in Distributed Generation (DG) such as photo-voltaic, wind turbine, diesel/gas generator, etc. in the grid. The traditional grid needs to be upgraded for handling these DGs. Decreasing prices of renewable generation has increased micro grid installations worldwide. Micro grids (MG) are seen as a solution for large scale integration of DG in the distribution systems [2]. Micro grids are often resorted to for supplying remote areas where gird cannot reach [3]. Micro grids are being increasingly used for electrification in developing countries, where many area still do not have access to reliable electricity.

Micro grid mainly consists of distribution network made up of electrical loads, DGs (predominantly solar and wind generation) and energy storage systems capable of operating autonomously in stand alone mode or with grid connection [4]. MG offer benefits including higher efficiency, enhanced compatibility, increased reliability, lower environmental impact and solution for growing demand [5]. MG management is a challenging task due to the intermittent nature of renewable generation and an increase in dynamic loads in system. For enabling maximum efficiency of energy utilization, renewable generation is normally controlled with maximum peak power tracking algorithms [6]. Also, it is a non-dispatchable generation due to the fast changing and uncontrollable nature of weather conditions. The presence of non-dispatchable generation and varying load demand lead to issues in active power balance of load, generation and energy storage. The operation of micro grid is thus controlled with an Energy Management System (EMS) which ensures its reliable, secure and economical operation in both grid-connected or stand-alone mode [4].

With advancement in information and communication technology grids are becoming smarter. Smart micro grid enables secure and optimal operation of potentially islanded system. But for implementing smart micro grid control strategies like EMS, there is a need of communication between components of micro grid [4]. A number of communication protocols are employed in a MG and components of MG must be equipped with required communication protocol for being compatible to operate in MG. Sensors send data to Intelligent Electronic Devices (IEDs) which issue control commands to components of MG. The MG controller may use Modbus compliant with IEC 61850 standard for communicating over Ethernet using TCP/IP. Employing internet communication protocol suite, a secure and reliable communication between components of MG can be ensured. Human Machine Interface (HMI) clients are also expected to be used for monitoring and controlling requirements, additionally a data logging server and an event recorder is also expected. Figure 1 illustrates the data flow between the microgrid components.

IEDs receive power system data from DGs, ESS and load which is transmitted as feedback to MG controller. This data is utilized by MG controller for issuing control signals and reference values of voltages, frequency, active and reactive power to IEDs. IEDs consequently provides control signals to DG, ESS and load.

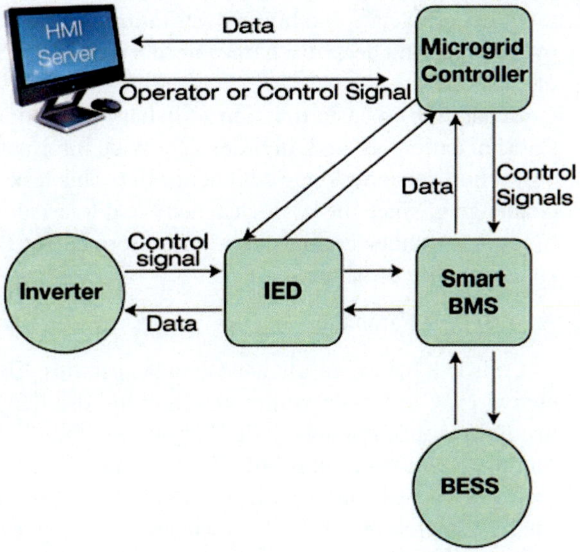

Fig. 1 The data flow between the microgrid components

Energy storage system (ESS) is an essential component of smart micro grid for compensating intermittent renewable generation and continuous power supply. It reduces need of diesel generation in micro grid and helps in optimizing the cost of operation. ESS is implemented with many different technologies like pumped hydro, fly wheels, batteries, capacitors etc. Battery energy storage systems have been found most suitable for micro-grid considering their efficiency, energy density, response time, discharge duration, depth of discharge, lifetime cycle capacity, etc. [7]. Batteries are made up of cells and each cell needs to operate within its safe operating limits for the battery to have long life. A Battery management system (BMS) ensures safe and optimal operation of batteries. In this paper a smart BMS is developed for using battery energy storage in a smart microgrid.

2 Battery Management System

The performance of battery depends on the chemicals inside the battery. With time and usage the chemicals in battery undergo degradation and the energy storage capacity of battery also reduces. The battery charging and discharging profile needs to be controlled under various load conditions for curtailing the battery depreciation process [8]. Operating conditions like frequent charge and deep discharge cycles, wide range of operating temperature and high current pulses on battery diminishes its life. Now-a-days, Li-ion chemistry is being preferred due to its good energy density, long life, efficiency and high power rating. Li-ion chemistry being very sensitive to overcharges and deep discharges need a proper BMS for safe and reliable operation of each cell.

In addition to basic function of battery protection, the BMS needs to determine status of battery as well, in order to provide information about its energy supply and absorption capacity to the MG controller. This task of determining battery status is challenging, since the usable capacity and internal resistance of battery varies over time [9]. Another crucial function performed by BMS is cell level balancing for enhancing life of battery.

A. Battery Parameter

Cells in a battery rarely have equal capacities. This causes a mismatch in state of charge (SoC) of cells while charging and discharging. Battery protection system needs to stop charging or discharging as soon as even a single cell reaches its minimum and maximum SoC limit. However, due to this all cells might not reach their full state of charge. It is often found that cells with higher voltage have higher SoC. Mismatch in cell SoC affects the performance of battery. The battery protection circuit stops battery charging before reaching its full charge voltage due to any one cell reaching its maximum SoC. Similarly, during discharging the battery stops discharging even before reaching its maximum discharge limit. Mismatch in cell capacities is the cause of underutilization of battery. It needs to be addressed with appropriate cell balancing technique for improving the utilization capacity of battery and its cycle life [10]. BMS performs the function of cell balancing along with protection of battery.

BMS needs majorly battery current, voltage and temperature measured over time as input. Using these inputs the BMS performs battery protection and estimation of battery state of charge (SoC), state of health (SoH) and state of function (SoF). Additionally, it also performs the tasks of controlling the heating/cooling subsystem and main power switch. It also ensures the isolation from high voltage when used in high voltage application by implementing isolated communication.

Another requirement from BMS is that of meeting accuracy and synchronization of current and voltage measurements of the battery pack and its cells. Accuracy targets for current measurement up to 140 A are typically 0.5–1% and 1–2 mV or 0.1% in case of cell and pack voltage measurements. Such stringent voltage accuracy demand is mostly driven by $LiFePO_4$ chemistry. Since, it has very flat voltage versus state-of-charge profile which make it difficult to estimate SoC in 80–20% range.

The accuracy of BMS measurement is affected by error sources such as variation in shunt resistance, amplifier gain and Analog-to-Digital Converter (ADC) reference over temperature and time. The BMS accuracy must be maintained since recalibration normally is not a feasible option. Accurate predictions over lifetime can be achieved with comprehensive qualification tests. Biased high temperature operating conditions which cause pre-aging in electronics also need to be considered for maximizing long term accuracy.

B. BMS architecture

BMS architecture depends on physical structure of battery used. High power application requires over one hundred cells to be connected in series. Normally modules consisting of 4–16 series connected cells are combined to form the higher voltage string of cells. Thus, battery can be viewed as three layer structure namely the elementary cell, the module and the overall pack. The inner most layer is of cell monitoring with Cell Monitoring Unit (CMU) for each cell in the module. The middle layer comprises of Module Management Unit (MMU) one for each module. The monitoring data from CMUs in module is used by MMU for providing services to the Pack Management Unit (PMU) whose function is to supervise all the modules. Each CMU can be connected to each MMU using a dedicated and custom bus. Normally communication between PMU and the MMUs is implemented using Serial Peripheral Interface (SPI) or a shared galvanic-isolated Controller Area Network (CAN) bus. Furthermore, the communication between battery and other control systems like inverters etc. is also normally using CAN bus.

Battery is protected against overcharge, deep discharge and over-temperature usually by breaking the battery current flowing through the Main Switch (MS) contactor/High power Relay. This is also controlled by the Smart BMS.

C. Battery Balancing

Battery balancing, one of the most important function of BMS, can be performed with a number of approaches. In passive balancing technique, excess energy in a cell with higher SoC is dissipated in bleeder resistor. This method is inefficient as it wastes energy in bleeder resistor which also causes increase in temperature. Active balancing technique on the other hand transfers energy from the cell with higher SoC to a cell with lower SoC which makes it more efficient than passive balancing. There are different methods of energy transfer between cells in active balancing techniques. However, for making active balancing preferable over passive balancing, a trade-off between complexity of active balancing circuit and its efficiency needs to be found. Passive balancing technique can be easily implemented using just one controlled switch and a resistor. MMU includes the hardware implementation of charge equalizer whereas supervising the overall balancing procedure is done by PMU. It estimates the SoC of each cell and controls the amount of charge to be stored in it. However, as battery capacity decreases with time, aging effect needs to be taken into account for accurate SoC estimation.

One of the method of estimating SoC is by coulomb counting assuming capacity of cells is known. Coulomb counting is performed by integrating the battery current over time. It is usually used in low power applications like portable consumer devices. Coulomb counting method is affected by measurement errors particularly in current sensor. Using relation between SoC and open circuit voltage (OCV) is another method of estimating SoC. In this method errors get introduced with error in OCV measurement due to its dependence on accurate OCV. Other methods like discharge tests, neural networks, internal resistance measurement are also available [11]. Model-based algorithm (like Kalman filters) are found to be more suitable for online SoC estimation [12]. Since it does not requires long tuning times. However, an accurate cell behavior model over its life time also needs to be developed for model-based methods. Temperature effects must also be considered while modeling the cell behavior [13].

3 Smart Battery Management System

The smart BMS developed in this work is used for monitoring 48 V battery containing 16 $LiFePO_4$ cells as shown in Fig. 2. It determines individual cell voltage, current, temperature, SoC and SoH. It protects the cells against overcharge, over discharge, over current, over temperature and short circuit. It is also possible to set depth of discharge and charge levels of the battery. It works as MMU in high voltage packs for up to 128 cells in series and one of the BMS acts as both MMU and PMU.

The smart BMS supports active cell balancing and is capable of balancing cells within a single module and also in different modules. It supports CAN bus communication which makes it compatible with inverters commercially available.

Fig. 2 Smart BMS connected to Li-Rack battery

4 Smart Microgrid in India

A smart MG is installed in Goa in India which comprises of 10 kWp solar generation, battery energy storage system (BESS) of 11.2 kWh, diesel generator of 10 kW, load and utility grid. The developed smart BMS is implemented in this MG successfully. The ethernet port on the smart BMS enables monitoring the battery energy storage and controlling its working as well. It can be paired to a SCADA system using MOBDUS protocol over ethernet port. Further, pairing to Wifi network though ethernet port is also possible. The MG controller can control the battery energy storage by communicating with the smart BMS. The inverter used in MG is also smart inverter and is capable of being controlled by MG controller and battery. Thus, using MG controller it is possible to implement optimal management of energy for minimizing operating costs of MG.

5 Energy Management System for Minimizing Operating Cost of Microgrid

The supply of grid is unreliable and battery energy storage or diesel generation is used to supply the load while it is unavailable. The cost of energy generated by a diesel generator is higher than grid supply. Hence, it is economical to use BESS during outage as compared to diesel generator. Also, solar generation is installed which further reduces the cost of operation. The presently used EMS is Rule-Based, wherein solar generation is priority and is used for supplying load with BESS and grid. Proportion of grid supply in total power to MG increases as solar generation reduces. In absence of solar generation load is supplied with utility grid. For backup, the battery SoC is maintained around 90%. During grid outages and absence of solar generation, BESS is used to supply the load and when BESS SoC reaches minimum limit, diesel generator supplies the load.

The battery is charged back to 90% when utility grid supply is available again. The data of MG operation on a random day are displayed in graphs in Figs. 3, 4, 5 and 6.

Readings obtained from smart MG reveals that the solar generation is wasted, if it is more than load and battery is also fully charged. Daily average energy which solar panel can generate is found to be 40 kWh of which an average of 22 kWh of energy is utilized. BESS needs to be at lower SoC for storing the excess solar generation. Whereas, for minimum use of diesel generation during grid outage, battery needs to be maintained at full SoC. Thus, EMS needs to maintain optimal SoC for minimizing use of diesel generation and maximizing use of solar generation. Optimal operation of MG is achieved by minimizing the consumption of energy from diesel generation and grid.

In this system, solar generation is assumed to be free of cost. Charges need to be paid for using energy from grid and diesel generator. A day-ahead forecasting of

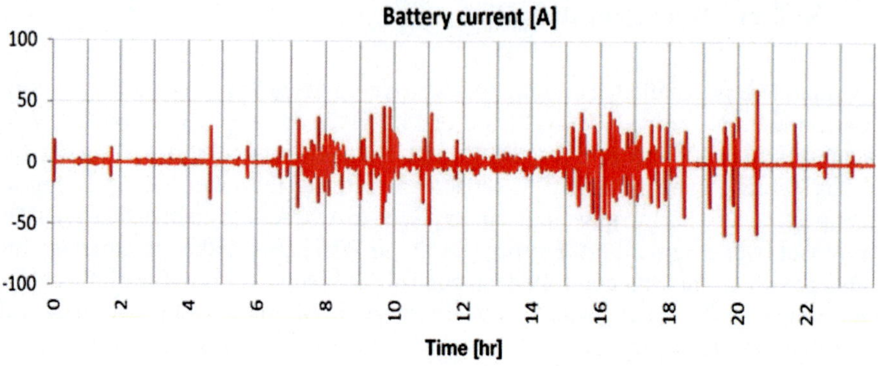

Fig. 3 Battery current

Fig. 4 Battery voltage and SoC

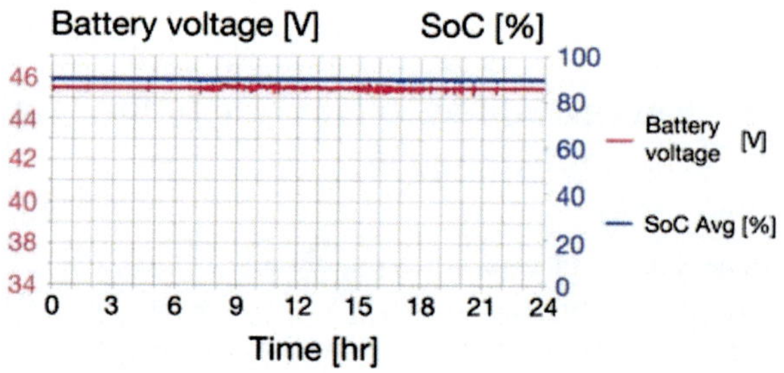

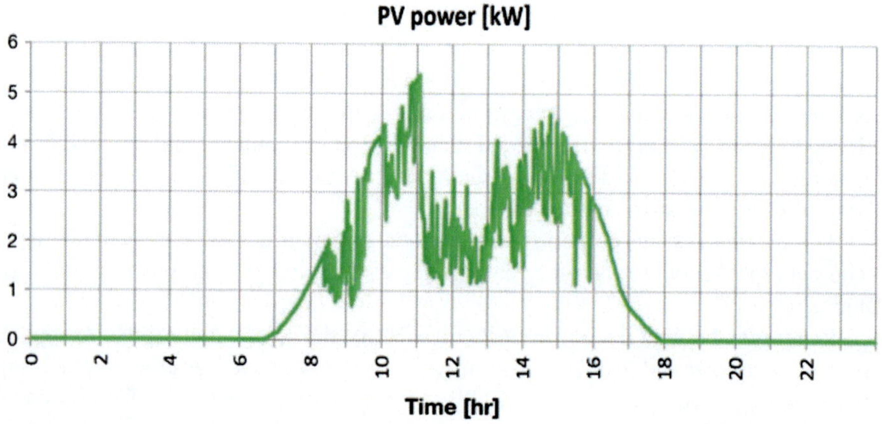

Fig. 5 PV power consumed by load

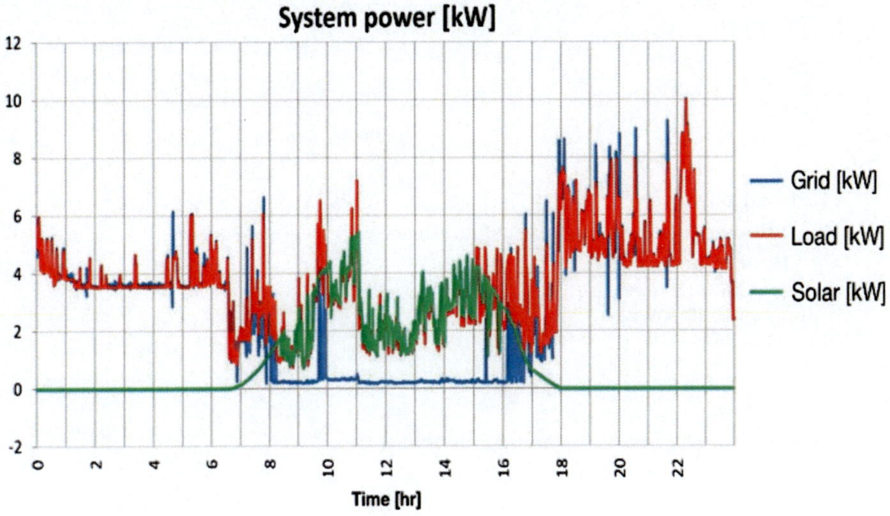

Fig. 6 System power

load and solar photovoltaic (PV) generation [14], is used for scheduling sources to supply load. The optimal energy management of MG can be achieved using Bellman algorithm by scheduling sources for minimizing cost of operation [15].

A. Objective Function

The total cost of operating MG for one day is denoted by Cost in (1) and the objective is to minimize the value of Cost.

$$\min(Cost) = \min \sum_{t=0}^{T} P_g(t) \cdot T_g + Dsl(t) \tag{1}$$

P_g : Utility grid power
T_g : Utility grid tariff
Dsl : Cost associated with probability of using diesel generator.

The discharging of battery increases probability of using diesel generation. $Dsl(t)$ can be calculated as product of probability of using diesel generation and cost of supplying average load with diesel generation. Thus, as battery SoC decreases value of $Dsl(t)$ increases.

The data obtained from the implemented smart MG is used for calculating probability of using diesel generation as function of time of day and battery SoC. The probability function will be updated with time to give more accurate results. Similarly, average load is also derived using the data and regularly updated with time.

B. *Constraints*

Power Balance constraints

$$P_l(t) = P_g(t) + P_B(t) + P_{PV}(t) \tag{2}$$

P_t : Load power
P_{PV} : Power supplied by solar generation
P_B : Battery power.

Battery output power

$$P_{B\min} < P_B < P_{B\max} \tag{3}$$

State of charge of battery can be calculated as

$$Soc = \frac{C(t)}{C_{ref}} \tag{4}$$

where $C(t)$ is instantaneous capacity of battery and C_{ref} is the reference capacity.
The SOC variation constraints are:

$$\Delta SoC_{\min} < \Delta SoC(t) < \Delta SoC_{\max} \tag{5}$$

SoC constraints are taken as

$$SoC_{\min} < SoC(t) < SoC_{\max} \tag{6}$$

Minimum SOC of 20% needs to be maintained for diesel generator to reach its full capacity while BESS supplies load. Maximum Solar generation can be of instantaneous load and BESS charging power and excess generation will be wasted.

$$0 < P_{PV}(t) < P_l(t) + P_B(t) \tag{7}$$

6 Optimisation in EMS

The MG optimization problem can be seen as Multi-Stage Decision Problem (MSDP) [16]. Each step of MSDP consists of number of system states which determine the value of current stage. On each state of a stage, a set of decision variables acts for generating new states which belong to next stage. The aim of using Bellman algorithm is to find the optimal path of state transition which minimizes summation of cost of all stages.

In the current optimization problem SoC of battery acts as decision variable. The Bellman algorithm method as proposed in [15] can be used for finding the sequence

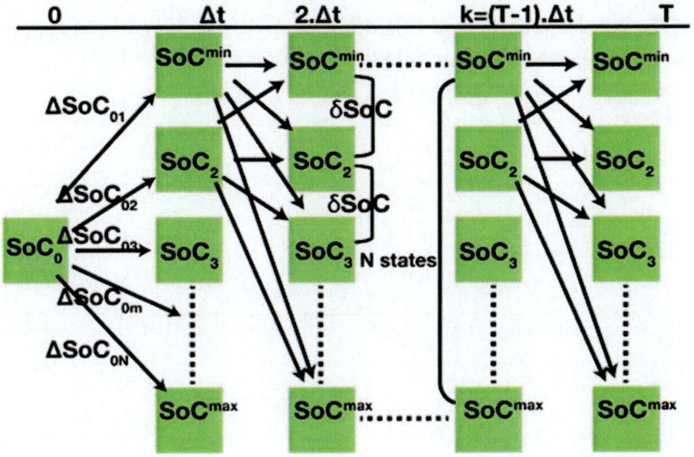

Fig. 7 Paths of SoC variation

of SoC variation in discrete steps such that Cost value is minimum. A set of variables determine the state of system at each time. The system is discretized with a time step of Δt and SoC is varied with a step size of ΔSoC [15]. At each stage, change in SoC is used for calculating the battery power (P_B) for next stage. Using calculated P_B and forecasted solar generation and load (P_l), maximum possible P_{PV} is determined. After which P_g required for maintaining power balance is inferred. And finally cost of each stage is calculated.

The initial value of SoC in the first stage is taken as the instantaneous value of SoC at the start of simulation. It is taken as source vertex for implementing Bellman algorithm. And with time step of one hour there are 24 stages remaining.

Each of the remaining stage has same number of N vertices with each vertex varying increasingly by step size δSoC from $SoC_{\min}$ to $SoC_{\max}$. The vertices of the 2nd stage are obtained from the source vertex by adding SoC variations ΔSoC (Fig. 7). The set of vertices $\psi(1)$ which satisfy the constraints in Eqs. (2)–(7) are used for finding vertices in next stage. The value of $Cost(i, t)$ corresponding to the vertices in $\psi(1)$ are also calculated and is taken as weight of path for reaching that vertex. Using each vertex in previous stage $\psi(i)$, new vertices are obtained by adding the SoC variations (ΔSoC) and verifying whether they satisfy Eqs. (2)–(7). In this way valid vertices are found in each stage and the cost calculated at each vertex is taken as weight for path of traversing from vertex in previous stage, by which it reached current vertex. Using Bellman algorithm, minimum weight path of reaching each vertex in the last stage is found. The path with minimum weight ($\min(Cost)$) is taken as optimal path. The SoC transition along this path gives optimal day ahead schedule of sources and storage components for minimum operating cost.

Using this schedule as guide, energy management is performed by MG controller for the day and the controller simultaneously calculates optimal schedule for the next day. The implementation of this optimal energy management is possible due to features of smart BMS which enable determining status of BESS and controlling it.

Simulation parameters

Name	Value	
T	24	H
Δt	1	H
δSoC	0.001	Pu
$SoC(t_0)$	0.5	Pu
$SoC_{\min}$	0.2	Pu
$SoC_{\max}$	0.9	Pu
$\Delta SoC_{\min}$	−0.7	Pu
$\Delta SoC_{\max}$	0.7	Pu
P_{Bmin}	−11.2	kW
P_{Bmax}	11.2	kW

Thus, scheduling of sources and storage components for optimal energy management is developed and being implemented in the smart MG using the smart BMS.

7 Conclusion

The smart BMS developed in this work accurately measures and calculates essential battery parameters like battery voltage, cell voltage, battery SoC etc. It communicates the calculated parameters using CAN, Ethernet, MODBUS and Wifi communication which makes it compatible with components of smart MG, and aids the decision making of MG controller by providing accurate and appropriate data as compared to the estimations by the other devices connected to the BESS system. It also ensures operation of BESS within safety margins. The smart BMS increases the efficiency of BESS with active balancing technique. It also increases the life and performance of BESS, thereby enhancing performance of MG as a whole. This BMS is successfully implemented in a smart MG in Goa. The data communicated by the smart BMS and other components of MG, form the basis for the MG controller to implement optimal energy management. The SoC of battery is controlled for implementing optimal energy management, which is possible only due to the accurate calculation and communication of battery SoC by the smart BMS.

References

1. Zia MF, Elbouchikhi E, Benbouzid M (2018) Microgrids energy management systems: a critical review on methods, solutions, and prospects. Appl Energy
2. Katiraei F, Iravani R, Hatziargyriou N, Dimeas A (2008) Microgrids management. IEEE Power Energy Mag 6(3):54–65
3. Battaiotto PE, Cendoya MG, Toccaceli GM, Vignoni RJ (2017) Stand-alone hybrid microgrid for remote areas topology and operation strategy. In: URUCON, 2017 IEEE, pp 1–4
4. Valencia F, Collado J, S´aez D, Mar´ın LG (2016) Robust energy management system for a microgrid based on a fuzzy prediction interval model. IEEE Trans Smart Grid 7(3):1486–1494
5. Xu Y, Zhang W, Liu W, Wang X, Ferrese F, Zang C, Yu H (2014) Distributed subgradient-based coordination of multiple renewable generators in a microgrid. IEEE Trans Power Syst 29(1):23–33
6. Xu Y, Shen X (2018) Optimal control based energy management of multiple energy storage systems in a microgrid. IEEE Access 29(6):32925–32934
7. Lawder MT, Suthar B, Northrop PW, De S, Hoff CM, Leitermann O, Crow ML, Santhanagopalan S, Subramanian VR (2014) Battery energy storage system (bess) and battery management system (bms) for grid-scale applications. Proc IEEE 102(6):1014–1030
8. Cheng KWE, Divakar B, Wu H, Ding K, Ho HF (2011) Battery management system (bms) and soc development for electrical vehicles. IEEE Trans Veh Technol 60(1):76–88
9. Brandl M, Gall H, Wenger M, Lorentz V, Giegerich M, Baronti F, Fantechi G, Fanucci L, Roncella R, Saletti R et al (2012) Batteries and battery management systems for electric vehicles. In: Design, Automation & Test in Europe Conference & Exhibition (DATE), 2012. IEEE, pp 971–976
10. Qi G, Li X, Yang D (2014) A control strategy for dynamic balancing of lithium iron phosphate battery based on the performance of cell voltage. In: Transportation Electrification Asia-Pacific (ITEC Asia-Pacific), 2014 IEEE Conference and Expo, pp 1–5
11. Piller S, Perrin M, Jossen A (2001) Methods for state-of-charge determination and their applications. J Power Sources 96(1):113–120
12. He H, Xiong R, Zhang X, Sun F, Fan J (2011) State-of-charge estimation of the lithium-ion battery using an adaptive extended kalman filter based on an improved thevenin model. IEEE Trans Veh Technol 60(4):1461–1469
13. Baronti F, Fantechi G, Fanucci L, Leonardi E, Roncella R, Saletti R, Saponara S (2011) State-of-charge estimation enhancing of lithium batteries through a temperature-dependent cell model. In: 2011 International Conference on Applied Electronics (AE), IEEE, pp 1–5
14. Dolara A, Leva S, Mussetta M, Ogliari E (2016) Pv hourly day-ahead power forecasting in a micro grid context. In: IEEE 16th International Conference on Environment and Electrical Engineering (EEEIC), pp 1–5
15. An, LN, Quoc-Tuan T (2015) Optimal energy management for grid connected microgrid by using dynamic programming method. In: Power and energy society general meeting, 2015 IEEE, pp 1–5
16. Zhao Z (2012) Optimal energy management for microgrids

Electric/H2 Vehicle with Renewable Energy Grid for Himalayan Region

Surendra Pal Sharma

Abstract India is a big country having huge renewable energy potential of 2600 GW, sufficient for world power requirement of today. Most of the power potential comes under solar, wind and hydro energy. Himalayan region is full of hydro power and small currents have been estimated to generate 20 GW of renewable power along with large hydro of 145 GW. The hydro power remains cheapest of all renewable power even today. As per small hydro power policy of India, any corporate or individual person can put up a small hydro power plant up to 25 MW after identifying the potential of any current/river suitable for the project. If the small hydro power plant in Himalayan region is coupled with hydrogen energy storage and electric vehicle charging stations along the mountain roads, it will give very economic viability of the project. The power cost of hydroelectric comes to merely INR 1.0 (USD 1.3 cents) per unit. The cheap power can be utilized to charge electric vehicles and to generate hydrogen. The hydrogen generated at the hydro power station will be quite cheap to replace the LPG cylinder for cooking and commercial purposes in the Himalayas. Additionally, the electric vehicle charging project will give another boost to clean transport in the region, reducing global warming effect in the area. The hybrid electric vehicles can also utilize Hydrogen for on the way recharging through hydrogen fuel cells. This will pave the way for 100% renewable economy in future.

Keywords Renewable · Energy · Global warming · Electric vehicle · Hydrogen · Hydro power · Energy storage

1 Introduction

India is crowned by beautiful Himalayas all through its northern range. Himalayas serve as a great storage of water and power. The beauty of Himalayan region is also reflected in the people living in hills having their own way of clean livelihood. The Himalayas have a history of longevity of saints who were used to worshiping there. Most of our Rishis (Saints) and ancient people have also lived and enjoyed Himalayas

S. P. Sharma (✉)
NTPC Ltd, Darlipali STPP, Sundargarh, Odisha, India

R. K. Pillai et al. (eds.), *ISUW 2019*, Lecture Notes in Electrical Engineering 764,
https://doi.org/10.1007/978-981-16-1299-2_6

Table 1 Renewable power potential of India [1]

Energy type	Potential (GW)	Installed (GW) 30.09.2018
Solar PV and roof top	750	24.0
Solar thermal power	200	0.4
Wind energy @ 100 m	302	34.6
Biomass gass. power	25	8.7
Solar PV floating	1000	0.01
Small hydro < 25 MW	20	4.5
Tidal/Ocean power	50	0.003
Waste to energy	100	0.14
Large hydro > 25 MW	145	45.5
Total	2592	117.853

as a source of medicinal plants and purity of water. But in modern days the hill people are running towards planes for employment due to which the potential of Himalayan region is not being fully utilized. With modern renewable technologies, we can utilize the full potential of Himalayas as a source of energy, Education, Innovation, medical tourism etc. and create a self sufficiency for the population to avoid shifting to planes. We can generate power, food and export energy while enjoying purest environment of Himalayas.

2 Renewable Power Potential of India

India is blessed with such a great potential of renewable power that it can feed to full of the world requirement of 2000 GW as on today. Table 1, gives an estimate of renewable power potential and installed power capacity in India as on 30th Sep 2018. The potential of large hydro and small hydro power of 165 GW is mostly in the regions of Himalayas. Although many more regions of Himalayan energy have remained unexplored so far. The total renewable energy potential in India is 2592 GW (excluding Off-Grid systems) out of which only 117.853 GW has been installed so far. Govt. of India has set a big target of installing 225 GW renewable energy by 2022. So all the efforts should be aimed at utilizing all possible technologies for generating renewable power in India to make a cool and pollution free country.

3 Himalayan Energy as Means of Livelihood

With the amount of energy potential available in Himalayan region, we can easily make it a power crown of the country. Govt. of India is already on the way of putting large hydro power plants in the region which will provide much needed boost to the development of the Hills. The hydro power still remains the cheapest source of power, more over it does not have any pollution and avoids floods, droughts and global warming. The small hydro is another method of allowing local population to generate power at the place of consumption. It is the right time to train people in small hydro power and techniques of utilization in various forms such as Hydrogen generation for cooking, transport, energy storage and commercial use, also electric vehicle recharging for multiple uses in Himalayan region. The new technologies shall avoid use of LPG, fire wood etc. and thus avoid global warming. The beautiful trees shall also be saved to give additional cover to Hills to maintain ecology in the area.

4 Technologies for Hills

There are various technologies available for producing power like solar PV, wind, small hydro and large hydro technology. The scope of this paper shall be to detail out only small hydro technologies coupled with energy storage in the form of Hydrogen and battery storage. The Himalayan region is full of perennial water currents. As per estimates 1 MW small hydro plants can be put up at every 1 km length of the small rivers and tributaries in Himalayan region depending on the river slope. Bulb Turbine and Screw Hydro turbine technologies are the simplest and require small investments. The rubber water dam technology can be used at various locations to make out a cheapest hydro plant system. The energy storage technologies of Hydrogen and battery are already fully developed for immediate use by Hill population.

5 Bulb (Propellor) Turbine Technology

The bulb or propeller turbine technology is quite popular in Europe. Norway has utilized this technology in a big way. They have installed these turbines for automatic round the clock operation even during the period when whole area is covered by ice. The turbines keep operating below the ice cover. There is no danger of flooding as they are already operating under water. Figure 1 shows a bulb turbine and its working in a small dam. The range of bulb turbines varies from 1 to 5 MW and operate with the head range of 1–10 m.

6 Archimedean (Screw) Hydro Turbine

The screw hydro turbine is based on the screw pump developed by Archimedes. In ancient times, this turbine technology was mainly used for Off-grid operations because of difficulty in grid frequency matching due to variable nature of water flow.

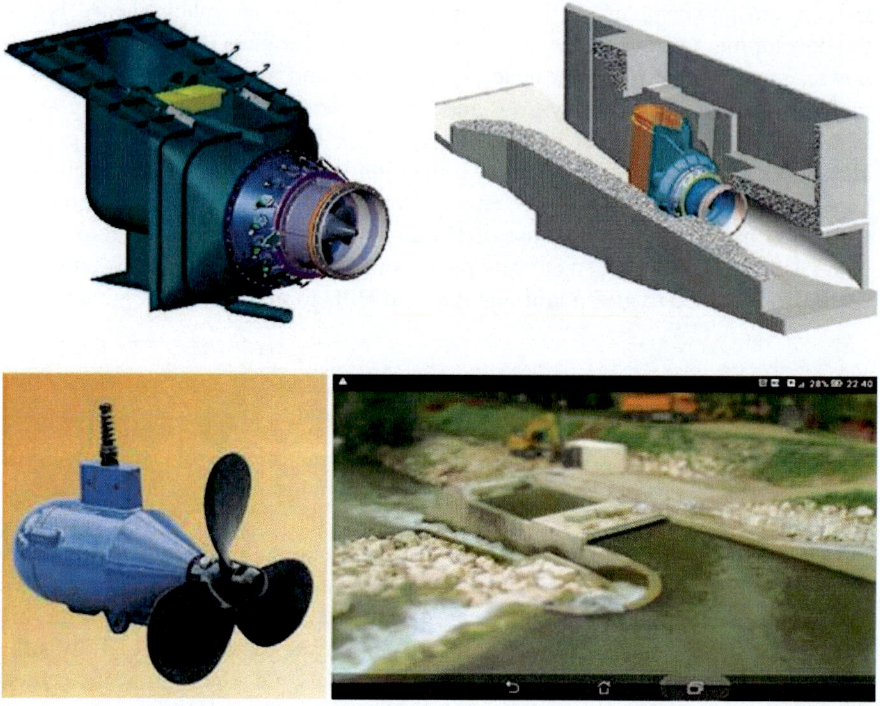

Fig. 1 The submersible bulb turbine technology

With the advent of electronic grid-tie inverters, this technology has a big potential of use in small hydro power plants. These turbines can be made with power ratings of 10–500 KW to operate with head range of 1–10 m. Figure 2 shows a design and installation of Screw Hydro turbine. The benefit of this technology is to extract small power from very small currents and at the point of consumption as per requirement, with very small investments (Fig. 3). These turbines can be put up in multiple quantity

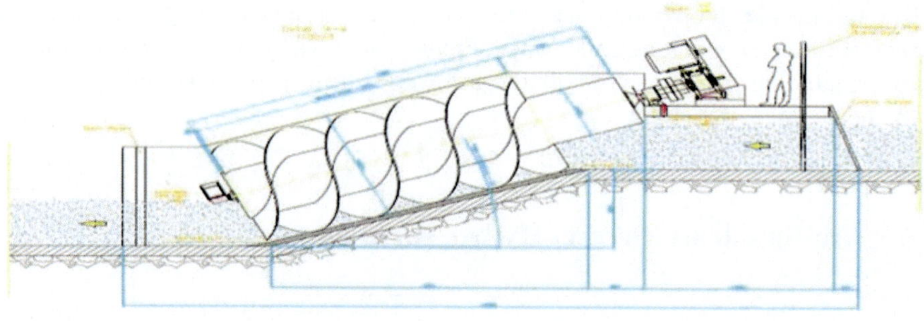

Fig. 2 The screw hydro turbine technology

Fig. 3 Screw hydro turbine water dam

to produce large power at required time with high range of variability see Fig. 4. The construction of turbine is very easy and small workshops can also manufacture with trained manpower.

Fig. 4 Screw hydro turbine multiple installation

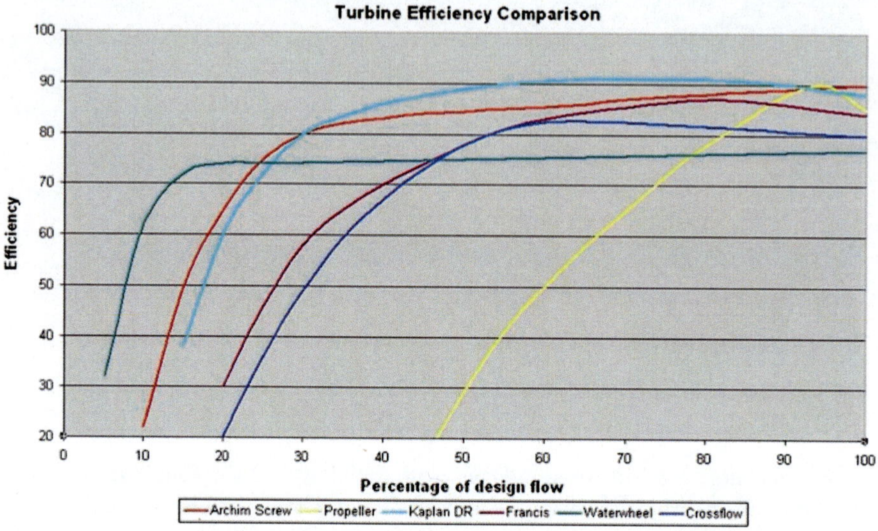

Fig. 5 Power efficiency of different hydro turbines

7 Efficiency Factor

The operating efficiency of bulb and Screw turbines is shown in Fig. 5. These turbines have a comparatively good efficiency factor at such low heads. The screw turbine has higher efficiency in comparison to Kaplan turbines if water flow is below 30% of design value, however it is little less than Kaplan Turbine but better than other types of turbines for water flows more than 30% of design value. The bulb (propeller) turbine has a best efficiency if water flow is around 95% of design value.

8 Inflated Rubber Water Dam

Inflated rubber water dam is a very cheap and convenient technology for putting up a bulb turbine or a screw turbine in a shortest possible time. It can be utilized at any location. In very cold areas where temperature may be lower than 0 °C air can be filled in the rubber pipe instead of water. The cost of rubber dam may be as low as 30% of the concrete dams, with additional benefit that the dam height can be varied by varying pressure inside rubber pipe, to allow flooded water to pass and natural desilting. Figure 6 shows a small rubber water dam. The rubber water dams are available in the world with a maximum height of 7 m and have a design life of 35 years (Fig. 7).

Fig. 6 Inflated rubber water dam

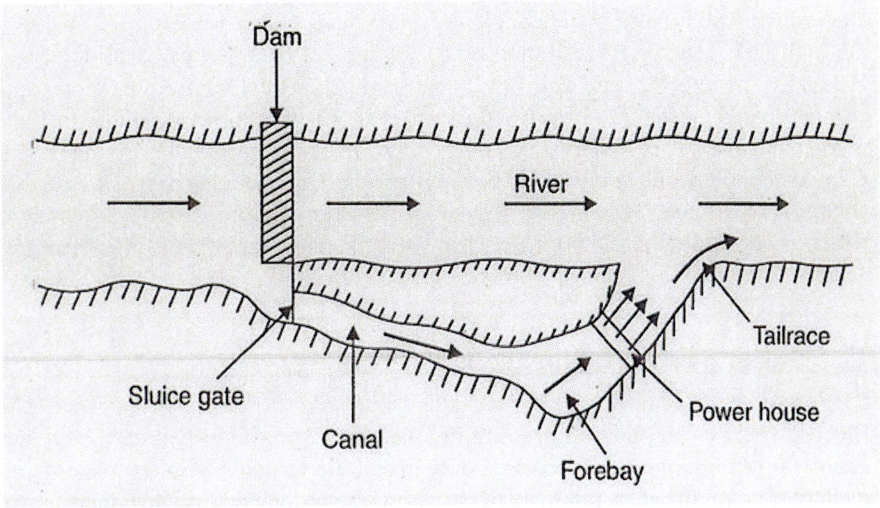

Fig. 7 General arrangement of water dam

9 Benefits of Rubber Inflated Dams Over Concrete Dams

The main benefit of rubber dam in place of concrete dams is its flexibility in maintaining the river ecology. It does not create sedimentation in upper stream and thus the water storage capacity in the dam remains intact all the years. Whereas the concrete dams create sedimentation and the water storage capacity is completely finished any time after 50 years depending upon the silt level in the river. After the storage capacity is finished, the concrete dams are to be destroyed to get the river flow characteristics restored. In the case of rubber dams, the dam height can be lowered to allow smooth river flow when silt is very high and height is again restored with the help of water/air pressure to increase storage capacity. Therefore the rubber water dam keeps the river ecology undisturbed and it keeps serving its purpose all through its life. The overall cost of rubber dams comes out to be very cheap based on its operability.

10 Hydrogen Energy Storage Technology

Hydrogen energy storage technology is best suited for Himalayan region. This can be coupled with small hydro power plants with off-grid systems. GES [2] high pressure technology can produce Hydrogen at 5000 psi without requirement of any further compressor to fill H2 cylinders. The cost of H2 generation comes out to be $2.85/gge H2 (Rs. 49/l gasoline eqvt) including electric component of $1.89 at the input power cost of $0.035/KWh (Rs. 2.28/KWh). This works out to Rs. 247/Kg of Hydrogen, when 7.3 kg H2 can replace 19 kg LPG commercial cylinder by heating value, thus costing Rs. 1803. Now if we couple H2 plant with hydro power at Rs. 1/- KWh, the hydrogen generation cost shall be $1.79/gge H2 (Rs. 31/l of gasoline eqvt. of H2 or Rs. 155/kg H2, and Rs. 1132/- per 19 kg LPG cylinder eqvt., shows quite competitive H2 generation even with respect to gasoline price available in Indian market and also the commercial LPG cylinder can be replaced by H2 cylinder. Thus, H2 plant coupled with small hydro power plant shall provide cheapest power to make cheapest Hydrogen. This idea will provide an alternative to Himalayan economy, which is consuming crude oil, LPG and wood for large population (Figs. 8 and 9).

11 Battery Energy Storage/EV Charging

Electric vehicles are the best preferred mode of transport in hilly areas due to their regenerative braking system possible only in electric vehicles. These vehicles can be recharged at night when power cost is lowest. Also, the small hydro plants can be coupled with EV recharging stations along the roads and in remote areas. This will also serve as local consumption of energy generated. The flow battery technology is the latest long-term high-density energy storage system best suited for hilly areas. Figure 10, shows Fe–Cl flow battery storage system for a MW scale. Such systems can store energy and can be used to recharge electric vehicles as and when required with a design life span of 25 years (Fig. 11).

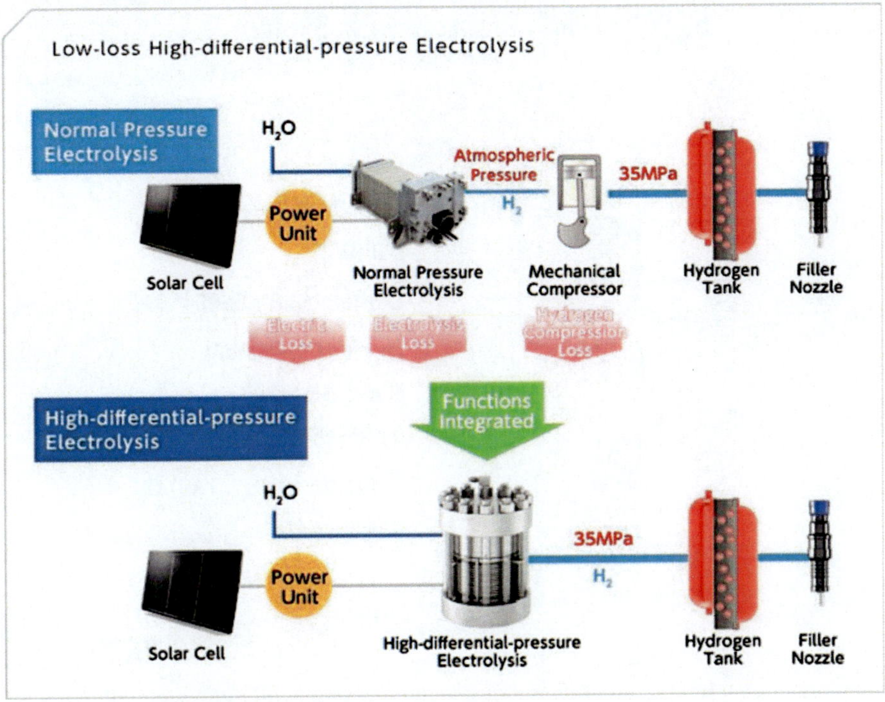

Fig. 8 High pressure H2 generation plant

12 Summary of Benefits

The benefits of small hydro power plant coupled with H2 generation and EV recharging system can boost the development and self-reliance of local areas in Himalayas. The area can also become energy exporter to planes. Following benefits are estimated to accrue by installing 1 MW SHP + H2 plant + EV charging station

i. Assuming energy requirement of 2 kw per household, 500 household can be fed energy by one power station of 1 MW.
ii. It can generate a direct/indirect employment for 100 people per 1 MW station.
iii. It can reduce 7000 tons of CO_2 per year, considering there are no tree cutting, no big construction activity for the small Hydro projects.
iv. The small hydro power systems do not inundate any area, so no population shifting.
v. The inflated rubber water dams have a facility of reducing to minimum height during rainy season allowing automatic desilting and thus storage capacity remains intact forever.
vi. Installation time is greatly reduced to 3–5 months, hence no cost escalation during construction period.

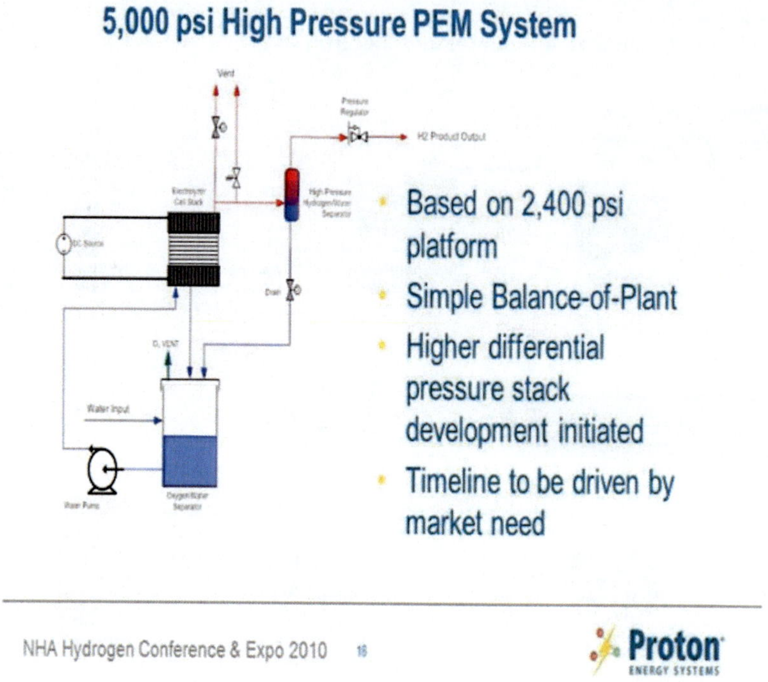

Fig. 9 Hydrogen generation at high pressure

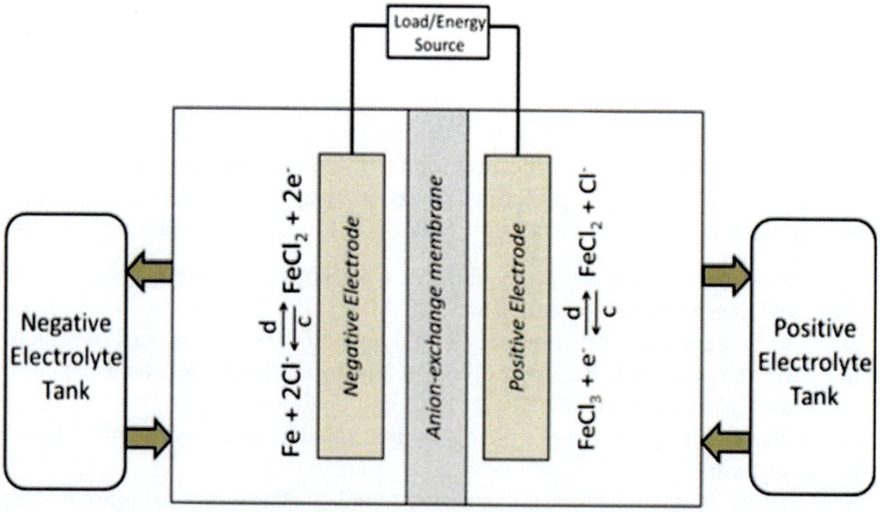

Fig. 10 Flow battery energy storage system

Fig. 11 Electric vehicle charging station

vii. Gestation period is very low, so expeditious output gives very good viability of the project.
viii. It will create a pollution free environment in Himalayan region.
ix. Upliftment of local areas turning into all round development in Hilly areas.
x. The system shall keep on working in case of landslides, natural calamities isolating the areas from rest of world.
xi. Being a distributed generation and consumption, it will not put any extra burden on the existing infrastructure in the area.

13 Popularising Technology

The Govt. of India in consultation with state Govt. should popularize the technology of Small Hydro Plant (SHP) coupled with H2 generation and Battery energy storage systems. The systems may be On-grid or Off-grid depending on feasibility. If system is on-grid, it will have an added advantage of reliability and also will serve as grid energy storage, in case grid power is cheaper in lean periods. The system can add to grid stability during day time due to heavy rush of solar energy. Presently in Germany, the power rates are zero or negative on Saturdays and Sundays, such a situation in India can be easily tackled by having energy storage systems already in place.

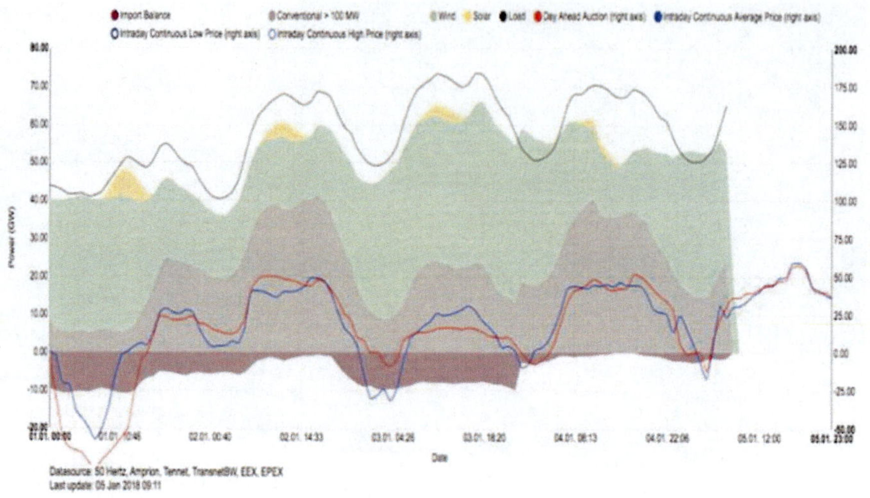

Fig. 12 Day ahead and real time power rates in Germany

14 Time of the Day Metering

It is right time to implement Time of the Day (TOD) metering of power in the country. Under this arrangement, power rates shall be given for 'Day Ahead' and 'Real Time'. With the day ahead rates the consumers will be able to plan their power consumption and energy storage taking benefit of the variable rates. The billing, however, shall be based on the real time power rates. Such arrangement is already in force in Germany where power rates become zero or negative during 24 h schedule and also on Saturdays and Sundays during day-time. A sample of TOD rates of Germany is given in Fig. 12. With the introduction of TOD rates various individuals shall start their own energy storage and a new area of business shall be opened giving new opportunities for employment. This step shall also provide optimum use of renewable energy in an efficient way.

15 Technoeconomic Viability

As per the estimates a 1 MW small hydro plant with H2 generation and EV charging stations will require an investment of Rs. 10 Cr. The system can produce H2 cylinders worth Rs. 2.5 Cr. per year. With employee cost Rs. 0.25 Cr, Interest @10%, net fund saving works out to be Rs. 1.25 Cr, thus giving a payback period of 6 years. The investment shall be quite investor friendly, but needs Govt. boost for such projects by giving single window clearance.

16 Conclusion

The renewable energy technologies and various energy storage technologies are available for implementation in Hilly areas. This will protect trees and stop firewood burning in the area, also will replace LPG cylinders transport from planes. The hill area economy shall improve and result in all round development in various activities giving new employment opportunities for local population. The Himalayan population will find suitable employment opportunities in the area and will not like to shift to bigger cities in planes. Pure air and healthy environment shall also be created for health tourism in the region.

References

1. http://www.cea.nic.in/reports/monthly/installedcapacity/2018/installed_capacity-09.pdf
2. https://www.hydrogen.energy.gov/pdfs/progress05/iv_h_1_cropley.pdf

Analysis of Impact of Electric Vehicles Penetration on Indian Distribution Network

Gajendra Malviya, Bharat Nandan, and Himanshu J. Bahirat

Abstract The recent advances in electric vehicles (EVs) technologies like batteries, increase in greenhouse gas emissions and increase in fossil fuel prices have caused the more interest in using EVs. But the charging of these EVs may put an additional requirement on the existing electricity grid. Therefore, the power grid must be ready for these challenges. In order to know the type of EVs charging infrastructure required for India, the distribution feeder ratings and different charging levels should be studied. The impact of number of EVs i.e. penetration impact on Indian distribution grid will be analysed. This will lead to identification of different issues due to charging. A typical Indian distribution network will be modelled in OpenDSS. The critical infrastructure issues like voltage profile at different nodes, transformer loading, peak load and power losses will be investigated for different penetration levels. Finally, the results will be analyzed to mitigate the impacts and to suggest various measures.

Keywords EVs charging infrastructures · Electric vehicle · Supply equipment · Loadshapes · OpenDSS

1 Introduction

THE transport sector currently relies on fossil fuels and therefore accounts for a significant part of greenhouse emissions. Issues associated with greenhouse gas emissions, reduction of fossil fuel resources, energy efficiency and security are among the factors that caused the increasing interest in using EVs. Therefore, one of the main future technologies to combat greenhouse gas emissions is the battery powered Electric Vehicle (EV) and Plug-in Hybrid Electric Vehicle (PHEV) [1]. A range of passenger electric vehicles are currently being developed by different manufacturers. Despite of the shortcomings associated with EVs like limited driving range, high cost

G. Malviya (✉) · B. Nandan · H. J. Bahirat
Department of Electrical Engineering, Indian Institute of Technology Bombay, Mumbai, India

H. J. Bahirat
e-mail: hjbahirat@ee.iitb.ac.in

R. K. Pillai et al. (eds.), *ISUW 2019*, Lecture Notes in Electrical Engineering 764,
https://doi.org/10.1007/978-981-16-1299-2_7

of battery, short life cycle of battery, high charging time, EVs are still gaining popularity. As the penetration of EVs in the market is perpetually escalating, thus, there is an urgent need of analyzing the impact of charging station loads on the power grid network. The increase in load due to EV charging may have adverse impacts like voltage deviations [2], transformer overloading [3], system losses and power quality issues etc. [4].

A residential EV charger in America provides a 120 V (Level-1) or a 240 V (Level-2) voltage supply to the connected EV through either a normal wall outlet or a dedicated charging circuit. Commercial chargers are generally high-powered, fast AC/DC chargers and installed in heavy traffic corridors and at public charging stations. However, because commercial chargers are still in the primary stages of deployment, EV owners typically charge their EVs overnight at residential charging stations primarily using Level-2 chargers. Unfortunately, the increasing number of residential EV chargers may cause several challenges for the distribution system. Therefore, both a system level analysis of the impacts of EV integration on the residential distribution circuit and solutions to address their impacts are needed.

A. *Background*

The impacts of EVs charging on the distribution grid are addressed in terms of different parameters in the literature. An impact study of EVs charging on the Hungarian low voltage distribution system is conducted in [2]. For this purpose, a typical Hungarian low voltage grid was modeled in DIgSILENT software. Several distribution system parameters like the transformer loading, feeders loading, voltage deviation and total system losses were assessed on the uncoordinated and delayed charging at different penetration levels of EVs.

The potential impacts of EVs charging in different penetration scenarios for Singapore with particular regard to distribution transformer rating is analyzed in [3]. This paper address the problems on the distribution side and also analyze the capacity based on a projected EV penetration and calculate and predict the overloading on the distribution transformer requirements in particular under the different EV penetration scenarios mentioned in the Electro mobility roadmap study for Singapore.

Evaluation of supply/demand matching and potential violations of statutory voltage limits, power quality and imbalance due to EVs charging are presented in [4].

The review paper [5] makes an attempt to provide a qualitative as well as quantitative review of the literatures in this area published in the past few years.

Some of the batteries ratings of 2 wheelers, 3 wheelers, electric cars and electric buses are also reviewed to understand EVs scenario in India. It was found that 2 wheelers have battery of around 2 kWh (Ather e-scooter), 3 wheelers have battery of around 5 kWh (Mahindra e-rickshaw), electric cars have battery of around 15–20 kWh (Mahindra, Tata EVs) and electric buses have battery of around 50–100 kWh (Goldstone BYD). The study conducted by India Smart Grid Forum

(ISGF) on the implementation plan for electrification of public transportation in Kolkata also recommends potential models for electrification of public transportation in Kolkata (buses, 3 wheelers etc.) [8]. It suggests ratings of batteries for 3 wheelers, Buses (AC and Non-AC) use for public transportation.

2 EVs Charging Infrastructures

EV charging is either provided using a normal wall outlet or a dedicated charging circuit (e.g. wall box or charge pole). Usually EV charging is provided by a 120 V (Level-1) or a 240 V (Level-2) voltage supply (see Table 1) in North America and a 230 V single-phase or 400 V tri-phase in most other countries worldwide.

Although the couplers are specified for up to 690 V AC and up to 250 A at 50–60 Hz, Level-1 (up to 16 A) and Level 2 (up to 32 A) are most commonly implemented [6]. Fast charging circuits, for example, CHAdeMO and the Combined Charging System (CCS), usually deployed close to highways or on parking sites, are also becoming popular.

The Society of Automotive Engineers (SAE) is responsible for the standardization of EV charging stations, it is the American standard for EV electrical connectors. SAE identifies three charging levels (see Table 1) depending upon the energy transfer rate. Note that Level-1 and Level-2 chargers are deployed at residential facilities while Level-3 chargers are used at commercial charging stations.

The EU distribution system maintains its three-phase characteristic all the way to the house service connections. This system provides a three-phase power supply with a separate neutral and earth. The line-to-line voltage is 400 V, and the line-to-neutral voltage is 230 V. For single-phase EV charging, EV load is connected between one of the phases and the neutral wire. The maximum charging power for the singlephase charging is restricted to 4.6 kW, 20 A on 230 V [6]. As for the three-phase charging, the power limitation is 44 kW, which equals 63 A at 400 V. Table 2 shows the charging levels for Europe.

Some common EVs in US that are charged from AC Level 2 charging at residence are mentioned in Table 3 with their battery sizes.

Table 1 EVs charging levels (America standards) [6]

Charging level type	Voltage level	Power level
Level-1	120 V AC	Up to 1.8 kW
Level-2	240 V AC	Up to 19.2 kW
Level-3 or DC charging	480 V DC	50–150 kW

Table 2 EVs charging levels (Europe standards) [6]

Charging level type	Voltage level	Power level
Level-1	Single phase 230 V AC	Up to 4.6 kW
Level-2	Tri-phase 400 V AC	Up to 44 kW
Level-3 or DC charging	480 V DC	50–150 kW

Table 3 Battery size of some common EVs [11]

EVs model	Battery size (kWh)
BMW i3	33.2
Chevrolet bolt EV	60
Nissan leaf	40
Fiat 500e	24
Ford focus electric	33.5
Kia soul EV	30

3 Modeling of Distribution System in Simulation Software

In this section, the test circuit IIT Bombay 10 node feeder, simulation software OpenDSS and different loadshapes curve were discussed that are used to model distribution network.

A. *IIT Bombay 10 node feeder*

IIT Bombay 10 node feeder taken as a typical Indian distribution network to do analysis. It is of 2.5 km length having primary voltage of 4.16 kV and peak load of 3.6 MVA. This small and highly loaded test feeder includes most of the common features that are used in actual networks like voltage regulators, shunt capacitor banks, overhead and underground lines, and unbalanced loads. This feeder provides a starting point to test power-flow convergence problems for a highly unbalanced system [7]. The substation transformer is of 5 MVA rating and it is 4.16 kV distribution network, which is common in US (Fig. 1).

B. *Simulation Software OpenDSS*

OpenDSS software was used to simulate IEEE 13 node feeder and do steady state analysis by adding EVs load on distribution network.

The Open Distribution System Simulator (OpenDSS) is a comprehensive electrical system simulation tool for electric utility distribution systems [12]. It can support all kinds of steady state analysis commonly performed for utility distribution systems. In addition, the most important advantage of OpenDSS is that it supports analysis with distributed generation integration and time series power flow.

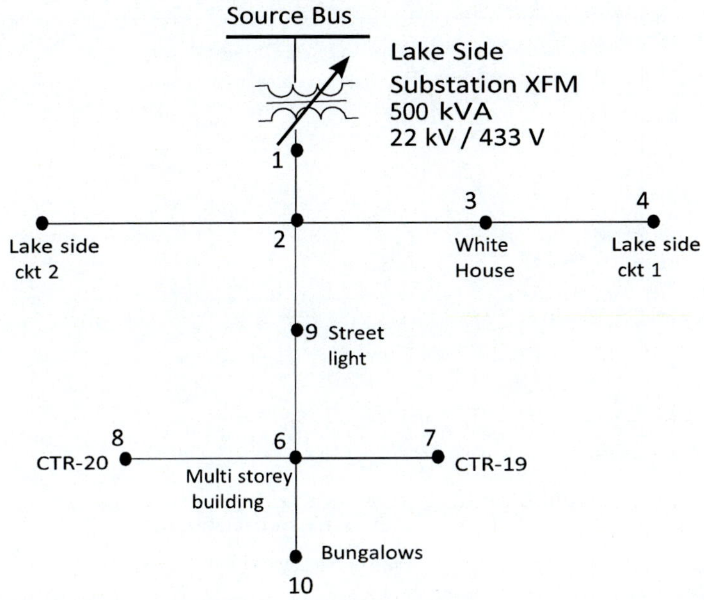

Fig. 1 IIT Bombay 10 node feeder

The iteration cycle is started by obtaining the current injections from all the power conversion elements and introducing them in the line vector. The sparse set of matrices is solved until the voltages converge to the specified tolerance. When performing daily or yearly simulations, the solution at the present time step is used as the starting point for the solution at the next time step. The solution typically converges in two iterations unless there is a large change in the load. After a converged solution, control iterations are performed to if control actions are needed.

C. *Different Loadshapes*

IIT Bombay residential loadshape is used to see AC Level 2 charging effect which was downloaded from Electric Power Research Institute (EPRI) website [10]. Residential loadshape of California is chosen as IEEE 13 node feeder network is a typical US distribution network, so the loadshape is also chosen of a state of US. Residential loads considered in loadshapes are lighting, central air conditioning (CAC), heating, refrigerator, TV & PC, clothes washer, clothes dryer, dishwasher and water heating. This residential loadshape is of an average weekday of peak season (summer) of California (Figs. 2 and 3).

EVs loadshape curve for AC Level 2 charging is chosen by taking some assumptions like there is facility for AC Level2 charging of EVs is available at all homes i.e. 240 V and 32 A (7.7 kW peak) charging facility is available at all homes. There are homes (as IEEE 13 node feeder is mainly residential) at each node of the feeder. The number of homes are counted by taking 4 kW as peak load of a home and it is

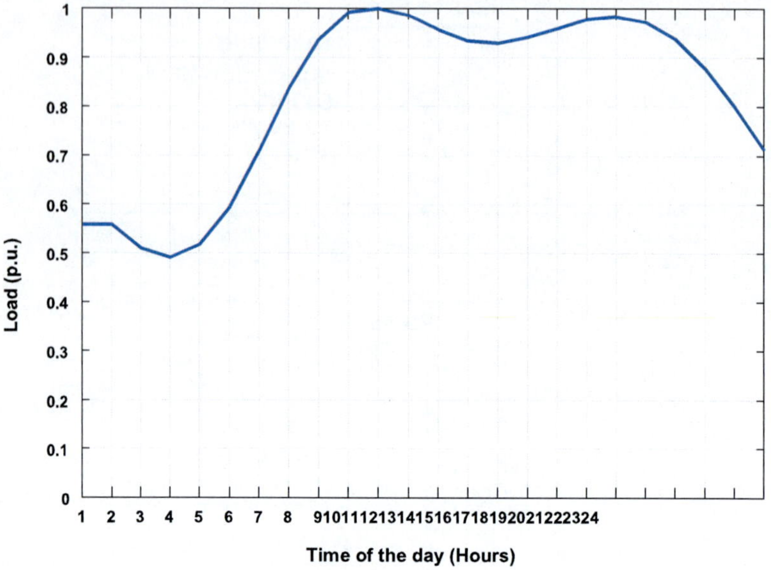

Fig. 2 IIT Bombay residential loadshape curve for Jan (winter)

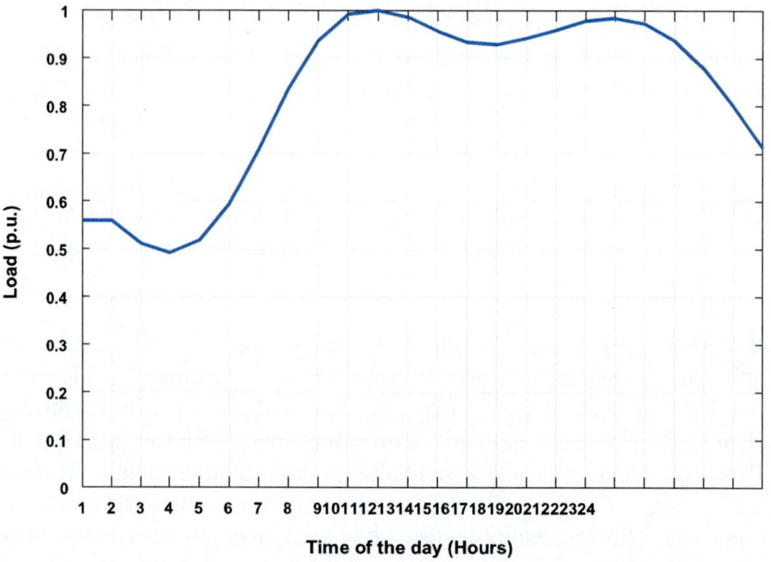

Fig. 3 IIT Bombay residential loadshape curve for May (summer)

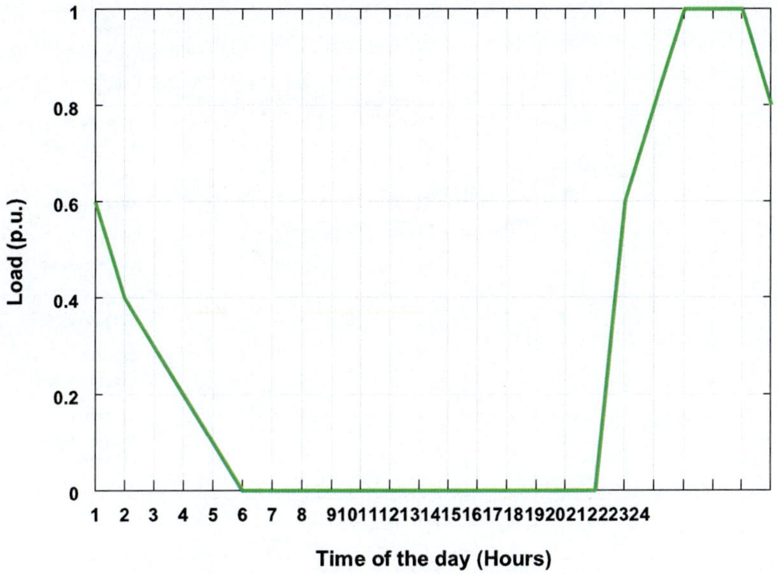

Fig. 4 EVs loadshape curve for AC level 2 charging

assumed that there are two vehicles in each home. Thus, accordingly EVs penetration level is given for study. In OpenDSS simulations, load of an electric vehicle is taken as 7 kW, thus EVs loadshape is generated considering 7–8 h of charging. Peak of loadshape is assumed at night (9–11 pm) considering large number of EVs are charging at that time (Fig. 4).

4 Impact of EVs Charging on Distribution Grid

In this, different impacts due to EVs charging on the distribution grid were studied using simulations.

A. *Voltage Drop*

As EVs charging load increases, it can draw more current through lines or cables that have impedances which results in voltage drop. There may be scenario where voltage may go below minimum required levels at certain homes in the network which are away from the transformer. Residence connected to highly loaded phase and most away from transformer is most vulnerable to the voltage drop. Due to this voltage drop electrical appliances may run inefficiently and there lifetimes can be reduced, thus it is important to have these voltages within limits (Fig. 5).

Node 671 is chosen for study of voltage drop as it is quite away from the substation and having load at all the phases. From the Fig. 1, Voltage profile of node 671 can be observed at different EVs penetration. V1, V2, V3 indicates the voltages of phases 1,

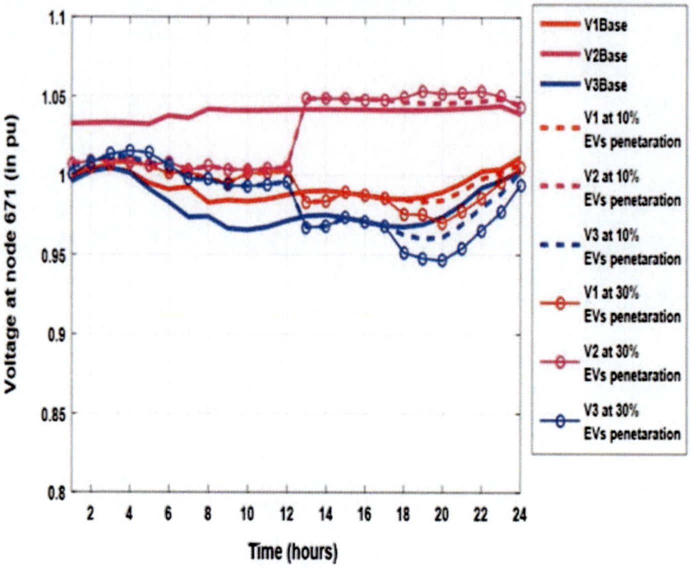

Fig. 5 Voltage profile of node 10 for different EVs penetration in Jan (winter)

2, 3 respectively. Minimum pu voltage at node at different EVs penetration can also be seen in Table 1. It can be observed that it is crossing the under voltage limits of American National Standards Institute (ANSI) at 30% penetration level as according to the American national standard the voltage deviation limit is ±5% of the nominal voltage [9].

Node 634 voltage profile was also plotted, it can be observed that its voltage remain within limits in different cases as it is not that farther from the substation. Node 634 is at 2500 ft. from substation transformer where as Node 671 is at 5000 ft. from substation transformer. Thus, it can observed from the Figs. 6 and 7, their is more voltage deviation at node which is farther from the substation and it may cross limits at higher penetration levels of EVs. These voltage profiles are found at unity power factor of an EV charger.

B. *Peak Load Demand*

Due to EVs charging, peak load demand will increase and there might be supply/demand mismatch. There should be sufficient power supply to cater EVs charging load. The increased load demand for EVs charging results in rise in the peak load demand of the grid which is accompanied by decrease in reserve margin. There can be scenario of increment in peak load demand as generally residential peak load demand is around 7–8 pm (lighting, AC) in US, when people come home from work and it may happen they just plug in their EVs as they reach home which can lead to huge increment in peak load demand.

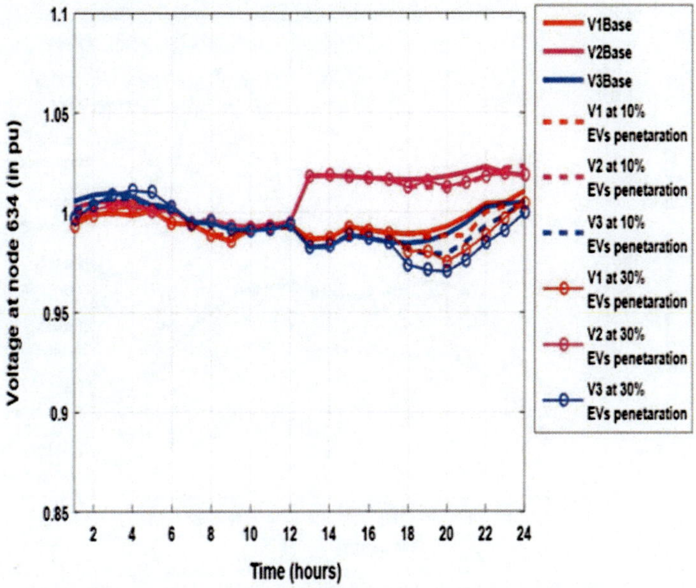

Fig. 6 Voltage profile of node 10 for different EVs penetration in May (summer)

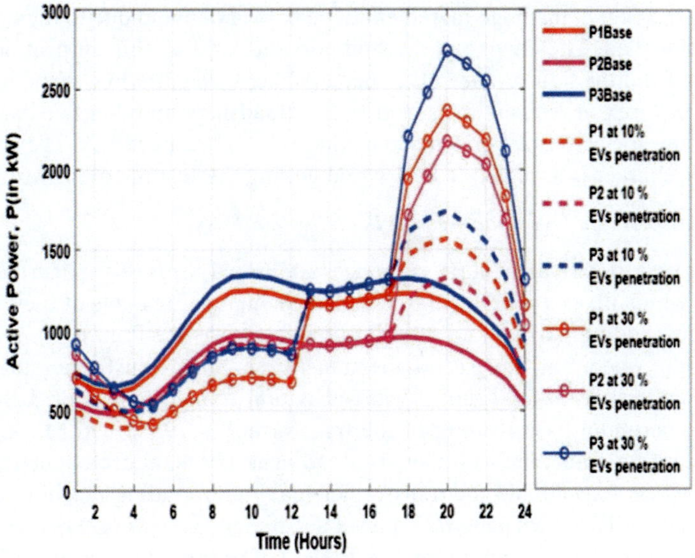

Fig. 7 Total active power demand of system at different EVs penetration in Jan (winter)

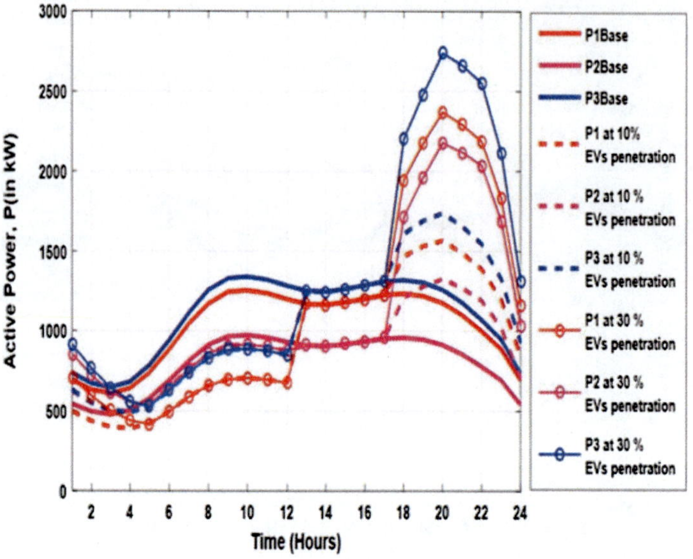

Fig. 8 Total active power demand of system at different EVs penetration in May (summer)

From the Fig. 8, the huge increment in peak load demand due to EVs penetration can be observed. The change in demand curve as well as shift in peak load can be observed from the figure. The curve shape is like residential loadshape in base case while it is a mix of residential as well as EVs loadshape in other two cases. P1, P2, P3 indicates the active power demand in phases 1, 2, 3 respectively. The power factor of an EV charger is assumed to be 1 while getting these simulations results.

C. *Transformers Overloading and Peak System Losses*

Large scale deployment of EVs produces additional stress on the distribution transformers which plays a prominent role in decreasing the life cycle of the transformer. Though exceeding normal ratings will not result in device failure [3], it effectively reduces the operation lifespan of the transformer thus transformer loading must be kept under permissible limits. The substation transformer is of 5 MVA rating and the maximum base loading of the transformer is 79% (3.96 MVA) and since PEVs charging coincides with the base load peak hours, at each penetration levels a new peak is formed and the transformer maximum loading reaching about 99% (4.95 MVA) at 10% EVs penetration level and about 152% (7.62 MVA) at 30% EVs penetration level which can be seen in Table 4. The power factor of an EV charger is assumed to be 1 while doing these simulations (Table 5).

EVs charging will result in increasing the load demand. With increasing the load demand, the current flowing in the feeders will increase which in turn will increase the system losses. In Table 4, it can be observed that by increasing EVs penetration, percentage peak active power losses increases.

Table 4 Simulation results at for EVs charging in winter

Case study [Jan (winter)]	Minimum node voltage (pu)	Peak active power demand (MW)	Peak MVA demand	Time of the day (Hour)	Peak active power losses (%)
Base case (No EVs)	0.961	3.57	3.96 MVA	10	3.15
10% EVs penetration	0.954	4.63	4.95 MVA	20	3.61
30% EVs penetration	0.937	7.23	7.62 MVA	20	5.18

Table 5 Simulation results at for EVs charging in summer

Case study [May (summer)]	Minimum node voltage (pu)	Peak active power demand (MW)	Peak MVA demand	Time of the day (Hour)	Peak active power losses (%)
Base case (No EVs)	0.961	3.57	3.96 MVA	10	3.15
10% EVs penetration	0.954	4.63	4.95 MVA	20	3.61
30% EVs penetration	0.937	7.23	7.62 MVA	20	5.18

5 Conclusion

From the case studies that were performed showed that large deployment of EVs may lead to potential problems for existing power distribution networks. It can be concluded that after a certain penetration levels of EVs, there is violation of statutory voltage limits. This increased EVs penetration may result in sustained secondary service under-voltage conditions, violation of under-voltage limits, and which would deteriorate the service voltage quality. The study also concludes that EV load charging may increase the system peak load demand. It can be speculated that if the charging infrastructure is not planned properly, the widespread adoption of EVs over the distribution network can significantly increase the substation load demand and could results in violation of supply/demand matching. Furthermore, the increased peak load demand due to EVs load charging may overload service transformers, resulting in transformer overheating, thus deteriorating the transformer's life.

The impacts like the transformer overloading and service voltage quality in the distribution network can be mitigated by some infrastructural upgrades like increasing the size of the service transformer and/or reconfiguring the distribution circuit using additional service transformers. Several methods like indirectly control EVs charging by utilities using Time of-Use (TOU) pricing and smart charging were proposed in literature to mitigate the impacts of EVs charging on the distribution grid. Integration of renewables (solar PVs) and storage devices in distribution network can also be studied to mitigate the impacts of EVs charging.

References

1. Yang Z, Wu Y (2012) Projection of automobile energy consumption and CO2 emissions with different propulsion/fuel system scenarios in Beijing. In: 2nd international conference on Remote Sensing, Environment and Transportation Engineering (RSETE), IEEE
2. Ramadan H, Ali A, Farkas C (2018) Assessment of plug-in electric vehicles charging impacts on residential low voltage distribution grid in Hungary. In: 2018 6th International Istanbul Smart Grids and Cities Congress and Fair (ICSG), IEEE
3. Vaisambhayana S, Tripathi A (2016) Study of electric vehicles penetration in Singapore and its potential impact on distribution grid. In: Asian Conference on Energy, Power and Transportation Electrification (ACEPT), IEEE
4. Putrus GA et al (2009) Impact of electric vehicles on power distribution networks. In: Vehicle Power and Propulsion Conference, VPPC'09, IEEE
5. Deb S, Kalita K, Mahanta P (2017) Review of impact of electric vehicle charging station on the power grid. In: 2017 International Conference on Technological Advancements in Power and Energy (TAP Energy), IEEE
6. Dubey A, Santoso S (2015) Electric vehicle charging on residential distribution systems: impacts and mitigations. IEEE Access 3:1871–1893
7. Marcos P, Fernando E et al (2017) A review of power distribution test feeders in the United States and the need for synthetic representative networks. Energies 10(11):1896
8. India Smart Grid Forum (ISGF) with Shakti Energy Foundation (2017) The implementation plan for electrification of public transportation in Kolkata, Report
9. National Electrical Manufacturers Association (2006) American national standard for electric power systems and equipment voltage ratings (60 Hertz), pp 19
10. California Residential Loadshape Curve [Online] Available: http://loadshape.epri.com/enduse. Accessed Oct 2018
11. Electric Vehicles Battery Ratings [Online] Available: https://insideevs.com/compare-plug-ins/. Accessed Oct 2018
12. OpenDSS Manual [Online] Available: http://smartgrid.epri.com/SimulationTool.aspx. Accessed Oct 2018

Global System of Record and Framework to Preserve Energy-Usage Data with Blockchain

Gideon Bandaru and Puneet Paneri

Abstract In the current scenario of the energy market, the availability of the energy-usage data of a consumer is limited. Utilities maintain the historical energy-usage data to some extent but the data is at the service address level and not at the consumer level. The historical data is available only for the period the consumer stays with the utility and the consumer cannot carry along when he moves to a different utility in a different region. The data format and the period of historical data maintained is inconsistent and is dependent on the utilities. In this paper, we propose a solution to maintain the energy-usage history of a consumer at a global level independent and irrespective of the utilities and regions using Blockchain technology with an emphasis on the data standards. The consumer who is the owner of the historical data can authorise the market participants to access and utilise the data for a meaningful purpose.

Keywords Energy consumption · Consumption history · Blockchain

1 Introduction

The energy-usage data of the consumer is of immense value for different entities operating at different levels of the market.

- Utilities—utilities use the energy-usage history to uncover the consumer's usage pattern, estimate load forecasting and can work towards energy efficiency
- Retail Suppliers—retail suppliers identify the consumer energy-usage pattern and propose best-suited plans, prepare purchase plan.
- Consumer—assess and identify their own energy-usage pattern, compare their usage with the peers and can take steps to bring down consumption and improve efficiency in usage

G. Bandaru (✉) · P. Paneri
Fluentgrid Limited, Visakhapatnam, India
e-mail: praveen.b@fluentgrid.com

P. Paneri
e-mail: puneet.p@fluentgrid.com

R. K. Pillai et al. (eds.), *ISUW 2019*, Lecture Notes in Electrical Engineering 764,
https://doi.org/10.1007/978-981-16-1299-2_8

- Third party organizations—third parties apply data analytics on the consumer energy-usage and provide useful insights to the consumers, utilities and other market participants.

The advantage is high but the availability of the data is low. In the current market, Utilities and Retails Suppliers maintain the consumption history but there are many shortfalls.

Utilities maintain history for a meter point or service address—Utilities maintain historical energy-usage data specific to a service address. At a particular address; consumers may move in and move out. Therefore, the history data is the consolidated consumption of all the customers who have resided in that particular address over a period and is not the record of an individual consumer (Fig. 1).

Retail Suppliers maintain history at the consumer level but for a limited period—In the deregulated market Retail Suppliers maintain the history at the consumer level but the data is available only for the period of their contract with the consumer. Once the customer switches to a new supplier, the retailers discontinue recording the data (Fig. 2).

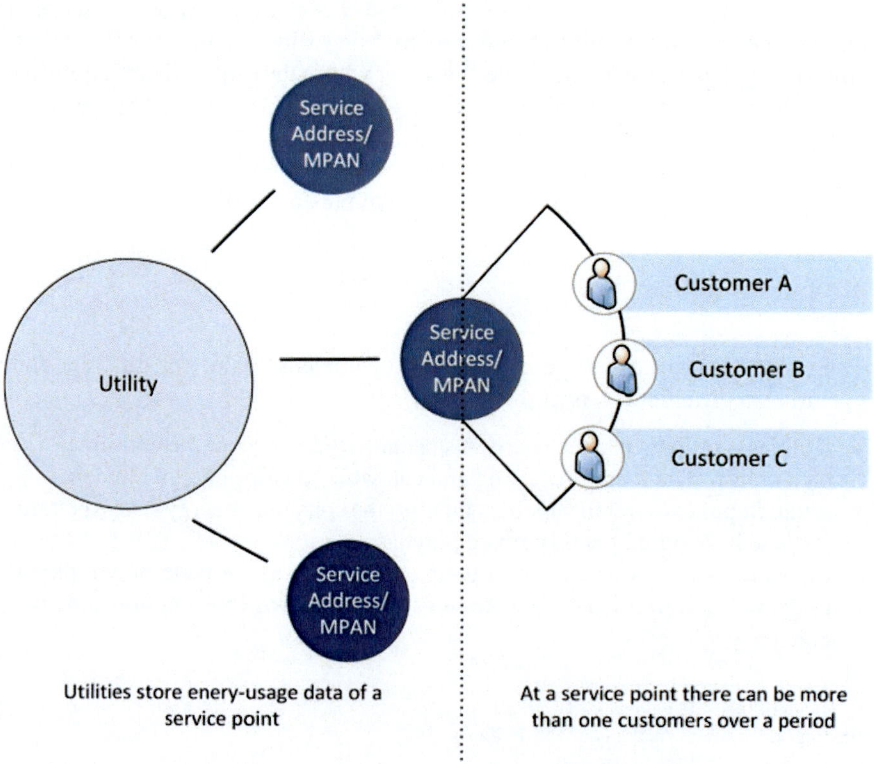

Fig. 1 Utilities maintain energy-usage history at the meter level and it can be a consolidation of multiple consumers resided in that address and consumed energy over a period

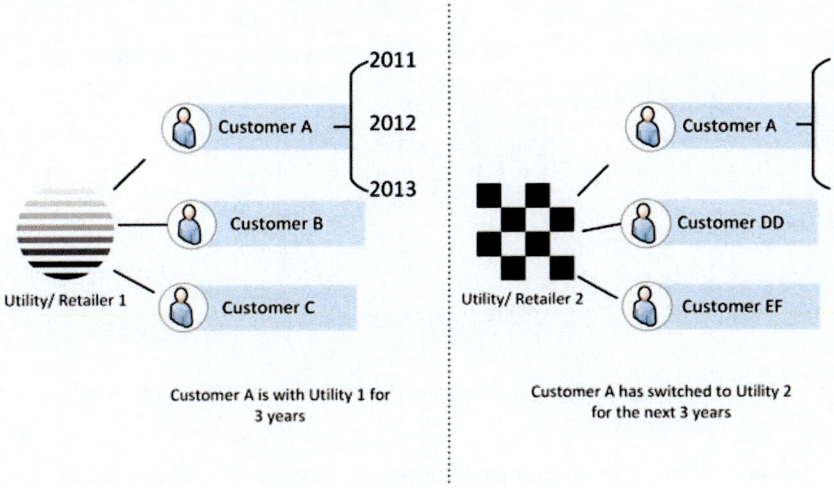

Fig. 2 Customer A is with the retailer AB energy for the years 2011, 2012, 2013, and switched to another retailer XY Energy for the years 2014, 2015 and 2016

No common global platform for the consumers to preserve their energy-usage data—There is no common global platform for the consumers to maintain and preserve their usage data received from the utilities or the retailers at one place.

2 Maintaining a Global Record of a Consumer Energy-Usage History

Considering the problem, we propose a global system of record and framework to store and preserve the energy-usage history of a consumer. This framework provides a platform for the consumer to record and store their energy-usage data for their entire energy journey (Fig. 3).

Using Blockchain to Preserve Consumer Energy-Usage History

The core of the system is preserving the consumer energy-usage history in a secure, efficient and transparent way. Out of the many choices, we propose Blockchain technology to satisfy the purpose.

What is Blockchain?

A Blockchain [1] is a growing list of records, called blocks, which are linked using cryptography. Each block contains a cryptographic hash of the previous block, a timestamp, and transaction data. By design, a Blockchain is resistant to modification of the data. It is "an open, distributed ledger that can record transactions between two parties efficiently and in a verifiable and permanent way". Once recorded, the data in any given block cannot be altered retroactively without alteration of all subsequent blocks, which requires consensus of the network majority. Blockchain already

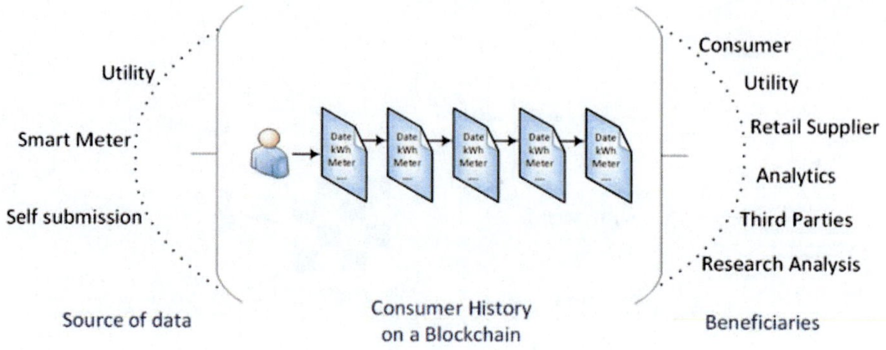

Fig. 3 Solution framework to record, preserve and share the energy-usage history of the consumer

have proven and meaningful implementations in areas of Crypto currencies, Smart contracts, Banks, Blockchain with video games.

Why Blockchain?

Blockchain has the potential to solve the following common business/IT pain points [2]:

- **Trust**: Blockchain eliminates the need for intermediaries and enables direct transactions among organizations. It establishes a distributed system that can be trusted inherently.
- **Security**: Blockchain ensures superior security for transactions compared with more traditional IT security mechanisms such as firewalls, encryption, intrusion detection systems and packet filters.
- **IT infrastructure overhead**: Because disparate systems today hold different copies of the same data, redundant effort is spent ensuring integration and reconciliation between IT systems. Blockchain-based solutions hold the potential to reduce this overhead significantly.
- **Integrated business process implementation challenges**: Blockchain could enable the next level of integrated business processes. For example, a metering solution based on Blockchain could ensure a meter-to-cash business process without third-party intervention.

3 Applying Blockchain

The functional components of the proposed framework constitute of three major areas in maintaining the historical data (Fig. 4).

1. Validation and Preparation of the data
2. Storing the data
3. Retrieval and Sharing the data.

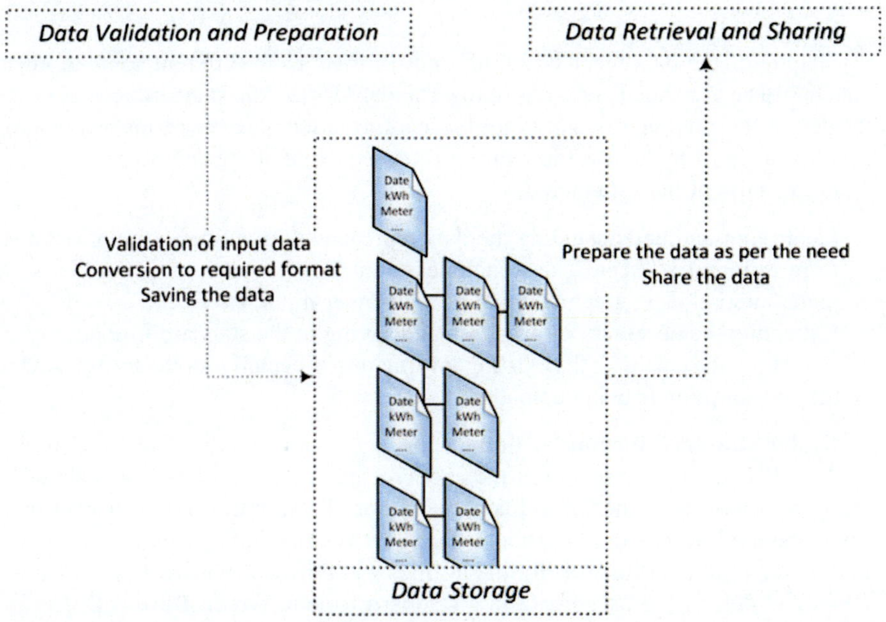

Fig. 4 Applying blockchain for validation, preparation, storage, retrieval and sharing of the energy-usage data

1. ***Validation and Preparation of data***

One of the core functional areas of the system is to acquire the energy-usage data of the consumer into the system. The energy-usage data can be available from different sources and in different formats. Validation and preparation of the data ensures that the incoming raw energy-usage data is stored in a consistent way irrespective of the format it is contained.

Sources of the energy-usage data

Few of the prominent sources of the energy-usage data are

- **Utility**—The utilities are the primary source and caretakers of the energy-usage data. The energy-usage data is recorded as a couple of fields per month for legacy meters and as a sequence of interval data for the smart meters.
  The data can be directly communicated by the utility to the proposed system using standard protocols and application interfaces.
- **Smart Meters**—Smart meters produce loads of interval data, which give important insights on the consumption pattern. The proposed system can have direct access to the smart meters and load the interval data into the system.
- **Self-submission**—The consumers submit the energy-usage data into the system. The consumers can capture the energy-usage data over a couple of fields, by uploading meter photographs or through the interval data files provided by a smart meter, utilities, retail suppliers, etc.

Preparation of the Data

In spite of the data received in different formats from different sources, all the data has to be translated and saved into a standard form. The standard format to save the data is designed in way that all the basic information required by the beneficiaries is constructed from the raw input provided irrespective of the data source.

Preparation of the data includes

- Identifying and understanding the data represented in multiple formats received from different sources. These include standard utility consumption file, smart meter interval data, green button data, consumer input, etc.
- Extracting the necessary information and saving in the standard format.
- Validating the data with the existing information especially for the dates for which the consumption is being uploaded.

Global standard format for data

A global standard to represent the energy-usage data is widely talked about and the impact will be immense and constructive if all the utilities deliver data in the same standard accepted across the globe. Few regions have taken a step forward in defining regional standards for the utilities and third party market participants to adopt and deliver the usage data in a standard format. **Green Button Data** [3] is one such initiative taken by the Green Button Alliance in US that is getting wide popularity. Most of the utilities are generating the energy-usage data in the Green Button format. Third party organizations are building applications around energy efficiency, forecasting and are effective in analyzing useful insights for the consumer and the utilities using analytics.

There is a strong need for a global standard across all the utilities of the globe, which helps in the efficient usage of the available information.

2. ***Storing the data***

The prepared data is stored on a Blockchain. Every new submission from a consumer is added to a new block linking it to the chain of blocks from the past submission (Fig. 5).

3. ***Retrieval and Sharing the data***

The trusted non-modifiable data can be retrieved and shared with the consent of the consumer. Depending on the search criteria, the data can be retrieved from multiple

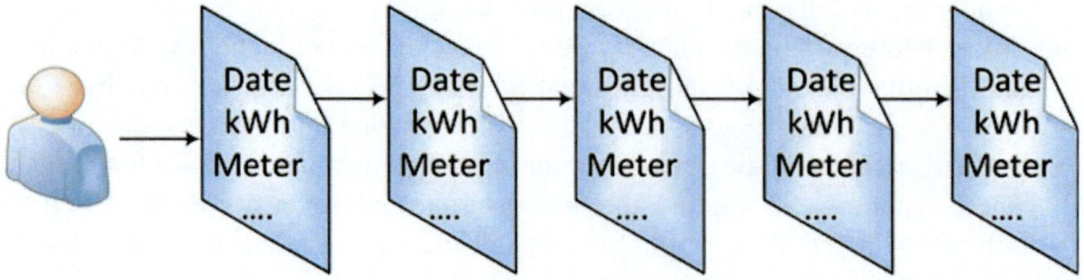

Fig. 5 Energy-usage details are added to the block linking to the chain of blocks from the past submission

blocks. The data can be shared by the consumers with utilities, retail suppliers and other third parties.

4 Benefits and Advantages

1. **Benefits of accessibility to the consumption data**

- **Utilities and Retail Suppliers**

The accessibility of the energy consumption data encourages the utilities and retail providers to facilitate evidence-based research for better understanding of electricity consumption pattern, improvement in energy consumption efficiency. The data is useful for energy model consumption model development, energy audit, load forecasting, energy management and tariff design.

- **Consumers**

The energy consumption data on the consumer side enables the consumer to identify the consumption pattern, strive towards the improvement of energy efficiency, compare with the peers, understand the bills paid and for cost savings if necessary change slabs or switch to other suppliers for better plans (in the case of deregulated markets).

- **Third Parties**

The third parties by applying new technologies on the energy-usage data provide meaningful insight and actionable tips for saving energy and money [4]. Provide Benchmarking, Energy Star scoring and compliance with transparency laws. Suggest cost savings by switching rate plans based on usage patterns.

1. ***Benefits of the Framework***

The availability of the energy-usage data of a consumer is definitely an advantage across the market participants.

- Availability of a ***common global platform***—the consumer has a platform to record, preserve and share his consumption data
- The energy-usage data is ***customer centric***—the consumption history is maintained at the consumer level and not at the meter/service address level
- The energy-usage data is ***continuous***—the energy-usage data of the customer is continuous irrespective of the consumer moving from one utility region to another or switches from one retailer to another in the case of a deregulated market
- The energy-usage data is ***secure***—unauthorized access to the energy-usage data is restricted
- The energy-usage data is ***owned by consumer***—consumer is the complete owner of the data

- The energy-usage data is ***shareable***—customer has the sole authority to share the data with any utility, retailer or any third party for the mutual benefit.

5 Conclusion

There is a definite need for a common platform for the consumers to record and preserve their energy-usage data. The global record of energy-usage data can clearly benefit consumers, utilities, retail suppliers, third party analytics enthusiasts. The security, transparency and integrity of the Blockchain adds advantage in offering this novel solution for empowering consumers in the energy sector. While there are few challenges in the standardization of data across the utilities, considering the advantage of availability of the historical energy-usage data, there will be a solution explored soon.

References

1. Wikipedia on Blockchain. https://en.wikipedia.org/wiki/Blockchain
2. Govindan G, Gosar V, Singh R (2018) Blockchain for power utilities: a view on capabilities and adoption. https://www.cognizant.com/whitepapers/blockchain-for-power-utilities-a-view-on-capabilities-and-adoption-codex3372.pdf
3. Burns MJ, Teeter JA, Wollman DA (2014) Green button: building an interoperable ecosystem
4. Murray M, Hawley J (2016) Got data? The value of energy data access to consumers

Impacts of Electric Vehicle Charger on the Power Grid

Jaya Aniruddha Mane, Minal Ramesh Rade, and Omkar N. Buwa

Abstract While the pollution is increasing resulting in green house gases and global warming, the amount of electric vehicle is predicted to increase. Most of the electric cars are design for urban use, and these vehicles need to be recharged in evening or during night, so the electric vehicles will interact with grid during this period. This process will impact the network voltage profiles and loading of grid elements. Existing Grids were designed several decades ago, so it becomes necessary to take a note of this that whether they will be able to support this increased loading, or is there a requirement of reconstruction to meet this demand. Most consumers will prefer charging during night time at their premises itself, which is LV system. And which operates at 230 V. So LV grid is taken in consideration to understand the impact of charging electric vehicles. As loads is connected to different phases, so there will be a phase asymmetry in the network. This asymmetry can be reduced by proper planning of the grid. Injection of electrical vehicle into the grid will also have an impact on power factor. As some consumers turn on or off their electric vehicles, the overall power factor changes. This has to be considered as this impact the reactive power consumption. With increase in the number of electric cars, the loading characteristics of transformer changes. For this work grid without electric vehicle and with electric vehicle is considered and Grid parameters are analysed.

Keywords Vehicle to Grid(V2G) · Electric Vehicle (EV) · Plugin Hybrid Electric Vehicle (PHEV) · Distribution Network · Power Quality

J. A. Mane (✉) · M. R. Rade · O. N. Buwa
Department of Electrical Engineering, K. K. Wagh Institute of Engineering Education and Research, Amrutdham, Panchavati Nashik, India
e-mail: jamane@kkwagh.edu.in

M. R. Rade
e-mail: mrrade@kkwagh.edu.in

O. N. Buwa
e-mail: onbuwa@kkwagh.edu.in

R. K. Pillai et al. (eds.), *ISUW 2019*, Lecture Notes in Electrical Engineering 764,
https://doi.org/10.1007/978-981-16-1299-2_9

1 Introduction

With the fast development of automobile industry and the increasing car ownership, air pollution, depletion of energy sources, resource are exhausting and other issues like global warming, acid rain and its hazardous health effects are arising from automobile industry. Hence it has now became a major concern and it is required to give more and more attention on all these issues all over the world. In order to protect the human living environment and to have sustainable energy supplies governments has to invest lots of manpower and material resources to find solutions for these problems [1].

There are many applications that contribute to air pollution and global warming. But among them transportation plays a major role. Hence it is required to replace the vehicles that run on conventional sources with the vehicles that utilizes clean energy without compromising with the efficiency. And Electric vehicles have these well known benefits of high efficiency, lower energy, low noise, zero emissions in comparison to conventional vehicles, so electric vehicles are now considered as one of the best solution to overcome all the environmental problems and it additionally results in the sustainable growth of entire nation.

So the conventional sources like petrol and diesel required for electric vehicle can be replaced with battery, Fuel Cell, Ultra capacitor. These sources can be used alone or in combination with conventional sources to have the maximum efficiency. Hence EV can be divided into pure battery electric vehicle (BEV), hybrid electric vehicle (HEV) and fuel cell electric vehicle (FCEV) [2].

As more electric vehicle penetrates into the market their will be a requirement for erection of charging station along the roadways. These charging stations will eventually take supply from grid and now will act as load. With the continuous increase in number of EV on road and hence the increased charging stations, vehicles are now taken as a new inductive load on power grid. This new load will have its own impact on grid, so it is required to understand the charging characteristics of electric vehicle various other issues such as load balance, power supply capacity, and power quality etc., need to be highlighted. So detail study of the impact of EV charging on the power grid is required.

The emerging EVs will not only act as load on grid but will also be a form of distributed generation in future and will play a major role in load sharing with grid. The plug-in hybrid electric vehicle (PHEV) has the ability to supply the grid when not in use, and can consume electricity when required, so it will act as load as well as a source. Therefore, the study of EVs charging effect on distribution network has great theoretical value and practical significance.

2 Development of EV

Increasing level of air pollution in Indian cities has been a cause of concern for policy makers. More than 25 Indian cities are within the 100 most polluted cities in the World [3]. The cause of growing air pollution in cities is related to a variety of sources however transport sector makes significant contribution [4]. It becomes important to reduce the emissions from transport sectors. The adverse effects of air pollution on human health and to the economy are well known and therefore to minimize the impact on the environment, policy makers are considering several options. Electric vehicles is a promising technology option and several national governments have successfully implemented policies to promote the technology. Indian government is keen to promote electric vehicles as a green mobility options and is also considering it as a viable solution to reduce air pollution in cities.

In India, electric 3 wheelers have been partly successful, however not much diffusion of electric vehicles has happened within 2 wheelers, 4 wheelers and city bus fleets.

Techno-economic assessments however show that electric two wheelers can become commercially viable by 2020 itself and electric four wheelers can be a major technology option by 2030 [5], if government provides incentives and infrastructures for charging are available. Government is actively promoting the generation and use of renewable energy and see it as an ideal starting point to improve air quality problem. It is also encouraging the vehicle aggregators in expanding their base in the state to propagate shared mobility concept along with EV fleet introduction to deal with vehicular pollution concerns. Various proposals for EV introduction in public transport are also at advance stages of planning. The government has proposal to set up 4–5 charging under PPP and will be operational by the end of 2017. More charging stations will be added on a periodic basis as the demand grows. In addition Telangana state has already exempted EV's from road tax and other incentives are under advanced stages of consideration.

3 Charging Infrastructure

After usage, the EV's battery should be charged. With the increasing of the individual battery rating and number of EV, the charging load cannot afford to be neglected from the grid point of view. From the standards, the EV/PHEV charging methods can be divided into 3 categories, listed in Table 1.

A. *Standard charging*
   (Mode 1, 6 h < charging time < 8 h)

For a power of 3.3 kW, standard charging for electric vehicle is employed which uses a socket outlet of 230 V and 16 A. This method of charging is mostly expanded in Europe and it requires a single phase AC charger. It takes a nominal time of

Table 1 Charging modes of EV/PHEV

Type	KVA	Charging time	Charging method
Slow/Normal	1–5	6 h	AC: 1 phase, 230 V, 16/32 A
Semi-fast/Medium	10–25	1–3 h	AC: 3 phase, 230 V, 32/63 A
Fast	180–400	5–15 min	Undetermined, DC off-board charging

around 6–8 h for charging a vehicle. Also it locates the charger inbuilt in the vehicle itself which is also known as on board charger. This feature of Mode 1 charging makes it a feasible option for getting their vehicles charged at their own home space or workspace. However, this system adds on various safety issues like its safety is dependent on to the breaking capacity of residual current circuit breaker connected on supply side. This safety standard is made mandatory in many countries which enforce the implementation of such device.

B. *Semi-fast charging*
(Mode 2, 1 h < charging time < 3 h)

For a power level of 7–22 kW semi-fast charging is employed. This corresponds to single phase 32 A or three phases 16 A for a 30 kWh battery. It gives the benefit of getting twice the power that can be availed. It allows a moderate rate of charging around two to six hours. This corresponds to Mode 2 charging. In this mode vehicle is connected directly with AC supply. It includes a control pilot conductor which provides the required protection to the equipment and the user as well.

C. *Fast charging*
(Mode 3, charging time < 1 h)

This mode 3 charging indicates charging of vehicle by off board charger working on a DC supply. This mode provides the benefit of fast charging, however this mode requires integration of fully developed technology and which is ready for deployment. Following are the available fast chargers:

- **Fast charger**
- **Super fast charger**
- **High AC charger.**

Fast charger—It is a charging system based on the power electronics. It converts AC power to controlled DC power in order to charge the EV battery. In Europe, fast charger station belongs to Mode 3. This mode is the most expensive. Its usage is limited only to public charging stations. In a fast charger station, the vehicle cannot be charged as fast as in a gas station. The charging time to full charge a battery is nearly 25–35 min. The power peak of this charger station is situated around 50–75 KW.

Super Fast charger—The goal of a Super Fast charger is to recharge a battery in the same time as required to refuel a conventional vehicle. The recharging time of this component is also comparable with a "battery swapping" established by the Renault project Better Place. Since the peak power is huge, it requires a special component for the high power.

High AC charger—Mode 3 contains high power AC charging up to 250 A. Today powerful AC sources are used to charge a traction battery via the traction inverter. In this case, adaption of voltage is performed by an off board mains transformer. As a consequence, this charger is used only for special applications.

4 Impact on the Distribution Network

Currently, the main research areas involved with grid power quality, economical operation and planning.

A. *Effect on Power Quality*

Integrating EV into grid results into various power quality issues like harmonic pollution, increased power losses, decreased voltage and three-phase voltage imbalance.

(i) *Harmonic Pollution*

Increased EV access will result in increased use of charging equipment, which includes a large number of highly nonlinear power electronic devices, and the DC link interconnecting the three-phase AC supply will generate harmonic, harmonic current may pollute grid, electrical components, and hence influence the power quality of distribution system.

(ii) *Voltage Drop*

EV technology is gradually getting mature and utilized in large scale. This cause the local load increase in power distribution network, large scale EVs charging will influence the node voltage, especially drop in end node voltage that ultimately affects the demand of users.

(iii) *Three-Phase Imbalance*

Considering a place for a fixed period of time, if there are less number of EV's getting charged then the charging gets reduced. And this results in a larger three phase unbalanced currents. However, charging of large number of electric vehicles cause unbalance current condition.

B. *Effect on Distribution Network Planning*

With increase in number of electric vehicles on road, demand for its charging facilities will be increased. And this necessitates the restructuring of distribution systems. Hence site selection, component sizing and system design becomes very important. If it is not optimally designed, it may result in notably increased power consumption and a significant drop in node voltage. Which further affects the traffic network layout thereby hampering the development of electric vehicles.

C. *Effect on Operation*

As far as economic operation of distribution network is concerned, it mainly reflects on the net loss, reduced life of cables, and life of distribution transformer.

(i) *Net Loss*

High permeability causes increase in EVs charging load rate, which corresponds to increased load loss rate.

(ii) *Cables*

The high harmonic currents impacts badly on the cable. This ultimately results in load losses and reduced life expectancy.

(iii) *Transformer*

Transformers form the weaker link in distribution network. However large-scale EVs charging will overloading of transformer, reduced life and even failure. This increased load cause increased burden on generation, transmission and distribution system and hence generator capacity and power transmission equipments will be correspondingly matched.

5 Key Technology and Challenge Involved of EV

A. *Vehicle-to-grid (V2G) Technology*

The electric vehicle and the grid complement each other as a systems for controlling not only energy but power as well. The power grid normally does not have any facility for energy storage. Not more than 2.2% of its total capacity as pumped storage plant. This is the main reason why it is important for controlling and continuously managing the electricity generation and transmission so as to match up the variable consumer demand. As electric vehicles are observed as electric drives that are powered by batteries, or hybrid sources, so they are compared with electrical system. Electric vehicles also provide an advantage that they can either generate or store electric energy in standstill condition. And if they are well connected with various auxiliaries in the system, then they can feed power in the grid. This kind of interconnection is known as V2G connection [6].

Hence the electric vehicle can be basically a battery or fuel cell powered vehicle or can have hybrid combination of these resources. But the most important job is to provide power to grid when it is in standstill or parked condition. As regulation, spinning reserves, and peak power are the most parameters in power markets [7].

Batteries have slow charging & discharging characteristics hence battery driven electric vehicle charge during low power demand and discharged when high power is demanded i.e. during acceleration or motoring mode. Whereas Fuel cell driven vehicles generate power from liquid or gaseous fuel. And Plug-in hybrid electric vehicles are the one that can function in either of these modes.

B. *Charging Control*

(i) *Harmonic Control*

Converters are among the important components of battery charging systems, which ultimately adds up the harmonics into the system. So, to control or minimize the harmonics, pulse width modulation technique can be implemented or multilevel operation of converter can be employed [8]. Increasing the number of pulses will result in great reduction of harmonics in RMS current. Also, methods for reactive power compensation can be implemented to sustain the unwanted harmonics. Harmonics will eventually lead to other power quality issues into the system which needs to be monitored continuously so that the unpredictable problems can be handled within time.

(ii) *Coordinated Charging*

Charging of large-scale electric vehicles can lead to concentrated charging and might affect regulation of grid. Therefore, coordinated charging is one of the way of regulating the grid and it is considered as the type of a controlled load. The main aim in this method is to optimize economics and to have minimum impact on grid. It mainly considers the state of grid, performance parameters of batteries and last but not the least, consumer requirements so that the complete charging process can be controlled. Also, it stabilizes the unwanted variations in load demand and avoids production of new peak load. In this way it can lead to improvement of reliability, power quality and the economy of the distribution network [9].

For successful implementation of coordinated charging, charging process needs to be scheduled and distributed EVs should be coordinated with grid. So, concept of middleman came into picture as it was difficult for grid to directly control the charging process. Multi agent technology can also be implemented for this purpose [10].

Integrating electric vehicles with smart grid automatically reduces the peak time of charging process.

6 Simulation and Results

Considering the V2G technology as a prominent solution, a micro grid is considered for the work. The micro grid mainly consists of four main equipment: A diesel generator which acts as the power generator; second a hybrid combination of PV farm and wind farm, which produces renewable energy; third a vehicle that is interconnected with grid which also acts as the load of the grid. This is similar to micro grid. The size of this micro grid depicts a community of a thousands of consumers. They considered to be operating during a low consumption day in spring or fall. Figure 1 shows the block diagram of grid when no electric vehicle is connected to the grid, and hence it is clear from Fig. 2 that the active and reactive power demand is fulfilled by hybrid combination of PV farm, wind farm, and diesel generator. Whereas, Fig. 3 indicates the block diagram of grid when electric vehicle is injected in to the grid. Figure 4

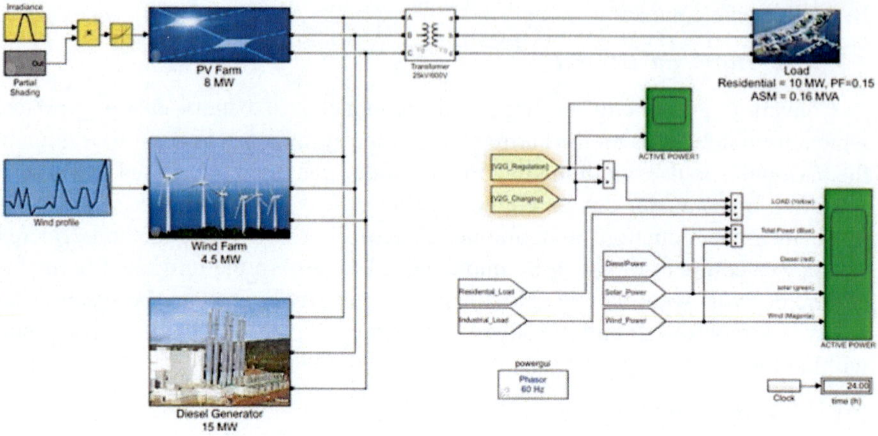

Fig. 1 Micro grid with no vehicle connected to it

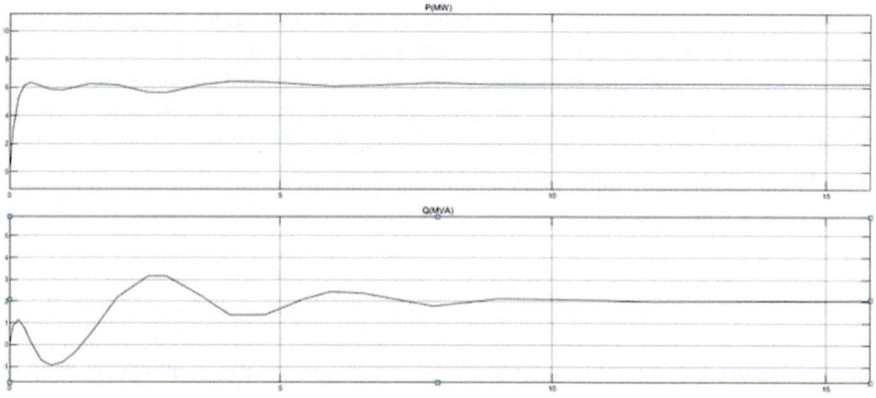

Fig. 2 Active and reactive power consumption when no vehicle connected to grid

clearly shows that electric vehicle not only acts as load on grid but also distributed generation.

The V2G interconnection has mainly two functions: firstly to control the charging state of the batteries which are connected to it and regulation of grid on the basis of power available during the day. The load is considered to be composed of mostly residential load and a few asynchronous machines that represents the impact of an industrial electric drive on the micro grid. For residential load, there is very rare scope of vehicle charging at work place with the desired power factor.

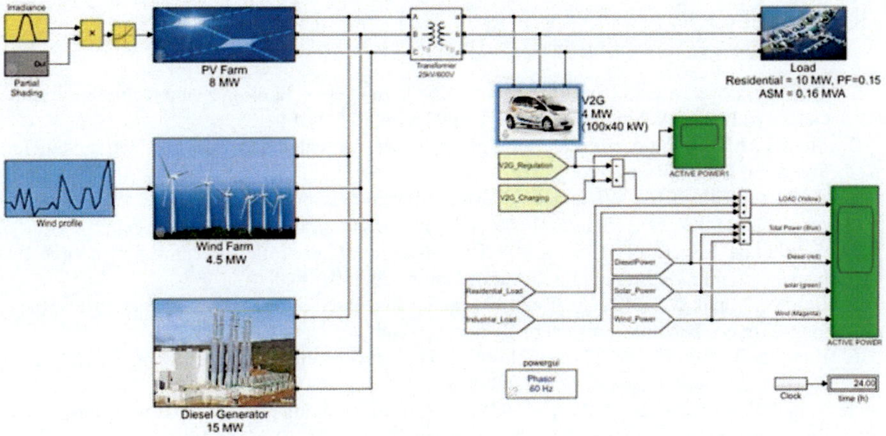

Fig. 3 Micro grid with electric vehicle connected to it

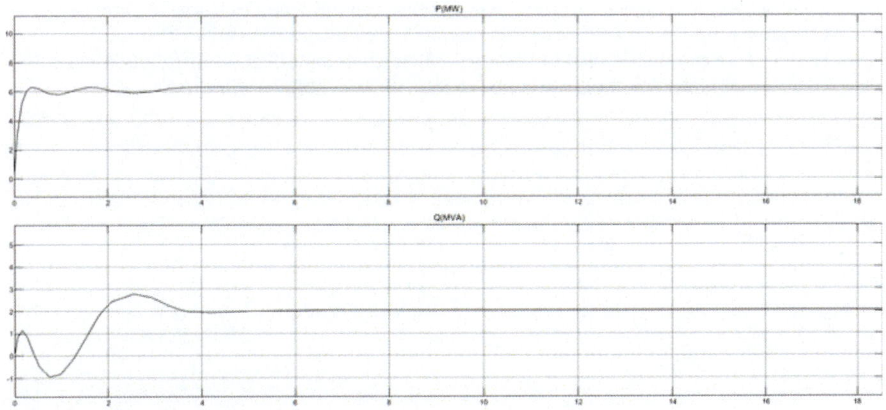

Fig. 4 Active and reactive power consumption when electric vehicle connected to grid

7 Conclusion

The integration of electric vehicle is now the inescapable trend in the expansion of distribution network. With increased usage of electric vehicles, potential problems for distribution system will get enhanced. With optimum configuration of electric vehicles, the adverse impacts on distribution system can be reduced to magnificently low level. There is a extensive research going on the various options of energy storage devices and the advanced features of vehicles with these new resources open ups the chances of prudent and profit making system. A complete change in present time transportation industry has raised the need for speedy evolution of electric vehicles and it will have a great impact on complete power system.

References

1. Qian L-j, Zhao H, Gao L-x (February 2002) On the key technologies and technology route of electric vehicle development. J Hefei Univ Technol 25(1):14–18
2. Yu Z-b (March 2004) Research status and development discussion of electric vehicles. J Shaoguan Univ 25(3):48–51
3. WHO (update 2016) WHO global urban ambient air pollution database WHO, [Online]. Available: http://www.who.int/phe/health_topics/outdoorair/databases/cities/en
4. Guttikunda SK, Goel R, Pant P (2014) Nature of air pollution, emission sources, and management in the Indian cities. Atmos Environ 95:501–510
5. Dhar S, Pathak MSP (2016) Electric vehicles and India's low carbon passenger transport: a long-term co- benefits assessment. J Cleaner Prod
6. Kempton W, Tomiü J (2005) Vehicle-to-grid power fundamentals: calculating capacity and net revenue. J Power Sources 144(1):268–279
7. Yang J, Wang M, Zhang Y, Wang DS, Ye FF (2010) Applying power battery of electric vehicles for regulating peak in grid. East China Electric Power 38(11):1685–1687
8. Peng Li, Chen Xinqi, Wentang Hu (2008) Analysis of the impact of electric vehicle charging station on the power grid harmonics. China Power 9:31–36
9. Zhou X, Zou L, Ma Y, Gao Z (2017) Research on impacts of the electric vehicles charging and discharging on power grid. In: 29th Chinese Control And Decision Conference (CCDC), Chongqing, pp 1398–1402, https://doi.org/10.1109/CCDC.2017.7978736
10. Huiling L, Xiaoming B (2011) Electric vehicle charging effect on distribution network and counter measures. Autom Electric Power Syst 17:38–43

Unmanned Substation—A Case Study

L. N. Mishra and Bharatkumar Soni

Abstract Just a few years ago it was extremely difficult to monitor and control the distributed substations. The reason for this was that the remote systems were either unable to communicate with a control center or communication involved a great deal of efforts and expense. The technological advancement in communication system and Information technology, the application of Remote Operation of all Substations can be adopted which in turn optimized the requirement of skilled manpower, virtual management of Substations and assets can be managed with better efficiency. Remote control technology describes the remote monitoring and control of physically separate system parts by means of data transmission. Measured values and control commands are transmitted over long distances and visualized, processed and stored in control center.

Keywords Remote control and monitoring · Substations · Data analytics

1 Introduction

Adani Transmission Limited (ATL) has established itself in Transmission Sector as India's largest Private Transmission Company in short span of time. ATL is operating various EHV Substations along with associated Transmission Lines with varied spectrum voltages in various region of India with best in class availability figures. ATL also owns and operates a 990 km long, ±500 kV HVDC Transmission system.

L. N. Mishra (✉) · B. Soni
Adani Transmission Limited, Ahmedabad, India
e-mail: ln.mishra@adani.com

B. Soni
e-mail: bharat.soni@adani.com

R. K. Pillai et al. (eds.), *ISUW 2019*, Lecture Notes in Electrical Engineering 764,
https://doi.org/10.1007/978-981-16-1299-2_10

2 Project Overview

Presently ATL is implementing 16 nos. Substations and 340 km of associated Transmission Lines in the state of Rajasthan. These 16 nos. Substations are as listed below:

1. 220/132/33 kV Ranpur SS
2. 132/33 kV Peeplu SS
3. 132/33 kV Chitri SS
4. 132/33 kV Bambora SS
5. 132/33 kV Khatoti SS
6. 132/33 kV Riyabari SS
7. 132/33 kV Baytu SS
8. 132/33 kV Ram Ji Gol SS
9. 132/33 kV Bar SS
10. 132/33 kV Ghumati SS
11. 132/33 kV Ahore SS
12. 132/33 kV Rajmatai SS
13. 132/33 kV Bengatikalan SS
14. 132/33 kV Shekhsar SS
15. 132/33 kV Ghamurwali SS
16. 132/33 kV Sorda SS.

SUBSTATION LOCATION:

The Substations and associated transmission lines are scattered and covers almost 85% of area geographically in Rajasthan. The average distance between two Substations is about 100 km. In the present configuration, it would require a large fleet of O&M staff and logistic supports to operate and maintain these substations, leading to high response time and operating cost. To overcome this challenge, a case was worked out to leverage the state of the art technology with respect to Remote Operation with minimal/no manpower.

3 Way Forward for Remote Operation

The transmission sector is currently facing multiple challenges like competitive bidding for transmission projects resulting in cost crunch and lean time schedules, scarcity of experienced manpower and stringent availability demands by regulators etc. In view of the latest developments in communication system and information technology, it is possible to create a remote control center for scattered Substations with an objective of better efficiency and optimization in skilled manpower. ATL decided to move towards the central operation of all 16 nos. Substations from Remote Control center at its existing Substation of Alwar and Deedwana in Rajasthan. Additionally, having information of entire assets at common location will help in better coordination and data analytics. Since complete set up of Operation is available at existing 400/220 kV Alwar and 400/220/132 kV Deedwana SS, ATL decided to build up the Alwar and Deedwana SS as Master Control Center. Further back up of Alwar Master Control Center can be built up at Deedwana SS and vice versa so that in case of any emergency, Operation of Substations can be managed through back up control Center.

4 Solution Overview

Entire Transmission lines under the projects are laid with OPGW for fiber optic telecommunication system, however all substations are not interconnected with each other; hence fiber network is not available among Substations under the projects. Since RRVPNL has planned the laying of OPGW in entire Transmission network of Rajasthan, ATL has tied up with RRVPNL for utilization of their network for connectivity of all Substations including Alwar and Deedwana also.

For remote SCADA system at each Master Control Center, ABB make 800xA and Micro SCADA system had been considered since all individual Substations have ABB make Micro SCADA system. This would result in better interfaces and compatibility. The remote SCADA system at each Control Center includes the following main functional equipment.

- Front end servers: Front end servers communicate with substation gateways for data collection on IEC104 protocol. It has been so organized that each server shall handle data from 5 substations keeping in mind reliability and scalability.
- Main Application Servers: Main servers will process the analog and digital data from front end servers for monitoring and controlling from the Operator Workplace.
- Historian Server: History servers shall store system generated real time data for long period which will be essential for data analysis.
- Data Recovery Server: The Data Recovery Server stores the image backup of the systems in local HDD. In case of breakdown of the servers/workstation image back up data can be used for recovery.

- SLDC Gateway Server: SLDC Gateway Server communicates with SLDC for sending the 16 Substations data on IEC 104 protocol on a single link.
- Corporate Connection Server: Corporate server host the smart client for the corporate network.
- LVS Workstation: LVS workstation is used to display the graphics on Large Video Screen.
- Operator workstation with dual Monitor: The Operator workstation is intended for operators from where the monitoring and controlling of Substation are done.
- Engineering Workstation: The Engineering workstation is intended for engineering of Database, display modification etc.
- DR Workstation: For DR evaluation which collects and stores the DR from Substation.
- Printers: For printing Graphics, reports and alarm events.

SYSTEM DIAGRAM

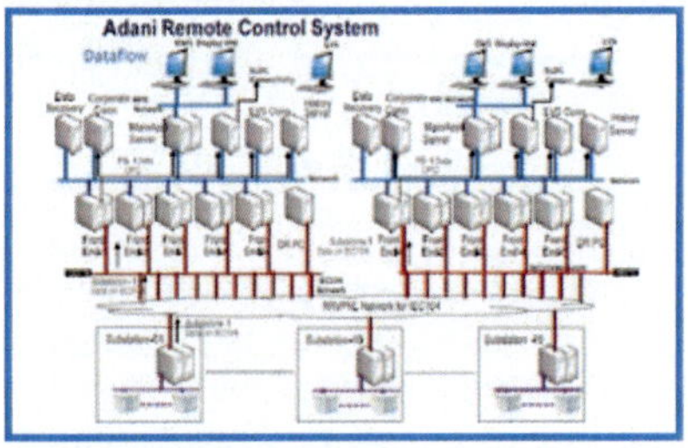

There are main two type of network forming the interconnection of various important SCADA devices;

1. Plant Network: This network interconnects all the Operator Workstations, Servers, printers through plant network switches of Remote SCADA System.
2. Control Network: This network interconnects all the IEC104 Gateways located at 16 Substations. This network shall be connected to IEC104 Front end servers. This network facilitates the transfer of status, commands and measurement data from 16 substations to remote SCADA system.

The SCADA servers and Front end servers with redundant configuration connected on dual LAN network along with Workstations, Engineering Workstations and printers. The Front end servers are communicating with Gateways available in Substations through OPGW network.

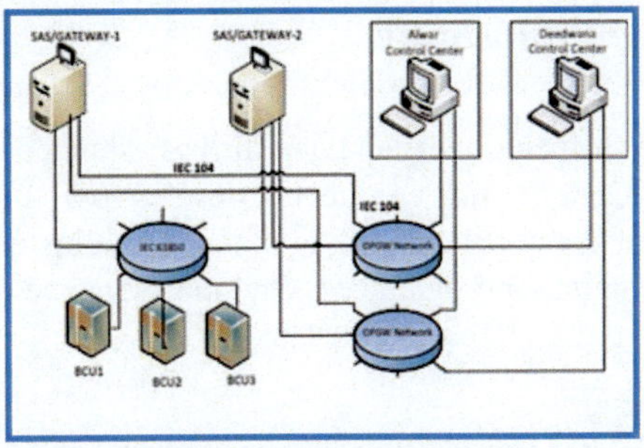

TYPICAL OVERALL COMMUNICATION N/W

5 Typical Overall Communication N/W

Remote SCADA system has been configured for redundancies at various levels of integration of the system like Main Servers, Front End servers, communication between Main servers and Front end servers, communication between Front end servers and Firewall and communication between Operator work station and servers.

ABB make 800xA system for Remote SCADA project is used for data acquisition and supervision of 220, 132 and 33 kV switchgear feeders. The system contains all information and means needed for the supervision of the process. It also includes tools for configuring the system and programming the process function.

The application creation involves the following function:

1. Graphic Displays
2. Event List
3. Alarm List
4. Status Monitoring
5. Trends
6. Reports
7. Time Synchronization with GPS
8. User Authorization.

Control Authority: System has been designed in such a way that 8 nos. Substations shall be controlled from Alwar SS and remaining 8 nos. shall be controlled from Deedwana SS, however control of all 16 nos. can be exchanged to either of Master i.e. Alwar or Deedwana con∞trol station if required during contingency.

For equipment like Circuit Breaker, Isolators, a reset control to operate from the Control Center has been provided for. Accordingly, respective devices have to be set in remote mode at local station.

In addition to installation of Remote SCADA system, Security Camera and Operation Camera system is also envisaged at each location for safe and reliable operations.

6 Operation and Maintenance Philosophy

As per Remote Operation planning, two shift engineers and one shift assistant considered at Alwar and Deedwana stations. Each of shift engineers operates four substations at a time. Further, maintenance of these substations is planned in clusters of 2 or 3 substations each, based on the interstation distances. One maintenance team will look after the maintenance of substations in particular cluster.

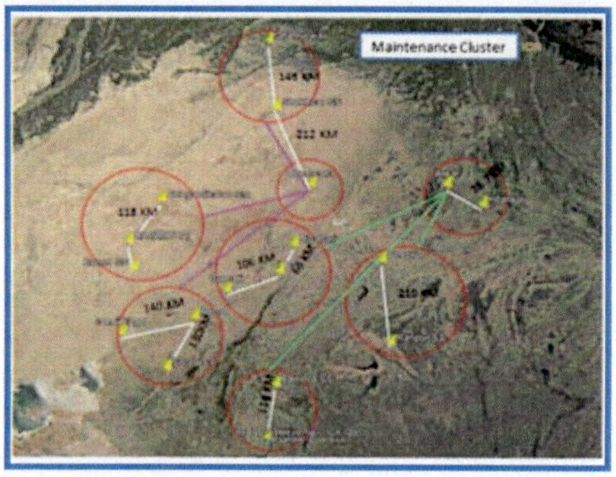

7 Cost Benefit Analysis

The implementation of Remote Control Center offers the reduction in operational manpower of around 70 personnel and also produce additional benefits like

- Reduction in site visit,
- Ability for engineers to access the entire systems at one location for better analytics,
- Quicker data access,
- increased end to end operations efficiency,
- Faster decision etc.

The total Capex and Opex requirements has been estimated to be around Rs.18 crore which includes cost of Remote SCADA system, IT infrastructure, Cameras, charges for third party OPGW network, Lease line charges for time being (2 year) etc. Considering the net reduction of operational cost, payback period is estimated to be around 3 years.

Performance Evaluation of Electric Vehicle Using Hybrid Energy Storage System

Minal R. Rade, J. A. Mane, and O. N. Buwa

Abstract The fuel efficiency and performance of novel vehicles with electric propulsion capability are largely limited by the performance of Energy Storage System. The battery system choice is a crucial item but no single type of energy storage element fulfils high energy density, high power delivery capacity, low cost per unit of storage, long cycle life, low leakage, and so on at the same time. One of the best solution is to use a Hybrid Energy Storage System. The main objective is to design of a hybrid electrical energy storage system which gives substantial benefits against battery issues such as reduction in battery stress. Also to maintain battery current as constant as possible during transients to limit battery stress. On the other hand supercapacitor has capability to charge as fast as possible without exceeding maximum current from regenerative breaking and to discharge most of its stored energy during acceleration. Adding supercapacitor bank will assist the battery during vehicle acceleration and hill climbing and with its quick recharge capability, it will assist the battery in capturing the regenerative breaking energy. This significant advantage a battery-supercapacitor energy storage system gained attention. Battery and supercapacitor sizing includes the most important and difficult steps is, the determination of the numbers of batteries and supercapacitor connected in series and parallel. The power management is essentially the optimal distribution of power between battery and supercapacitor. With supercapacitor added into the hybrid energy storage system, battery workload is reduced, which leads to significant extension of battery life.

Keywords Electric vehicle · Hybrid energy storage system · Supercapacitor · Battery · DC-DC converter

M. R. Rade (✉) · J. A. Mane · O. N. Buwa
Department of Electrical Engineering, K. K. Wagh Institute of Engineering Education and Research, Amrutdham, Panchavati, Nashik 422003, India

R. K. Pillai et al. (eds.), *ISUW 2019*, Lecture Notes in Electrical Engineering 764,
https://doi.org/10.1007/978-981-16-1299-2_11

1 Introduction

Nowadays, world is changing due to advent of numerous new technologies and innovations in EV. By 2030 all conventional vehicles will be fully electric. In Electric Vehicle energy storage system is a key ingredient as it affects the efficiency and driving performance [1]. The battery is the main power source available in the market. As vehicle is subjected to different time varying power demands battery has to supply large current which affects the battery performance and life. Also battery has limitations of power density and limited driving range [2]. Therefore no single element (battery) can fulfil the all desirable characteristics. Increasing size of battery to fulfil required power it will cause increase in cost and weight. The midway is hybridization that allows two different energy storage elements of different characteristics such as high energy density and high power density i.e. battery and supercapacitor (SC) respectively. So it gives efficient Hybrid Energy Storage System (HESS) [3–7]. In that battery is used to supply low and average power and supercapacitor is for peak power so as to battery current maintained as constant as possible. Thus efficiency and performance of an EV can be improved [8]. EV requires high power in dynamic condition which requires high power density and energy density source. So proper distribution of power between battery and supercapacitor requires more attention. To handle power split between two energy sources power electronic converter are used. For enhancing performance of energy sources DC/DC converters are used [9]. By using hybrid energy source not only workload on battery reduces but also overall performance improves [10]. Second another important point is proper sizing of energy storage system. It is critical as it related to cost of EV. This paper presents optimal sizing of battery and supercapacitor for pure EV based on theoretical formulas. The paper contents are as follows. The Sect. 2 Summarizes Hybrid Energy Storage System. Section 3 gives HESS Sizing. The Sect. 4 describes Overall System Structure, The Sects. 5, 6 and 7 estimates Power management algorithm, Simulation results and conclusion respectively.

2 Hybrid Energy Storage System

Figure 1 shows the Regon plot of specific energy verses specific power for all Hybrid energy technologies. Li–S have high energy density than the other technology also it have higher thermal runway onset temperature. This makes it suitable for EV applications. On the other hand supercapacitor has high power density than other. Combining these two technologies gives efficient energy storage System [11].

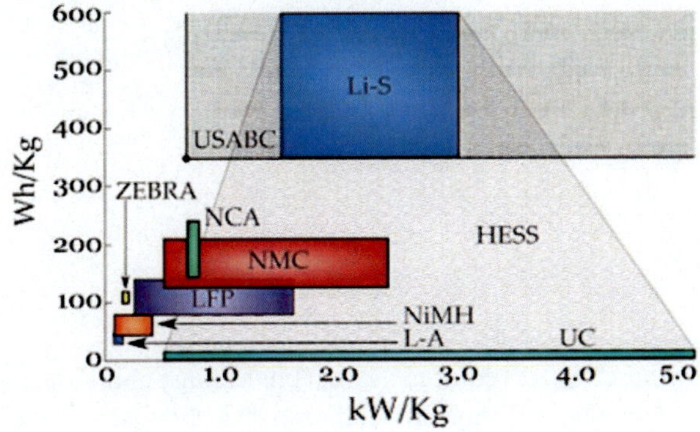

Fig. 1 Regon plot [11]

2.1 *Battery*

There are various batteries available as energy storage. Due to advancement in Li-based battery technology these are most popular in case of electric drive vehicles as it has high energy density and allows EV to have longer driving cycle. Conventional Li-ion battery has theoretical specific energy of 387 Wh/kg but commercially it is only up to 240 Wh/kg. Li Air (Li_2O_2) and $Li-S$ based technology have been gaining very high theoretical specific energy of 3582 and 2567 Wh/kg respectively [12, 13].

2.2 *Supercapacitor*

Supercapacitor is a double layer capacitor. Its electrical Characteristics are stable and operates on wide range of temperature. It have fast charging and discharging, also SC can meet other USABC goals such as specific power and life cycle which is suitable for EV applications. According to U.C Davis Institute study SC can maintain a specific energy and specific power of 30 and 10 Wh/kg and 3395 and 2540 W/kg respectively [11].

3 HESS Sizing

In pure electric vehicle sizing of energy storage system is the key point. Sizing should be such that it will meet all vehicle dynamics. Mainly two parameters have to consider namely nominal voltage and Ah rating of battery and nominal voltage and capacitance in case of SC. These specifications can be meet by reconfiguring series–parallel combinations. For desired bus voltage number of cells are connected in series and for desired energy number of cells are in parallel. Number of series connected cells N_{Bat_S} and $N_{SC}-s$ can be calculated by Eqs. 1 and 2 [14].

$$N_{Bat\_S} = \frac{V_{bus}}{V_{celbat}} \tag{1}$$

$$N_{SC\_S} = \frac{V_{bus}}{V_{SC}} \tag{2}$$

where V_{bus} is required system bus voltage and V_{celbat} and V_{SC} are the nominal voltage of battery cell and SC cell respectively. Capacitance of supercapacitor is given by Eq. 3

$$C_{series} = \frac{C_{sc}}{N_{SC}} \tag{3}$$

where C_{series} equivalent capacitance of series connected supercapacitor.

Therefore required No. of supercapacitors in parallel is given by

$$N_{SC\_P} = \frac{C_{defined}}{C_{series}} \tag{4}$$

where $C_{defined}$ is required capacitance to fulfil load profile. Energy and Power for both energy storage elements can be calculated

$$P_{te} = F_{te} * v_{vehicle} \tag{5}$$

$$Power_{SC} = \frac{V_{SC}^2}{4ESR} \tag{6}$$

$$BatteryAhrating = \frac{E}{V_{Nominal}} \tag{7}$$

$$E = \frac{P_{te}}{\eta_{DT}} * \left(\frac{ZEV\,Range}{3600^* v_{vehicle}}\right) \tag{8}$$

ZEV Range is Zero Emission Vehicle. The desired ZEV is 60 miles on a level road.

$$E_{SC} = 1/2CV^2 \tag{9}$$

where

ESR is the Equivalent Series Resistance of supercapacitor

F_{te} = Tractive force (N) P_{te} = Tractive power (W) P_{SC} = SC power (W)

$v_{vehicle}$ = Vehicle velocity (m/s)

η_{DT} = Drivetrain efficiency
E_{SC} = Energy required (Wh)
$V_{Nominal}$ = Battery Nominal Voltage (Volt).

4 Overall System Structure

Figure 2 presents the structure of overall system which is considered for study. The multiple converter topology is used. It allows complete control over both energy sources. It composed of two energy storage elements: Battery and Supercapacitor. The DC/DC Unidirectional boost converter for battery and DC/DC bidirectional buck/boost converter for Supercapacitor. DC link voltage value is kept constant as its output. The overall system is controlled using PID controller. Figures 3 and 4 shows unidirectional boost converter used for battery and bidirectional buck/boost converter for Supercapacitor. PID Controllers consist of gains, differentiation and

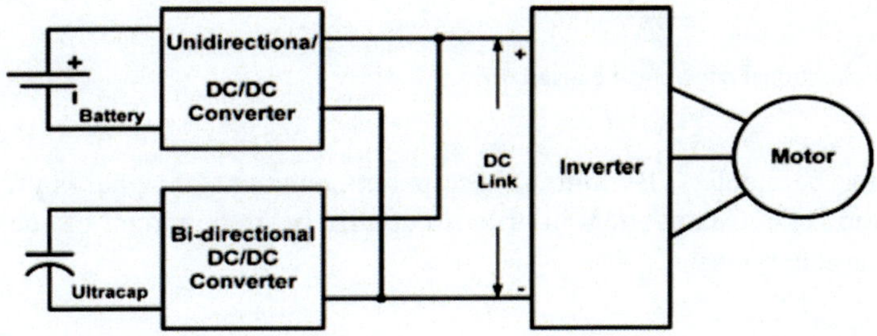

Fig. 2 Overall structure of system

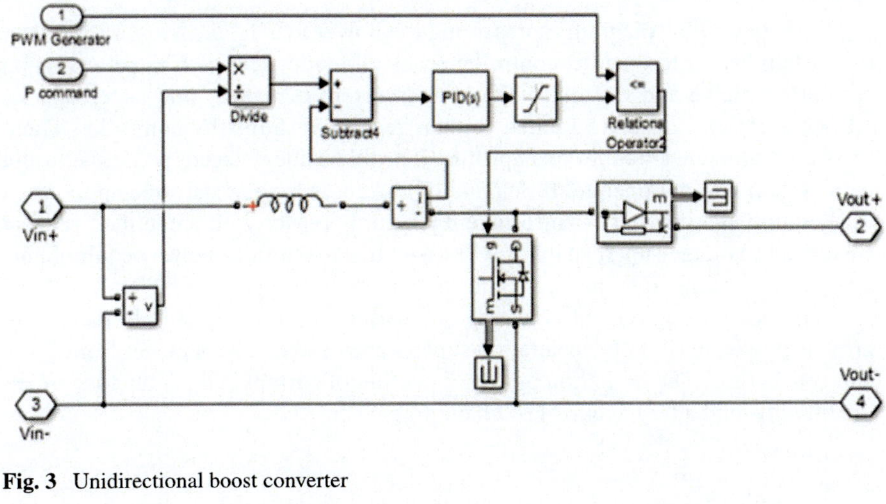

Fig. 3 Unidirectional boost converter

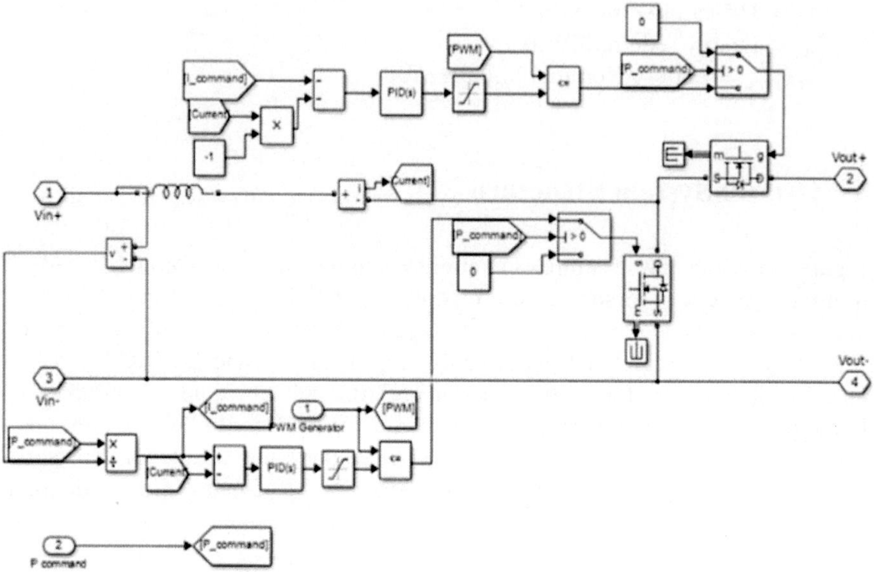

Fig. 4 Bidirectional buck/boost converter

integration parameters. By tuning these parameters control signal is produced and sent to the DC/DC converters in order to control desired amount of energy needed for the vehicle range.

5 Power Management Algorithm

Figure 5 shows flow diagram representing the power management algorithm. When the system begins to operate, controller reads all parameters such as battery, supercapacitor voltage and DC link voltage and current of battery and supercapacitor. All the variable values are limited to their respective limits by controller. Then it compares with power requirement profile. If profile values exceeds predefined values then system will not operate. If demand of current is high at starting and in case of acceleration and if battery is insufficient to supply power with predefined values of current and voltage supercapacitor will assist battery to fulfil power requirement at that particular time to reduce high current extraction from battery.

If the power requirement is average or within the limits then only battery will provide power. In case of deceleration supercapacitor will accept power from system and get charged. The limitation of rate of rise/fall of current in the inductors protects the lithium-ion battery against fast changing power.

Fig. 5 Power management algorithm

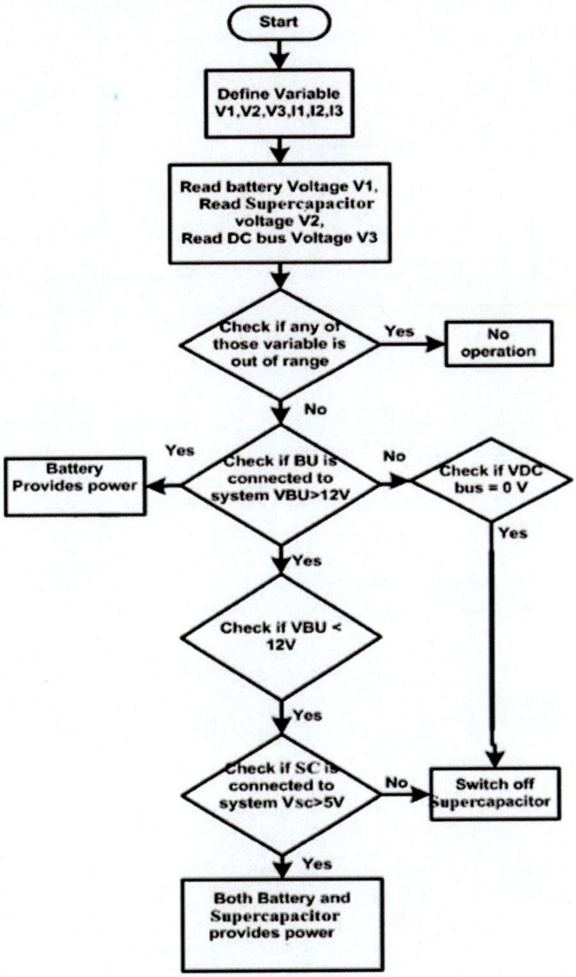

6 Simulation Result

In order to evaluate the behaviour of the studied system here directly its power requirement profile is considered is as shown in Fig. 6 MATLAB Simulink model of HESS is as shown in Fig. 7 and Table 1.

Figures 8 and 9 Shows the voltage, current and % SOC of Battery and supercapacitor respectively. Figure 10 shows the power supplied by Battery and supercapacitor respectively. It is observed that when the power requirement is high at starting almost up to 3 s battery is insufficient to provide power alone to the system, so in that case supercapacitor provides power with battery. After that from 3 to 6 s power requirement is moderate in that case battery is able to provide power alone without extracting high current. For that period power supplied by supercapacitor is zero. For the period

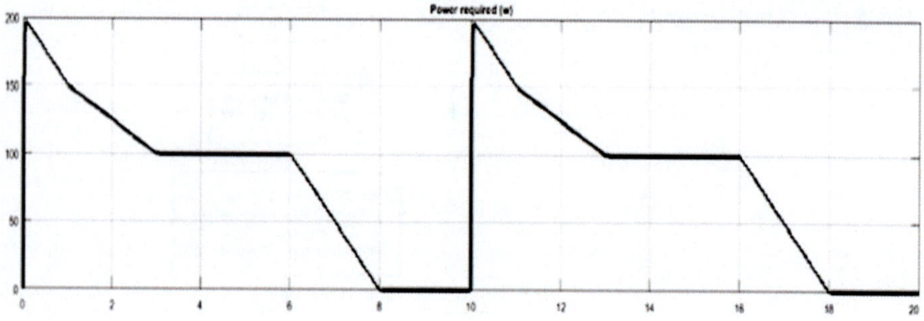

Fig. 6 Power requirement profile

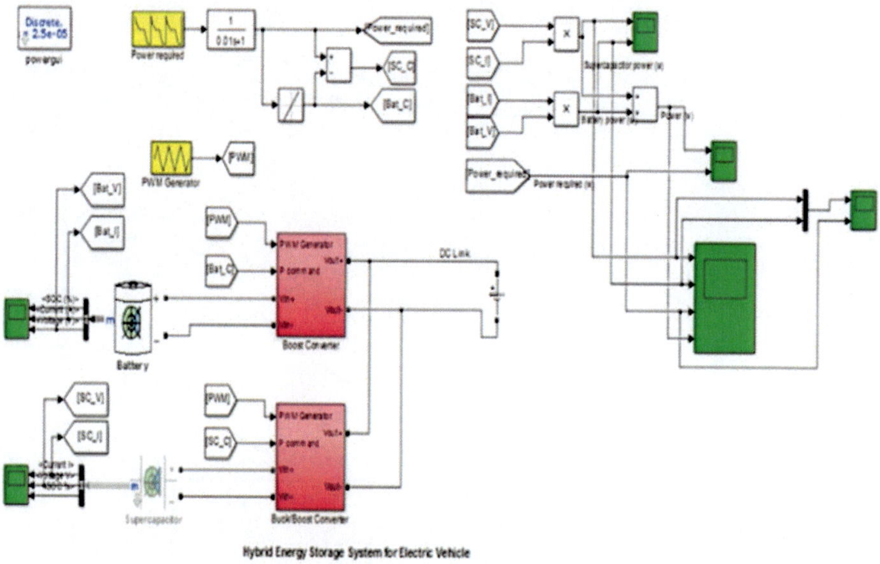

Fig. 7 MATLAB simulink model of HESS

Table 1 Specifications of energy storage device used

Device	Specifications
Supercapacitor module	5.4 V, 100 F
Battery Pack	7.4 V, 2.2 Ah @ 100% SOC
DC link voltage	12 V

from 6 to 8 s there is deceleration in that case regenerative power is recovered in supercapacitor and it get charged. For the period 8–10 s there is no power requirement from the system, so supplied by both battery and supercapacitor is zero. Also

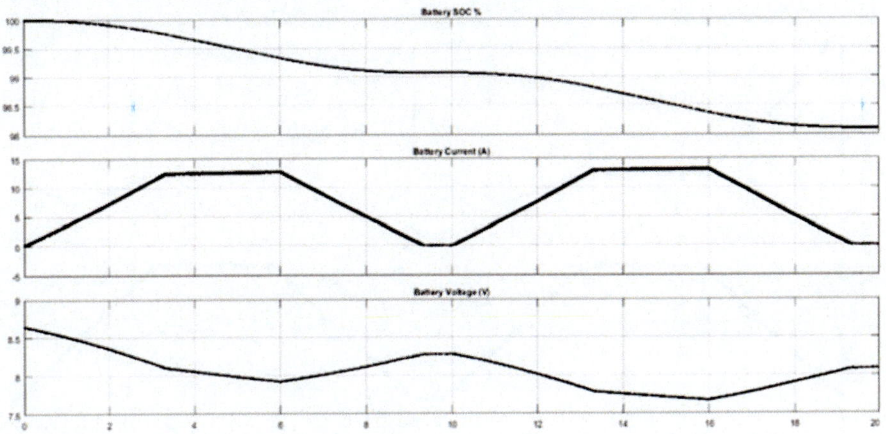

Fig. 8 Voltage, current and % SOC of battery

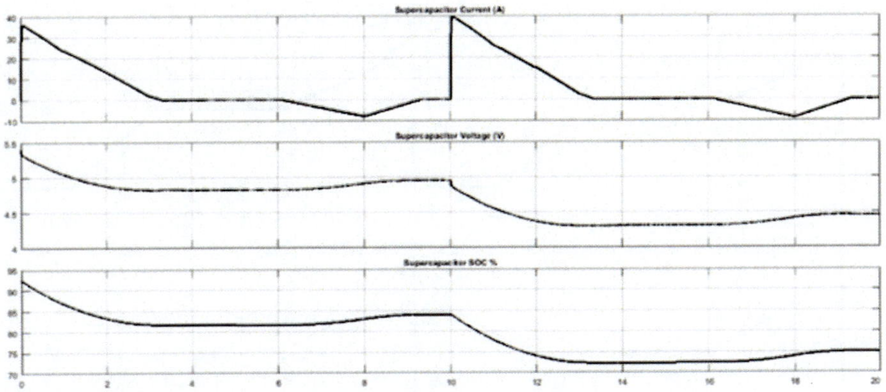

Fig. 9 Voltage, current and % SOC of supercapacitor

it is observed that battery is current is maintained as constant as possible in case of vehicle dynamics which is more important to extend battery life.

Figure 10 Shows the Power supplied by battery and Supercapacitor and Fig. 11 shows the combined power supplied by both supercapacitor and battery. From this it is observed that power requirement from the system is exactly fulfil by using hybrid energy storage elements i.e. battery and supercapacitor

7 Conclusion

From the MATLAB Simulink model of the HESS it is concluded that sizing of hybrid energy storage system for required power plays important role in pure electric

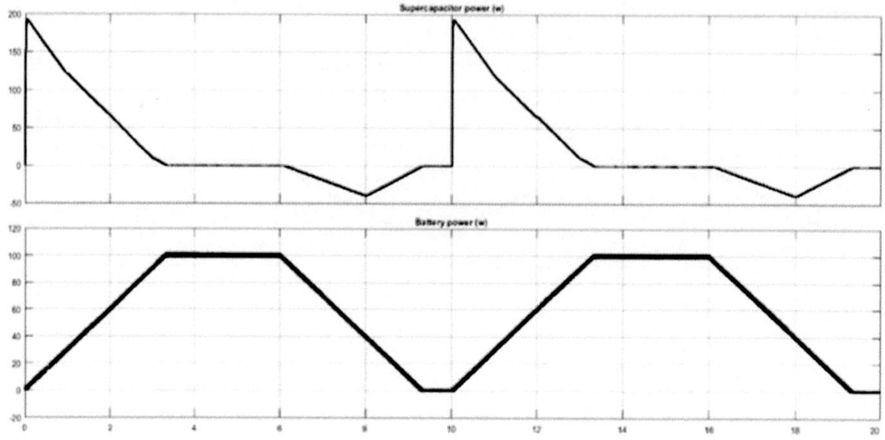

Fig. 10 Power supplied by battery and supercapacitor

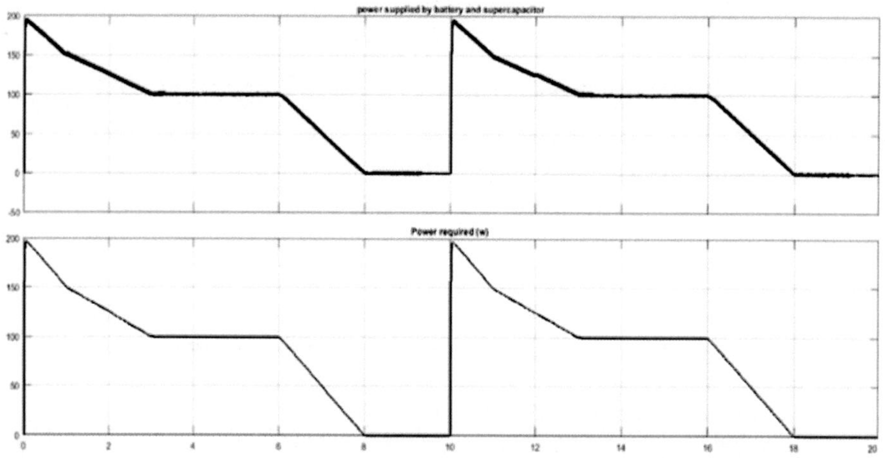

Fig. 11 Combine power supplied by battery and SC and power required

vehicle. The main challenge is to calculate number of series and parallel combination of individual energy storage element. Power required by the system is exactly fulfil by using Hybrid Energy storage system and DC/DC converter discussed in this paper. The battery current is maintained as constant as possible as peak current is supplied by auxiliary energy storage element i.e. Supercapacitor so battery life is significantly increased.

Future Scope

Design of MATLAB Simulink model with BLDC motor which is not considered in this paper. Design of On-board battery charger for the given system.

References

1. Zhang L, Hu X, Wang Z, Sun F, Deng J, Dorrell DG (2018) Multi objective optimal sizing of hybrid energy storage system for electric vehicles. IEEE Trans Veh Technol 67:1027–1035
2. Kachhwaha A, Shah VA, Shimin VV (2016) Integration methodology of ultracapacitor-battery based hybrid energy storage system for electrical vehicle power management. In: 2016 IEEE 7th power india international conference (PIICON), pp 1–6
3. Lahyani (2013) Battery/supercapacitors combination in uninterruptible power supply (UPS). IEEE Trans Power Electron 28(4)
4. Moungngam W, Phatrapornnant T (2014) Applied supercapacitor to energy storage during regenerative brake state in electric vehicle. IEEE Trans Power Electron 978-1-4799-2993–1/14
5. Badawy MO, Sozer Y (2015) A partial power processing of battery/ultracapacitor hybrid energy storage system for electric vehicles. In: 2015 IEEE applied power electronics conference and exposition (APEC). https://doi.org/10.1109/APEC.2015.7104804
6. Saikong W, Kulworawanichpong T (2017) Comparative study of energy consumption for electric vehicles with various on-board energy storage systems. In: 2017 international conference on alternative energy in developing countries and emerging economies 2017 AEDCEE, Bangkok, Thailand
7. Marzougui H, Kadri A, Amari M, Bacha F (2017) Improvement of energy management algorithm for fuel cell electrical vehicle with fuzzy logic. In: 2017 18th international conference on sciences and techniques of automatic control and computer engineering (STA), Monastir, pp 212–217
8. Herath N, Binduhewa P, Samaranayake L, Ekanayake J, Longo S (2017) Design of a dual energy storage power converter for a small electric vehicle. In: IEEE international conference on industrial and information systems (ICIIS), Peradeniya, pp 1–6
9. Kong Z, Cui N, Li P (2017) Energy management strategy coordinating lithium-ion battery and ultra-capacitor for electric vehicle. In: 36th Chinese control conference (CCC), Dalian, pp 9291–9296
10. Chemali E, Preindl M, Malysz P, Emadi A (2016) Electrochemical and electrostatic energy storage and management systems for electric drive vehicles: state-of-the-art review and future trends. IEEE J Emerg Select Top Power Electron 4(3):1117–1134
11. Bruce PG, Freunberger SA, Hardwick LJ, Tarascon J-M (2012) Li-O2 and Li-S batteries with high energy storage. Nature Mater 11(1):19–29
12. Bruce PG, Hardwick LJ, Abraham KM (2011) Lithium-air and lithium-sulfur batteries. MRS Bull 36(7):506–512
13. Schaltz E (2008) Electrical vehicle design and modelling. In: Soylu S (ed) Electric vehicles—modelling and simulations, pp 114
14. Ehsani SEGM, Gao Y, Emadi A (2004) Modern electric, hybrid electric, and fuel cell vehicles: fundamentals, theory, and design, vol 2. CRC Press, Boca Raton, FL

Performance Evaluation of Electric Vehicle Using Hybrid Energy Storage System

Minal R. Rade, J. A. Mane, and O. N. Buwa

Abstract The fuel efficiency and performance of novel vehicles with electric propulsion capability are largely limited by the performance of Energy Storage System. The battery system choice is a crucial item but no single type of energy storage element fulfils high energy density, high power delivery capacity, low cost per unit of storage, long cycle life, low leakage, and so on at the same time. One of the best solution is to use a Hybrid Energy Storage System. The main objective is to design of a hybrid electrical energy storage system which gives substantial benefits against battery issues such as reduction in battery stress. Also to maintain battery current as constant as possible during transients to limit battery stress. On the other hand supercapacitor has capability to charge as fast as possible without exceeding maximum current from regenerative breaking and to discharge most of its stored energy during acceleration. Adding supercapacitor bank will assist the battery during vehicle acceleration and hill climbing and with its quick recharge capability, it will assist the battery in capturing the regenerative breaking energy. This significant advantage a battery-supercapacitor energy storage system gained attention. Battery and supercapacitor sizing includes the most important and difficult steps is, the determination of the numbers of batteries and supercapacitor connected in series and parallel. The power management is essentially the optimal distribution of power between battery and supercapacitor. With supercapacitor added into the hybrid energy storage system, battery workload is reduced, which leads to significant extension of battery life.

Keywords Electric vehicle · Hybrid energy storage system · Supercapacitor · Battery · DC-DC converter

---

M. R. Rade (✉) · J. A. Mane · O. N. Buwa
Department of Electrical Engineering, K. K. Wagh Institute of Engineering Education and Research, Amrutdham, Panchavati, Nashik 422003, India

R. K. Pillai et al. (eds.), *ISUW 2019*, Lecture Notes in Electrical Engineering 764,
https://doi.org/10.1007/978-981-16-1299-2_11

1 Introduction

Nowadays, world is changing due to advent of numerous new technologies and innovations in EV. By 2030 all conventional vehicles will be fully electric. In Electric Vehicle energy storage system is a key ingredient as it affects the efficiency and driving performance [1]. The battery is the main power source available in the market. As vehicle is subjected to different time varying power demands battery has to supply large current which affects the battery performance and life. Also battery has limitations of power density and limited driving range [2]. Therefore no single element (battery) can fulfil the all desirable characteristics. Increasing size of battery to fulfil required power it will cause increase in cost and weight. The midway is hybridization that allows two different energy storage elements of different characteristics such as high energy density and high power density i.e. battery and supercapacitor (SC) respectively. So it gives efficient Hybrid Energy Storage System (HESS) [3–7]. In that battery is used to supply low and average power and supercapacitor is for peak power so as to battery current maintained as constant as possible. Thus efficiency and performance of an EV can be improved [8]. EV requires high power in dynamic condition which requires high power density and energy density source. So proper distribution of power between battery and supercapacitor requires more attention. To handle power split between two energy sources power electronic converter are used. For enhancing performance of energy sources DC/DC converters are used [9]. By using hybrid energy source not only workload on battery reduces but also overall performance improves [10]. Second another important point is proper sizing of energy storage system. It is critical as it related to cost of EV. This paper presents optimal sizing of battery and supercapacitor for pure EV based on theoretical formulas. The paper contents are as follows. The Sect. 2 Summarizes Hybrid Energy Storage System. Section 3 gives HESS Sizing. The Sect. 4 describes Overall System Structure, The Sects. 5, 6 and 7 estimates Power management algorithm, Simulation results and conclusion respectively.

2 Hybrid Energy Storage System

Figure 1 shows the Regon plot of specific energy verses specific power for all Hybrid energy technologies. Li–S have high energy density than the other technology also it have higher thermal runway onset temperature. This makes it suitable for EV applications. On the other hand supercapacitor has high power density than other. Combining these two technologies gives efficient energy storage System [11].

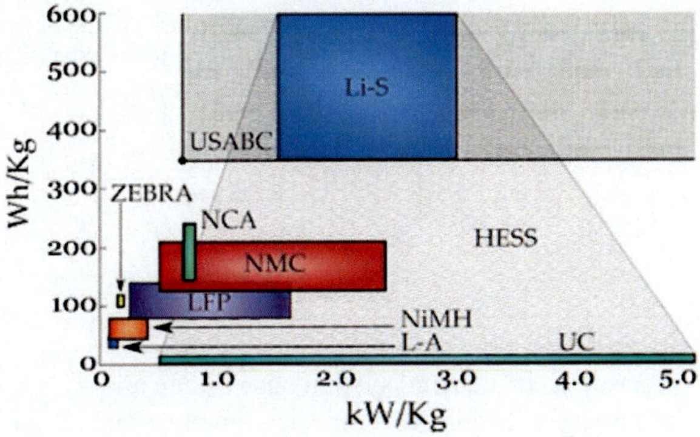

Fig. 1 Regon plot [11]

2.1 *Battery*

There are various batteries available as energy storage. Due to advancement in Li-based battery technology these are most popular in case of electric drive vehicles as it has high energy density and allows EV to have longer driving cycle. Conventional Li-ion battery has theoretical specific energy of 387 Wh/kg but commercially it is only up to 240 Wh/kg. Li Air (Li_2O_2) and $Li{-}S$ based technology have been gaining very high theoretical specific energy of 3582 and 2567 Wh/kg respectively [12, 13].

2.2 *Supercapacitor*

Supercapacitor is a double layer capacitor. Its electrical Characteristics are stable and operates on wide range of temperature. It have fast charging and discharging, also SC can meet other USABC goals such as specific power and life cycle which is suitable for EV applications. According to U.C Davis Institute study SC can maintain a specific energy and specific power of 30 and 10 Wh/kg and 3395 and 2540 W/kg respectively [11].

3 HESS Sizing

In pure electric vehicle sizing of energy storage system is the key point. Sizing should be such that it will meet all vehicle dynamics. Mainly two parameters have to consider namely nominal voltage and Ah rating of battery and nominal voltage and capacitance in case of SC. These specifications can be meet by reconfiguring series–parallel combinations. For desired bus voltage number of cells are connected in series and for desired energy number of cells are in parallel. Number of series connected cells N_{Bat_S} and $N_{SC}-s$ can be calculated by Eqs. 1 and 2 [14].

$$N_{Bat\_S} = \frac{V_{bus}}{V_{celbat}} \quad (1)$$

$$N_{SC\_S} = \frac{V_{bus}}{V_{SC}} \quad (2)$$

where V_{bus} is required system bus voltage and V_{celbat} and V_{SC} are the nominal voltage of battery cell and SC cell respectively. Capacitance of supercapacitor is given by Eq. 3

$$C_{series} = \frac{C_{sc}}{N_{SC}} \quad (3)$$

where C_{series} equivalent capacitance of series connected supercapacitor.

Therefore required No. of supercapacitors in parallel is given by

$$N_{SC\_P} = \frac{C_{defined}}{C_{series}} \quad (4)$$

where $C_{defined}$ is required capacitance to fulfil load profile. Energy and Power for both energy storage elements can be calculated

$$P_{te} = F_{te} * v_{vehicle} \quad (5)$$

$$Power_{SC} = \frac{V_{SC}^2}{4ESR} \quad (6)$$

$$BatteryAhrating = \frac{E}{V_{Nominal}} \quad (7)$$

$$E = \frac{P_{te}}{\eta_{DT}} * \left(\frac{ZEV\,Range}{3600^* v_{vehicle}}\right) \quad (8)$$

ZEV Range is Zero Emission Vehicle. The desired ZEV is 60 miles on a level road.

$$E_{SC} = 1/2CV^2 \quad (9)$$

where

ESR is the Equivalent Series Resistance of supercapacitor

F_{te} = Tractive force (N) P_{te} = Tractive power (W) P_{SC} = SC power (W)

$v_{vehicle}$ = Vehicle velocity (m/s)

η_{DT} = Drivetrain efficiency
E_{SC} = Energy required (Wh)
$V_{Nominal}$ = Battery Nominal Voltage (Volt).

4 Overall System Structure

Figure 2 presents the structure of overall system which is considered for study. The multiple converter topology is used. It allows complete control over both energy sources. It composed of two energy storage elements: Battery and Supercapacitor. The DC/DC Unidirectional boost converter for battery and DC/DC bidirectional buck/boost converter for Supercapacitor. DC link voltage value is kept constant as its output. The overall system is controlled using PID controller. Figures 3 and 4 shows unidirectional boost converter used for battery and bidirectional buck/boost converter for Supercapacitor. PID Controllers consist of gains, differentiation and

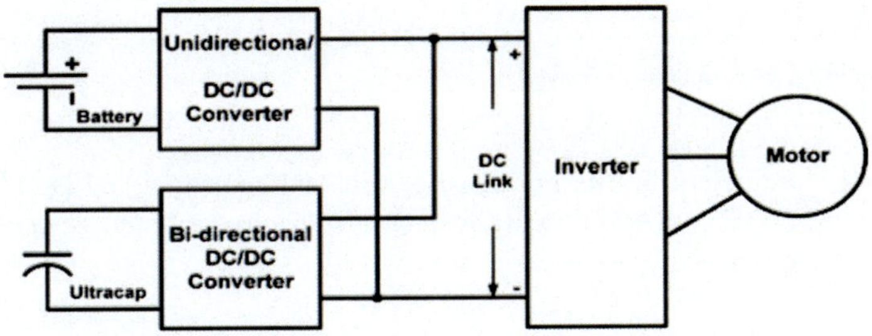

Fig. 2 Overall structure of system

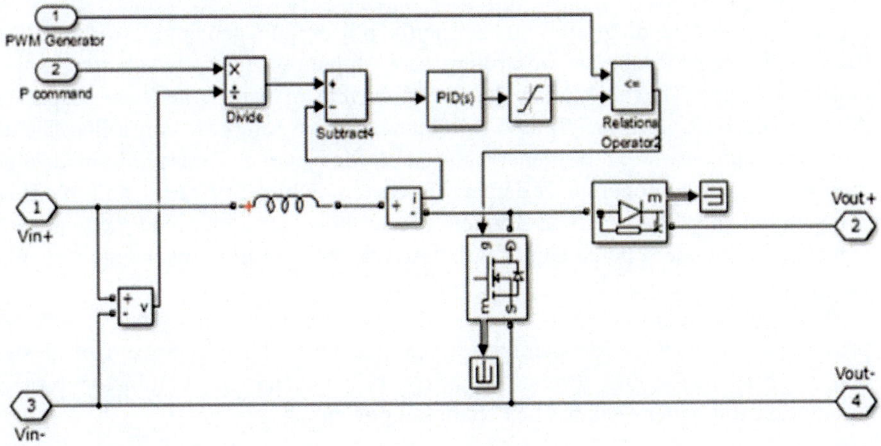

Fig. 3 Unidirectional boost converter

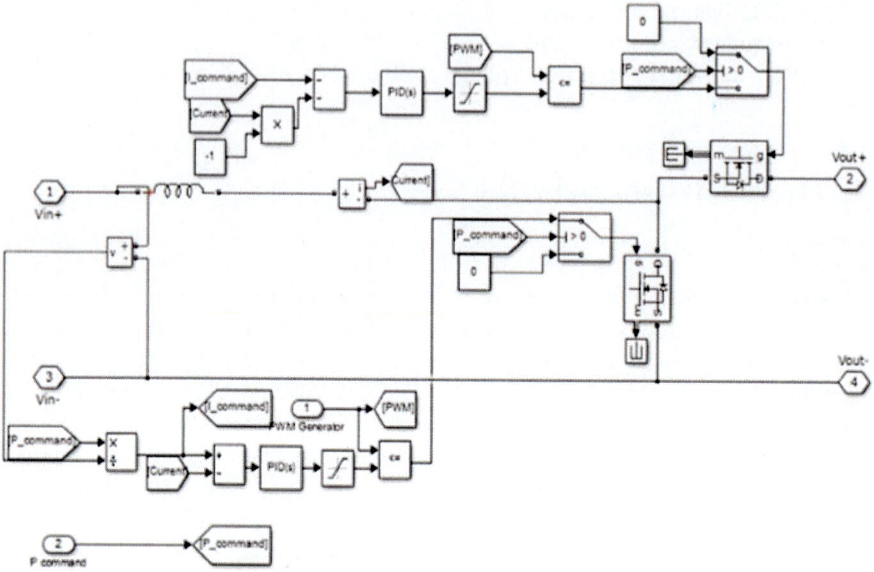

Fig. 4 Bidirectional buck/boost converter

integration parameters. By tuning these parameters control signal is produced and sent to the DC/DC converters in order to control desired amount of energy needed for the vehicle range.

5 Power Management Algorithm

Figure 5 shows flow diagram representing the power management algorithm. When the system begins to operate, controller reads all parameters such as battery, supercapacitor voltage and DC link voltage and current of battery and supercapacitor. All the variable values are limited to their respective limits by controller. Then it compares with power requirement profile. If profile values exceeds predefined values then system will not operate. If demand of current is high at starting and in case of acceleration and if battery is insufficient to supply power with predefined values of current and voltage supercapacitor will assist battery to fulfil power requirement at that particular time to reduce high current extraction from battery.

If the power requirement is average or within the limits then only battery will provide power. In case of deceleration supercapacitor will accept power from system and get charged. The limitation of rate of rise/fall of current in the inductors protects the lithium-ion battery against fast changing power.

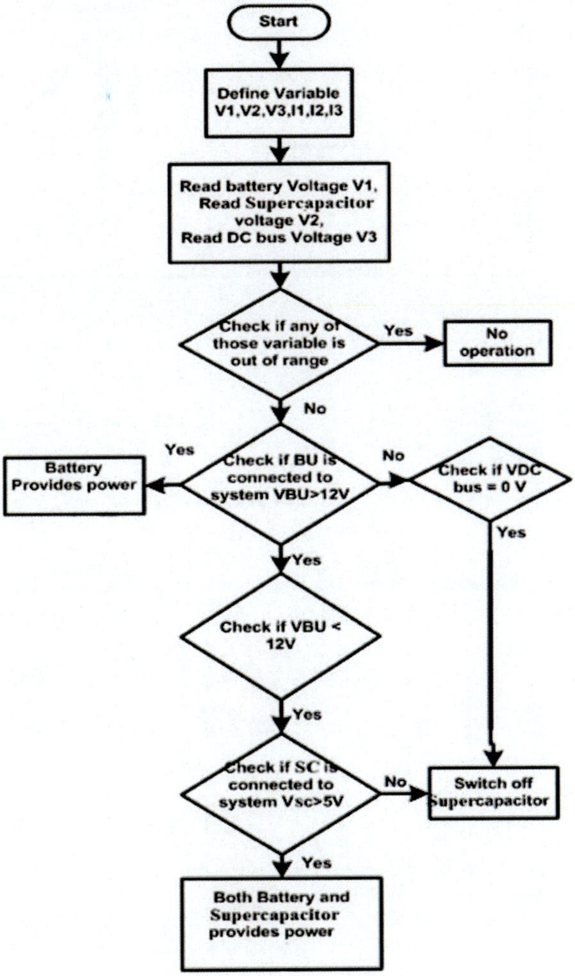

Fig. 5 Power management algorithm

6 Simulation Result

In order to evaluate the behaviour of the studied system here directly its power requirement profile is considered is as shown in Fig. 6 MATLAB Simulink model of HESS is as shown in Fig. 7 and Table 1.

Figures 8 and 9 Shows the voltage, current and % SOC of Battery and supercapacitor respectively. Figure 10 shows the power supplied by Battery and supercapacitor respectively. It is observed that when the power requirement is high at starting almost up to 3 s battery is insufficient to provide power alone to the system, so in that case supercapacitor provides power with battery. After that from 3 to 6 s power requirement is moderate in that case battery is able to provide power alone without extracting high current. For that period power supplied by supercapacitor is zero. For the period

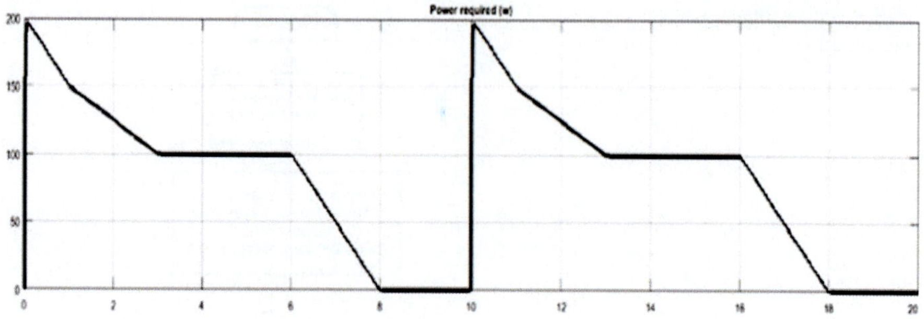

Fig. 6 Power requirement profile

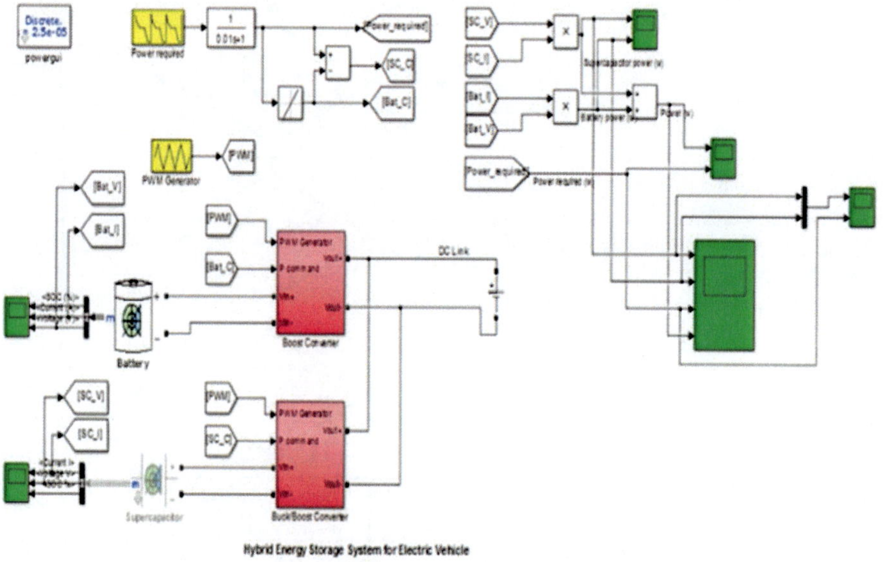

Fig. 7 MATLAB simulink model of HESS

Table 1 Specifications of energy storage device used

Device	Specifications
Supercapacitor module	5.4 V, 100 F
Battery Pack	7.4 V, 2.2 Ah @ 100% SOC
DC link voltage	12 V

from 6 to 8 s there is deceleration in that case regenerative power is recovered in supercapacitor and it get charged. For the period 8–10 s there is no power requirement from the system, so supplied by both battery and supercapacitor is zero. Also

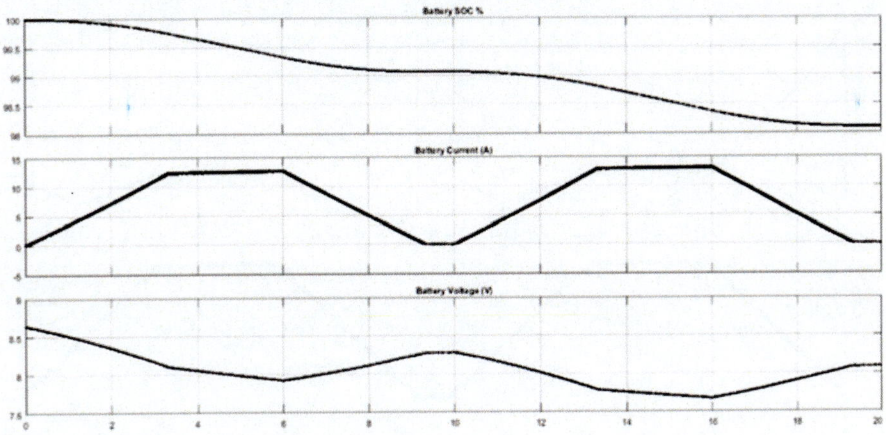

Fig. 8 Voltage, current and % SOC of battery

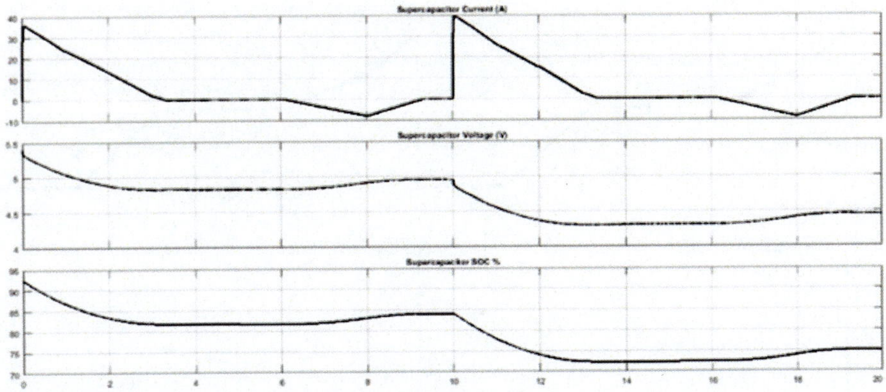

Fig. 9 Voltage, current and % SOC of supercapacitor

it is observed that battery is current is maintained as constant as possible in case of vehicle dynamics which is more important to extend battery life.

Figure 10 Shows the Power supplied by battery and Supercapacitor and Fig. 11 shows the combined power supplied by both supercapacitor and battery. From this it is observed that power requirement from the system is exactly fulfil by using hybrid energy storage elements i.e. battery and supercapacitor

7 Conclusion

From the MATLAB Simulink model of the HESS it is concluded that sizing of hybrid energy storage system for required power plays important role in pure electric

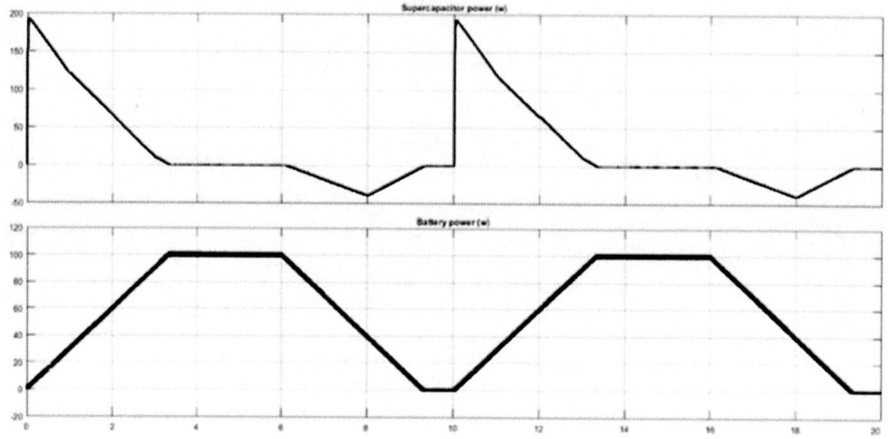

Fig. 10 Power supplied by battery and supercapacitor

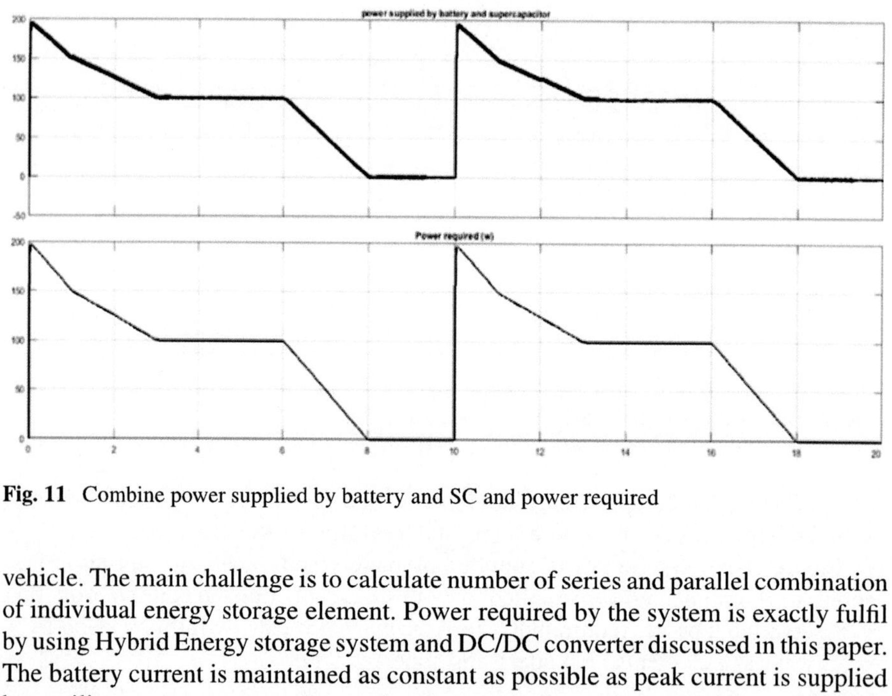

Fig. 11 Combine power supplied by battery and SC and power required

vehicle. The main challenge is to calculate number of series and parallel combination of individual energy storage element. Power required by the system is exactly fulfil by using Hybrid Energy storage system and DC/DC converter discussed in this paper. The battery current is maintained as constant as possible as peak current is supplied by auxiliary energy storage element i.e. Supercapacitor so battery life is significantly increased.

Future Scope

Design of MATLAB Simulink model with BLDC motor which is not considered in this paper. Design of On-board battery charger for the given system.

References

1. Zhang L, Hu X, Wang Z, Sun F, Deng J, Dorrell DG (2018) Multi objective optimal sizing of hybrid energy storage system for electric vehicles. IEEE Trans Veh Technol 67:1027–1035
2. Kachhwaha A, Shah VA, Shimin VV (2016) Integration methodology of ultracapacitor-battery based hybrid energy storage system for electrical vehicle power management. In: 2016 IEEE 7th power india international conference (PIICON), pp 1–6
3. Lahyani (2013) Battery/supercapacitors combination in uninterruptible power supply (UPS). IEEE Trans Power Electron 28(4)
4. Moungngam W, Phatrapornnant T (2014) Applied supercapacitor to energy storage during regenerative brake state in electric vehicle. IEEE Trans Power Electron 978-1-4799-2993–1/14
5. Badawy MO, Sozer Y (2015) A partial power processing of battery/ultracapacitor hybrid energy storage system for electric vehicles. In: 2015 IEEE applied power electronics conference and exposition (APEC). https://doi.org/10.1109/APEC.2015.7104804
6. Saikong W, Kulworawanichpong T (2017) Comparative study of energy consumption for electric vehicles with various on-board energy storage systems. In: 2017 international conference on alternative energy in developing countries and emerging economies 2017 AEDCEE, Bangkok, Thailand
7. Marzougui H, Kadri A, Amari M, Bacha F (2017) Improvement of energy management algorithm for fuel cell electrical vehicle with fuzzy logic. In: 2017 18th international conference on sciences and techniques of automatic control and computer engineering (STA), Monastir, pp 212–217
8. Herath N, Binduhewa P, Samaranayake L, Ekanayake J, Longo S (2017) Design of a dual energy storage power converter for a small electric vehicle. In: IEEE international conference on industrial and information systems (ICIIS), Peradeniya, pp 1–6
9. Kong Z, Cui N, Li P (2017) Energy management strategy coordinating lithium-ion battery and ultra-capacitor for electric vehicle. In: 36th Chinese control conference (CCC), Dalian, pp 9291–9296
10. Chemali E, Preindl M, Malysz P, Emadi A (2016) Electrochemical and electrostatic energy storage and management systems for electric drive vehicles: state-of-the-art review and future trends. IEEE J Emerg Select Top Power Electron 4(3):1117–1134
11. Bruce PG, Freunberger SA, Hardwick LJ, Tarascon J-M (2012) Li-O2 and Li-S batteries with high energy storage. Nature Mater 11(1):19–29
12. Bruce PG, Hardwick LJ, Abraham KM (2011) Lithium-air and lithium-sulfur batteries. MRS Bull 36(7):506–512
13. Schaltz E (2008) Electrical vehicle design and modelling. In: Soylu S (ed) Electric vehicles—modelling and simulations, pp 114
14. Ehsani SEGM, Gao Y, Emadi A (2004) Modern electric, hybrid electric, and fuel cell vehicles: fundamentals, theory, and design, vol 2. CRC Press, Boca Raton, FL

A Proposed System for Electric Charging Vehicle Infrastructure

Moreshwar Salpekar

Abstract Electric vehicle development has started in India. The Government of India has announced policies to promote electric vehicles. Electric vehicles are no different from other vehicles except they require charging. The charging is mostly controlled using microcontroller based systems in both vehicle and charging station. The paper proceeds to briefly describe how the design of software and firmware. It also gives the challenges that are perceived in design and briefly describes some reasons for the challenges and issues. It then proposes a design for the system that uses machine learning system to improvise the charging and generate billing accurately. The proposed system will give a generic software stack and how machine learning system will interact with this stack. This is expected to help companies build an efficient software for both charging and billing.

Keywords Electric vehicle · Battery management system · Firmware · Security

1 Introduction

The Government of India announced the FAME policy in 2018 to promote electric vehicles. This was followed by policies by states last one being Delhi which will set up electric vehicle charging stations every 3 km. Electric Vehicle charging infrastructure is already coming up in states all over India. Organisations are developing both vehicles and charging infrastructure both of which are controlled and monitored by firmware with respect to charging and on road operation.

The firmware residing in either charging station infrastructure or electric vehicle is usually modular and is in form of stack. Quite a few parts of the firmware modules or firmware modules like State of Charge Estimation have been patented even but still stack is generic. The stack even includes capability to send data over network for further processing and analysis like billing and data analytics.

M. Salpekar (✉)
Boson Motors Pvt. Ltd., Pune, India

R. K. Pillai et al. (eds.), *ISUW 2019*, Lecture Notes in Electrical Engineering 764,
https://doi.org/10.1007/978-981-16-1299-2_12

Communication networks have evolved to allow huge amount of data to be sent over network. Clouding computing also has evolved now to allow this data to be processed.

Machine Learning residing in the cloud allows data analytics to be carried out on data allowing automation of some operations that earlier required manual involvement.

A generic stack can be developed which incorporates, the necessary functionality while adding data communication with cloud and data analytics so to allow better monitoring and control of the system (vehicle and charging station).

2 The Electric Vehicle Charging Environment

The electric vehicle charging infrastructure firmware can be divided into two parts.

a. Software that resides in electric vehicles and controls and monitors the operation of battery including charging. It consists of Battery Management System, Motor Drive, Electronic Vehicle Charger Controller (EVCC) and Electronic Control Unit (ECU which is the master).
b. Software that resides in charging station and controls monitoring of the charging operation.

The data analytics system residing in the cloud has a software running on a server and caters to the analysis and processing of data.

2.1 *The Electric Vehicle Firmware*

The block diagram of this is shown in Fig. 1. It resides in the electric vehicle and is responsible for electric vehicle charging and operation. Three subsystems are identified in this firmware

1. Battery Management System
2. Motor Control System
3. Electronic Control Unit (Main Controller).

Each of the above are structured according to the generic software stack is given in Fig. 2 (based on Autosar Classic platform 4.4) [5]

Following vertical software modules relevant to this paper are identified for each of the above. It should be noted that firmware in each of the hardware module confirms to the below vertical module sub-divisions.

1. Control Module
2. Communication Module.

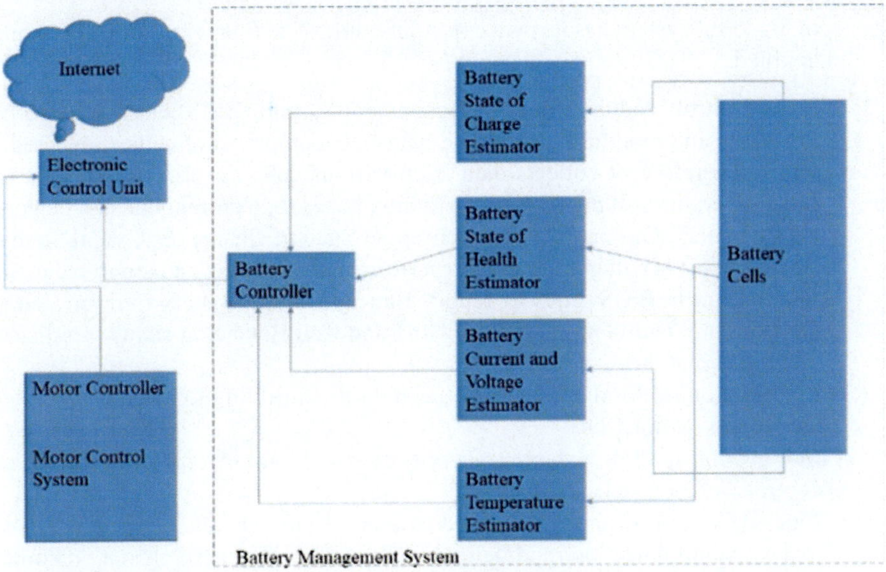

Fig. 1 The electric vehicle subsystems (derived from [1–4])

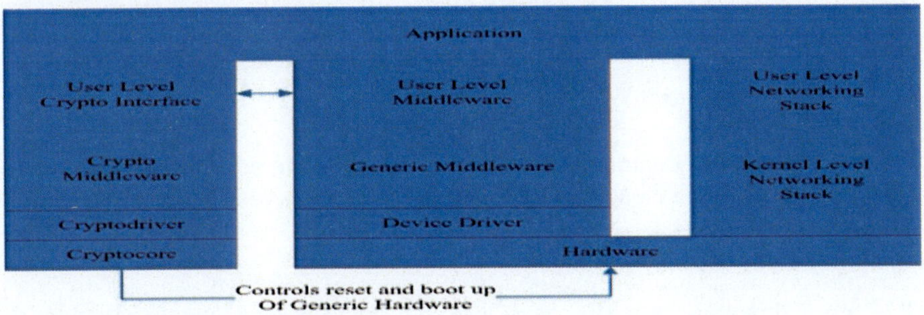

Fig. 2 Software stack (derived from [5])

Each of them is controlled by a thread, with appropriate interrupts to handle incoming and outgoing data. The threads communicate with each other using messages and shared memory.

Each of this is given in sections below.

2.1.1 Control Module

This module is responsible for controlling the hardware it is responsible for. It includes hardware initialization and getting measurements of parameters from system. Each subsystem has its own control module. The following control modules are identified

1. BMS Control Module: Battery Management System (BMS) has responsibility to control and monitor battery. The battery is comprised of cells connected in series. Therefore, it collects data from various cells in form of cell voltages and temperatures. It then computes total voltages total current that can be given out by battery. Optionally, it may compute State of Charge (SoC) and State of Health (SoH) of battery. It may also perform cell balancing and report balancing status. Thus the BMS firmware reports Battery Voltage, Battery Current, Battery Temperature, and Optionally SoC, SoH and Cell Balancing status (see [6] for more information).
2. Display Control Module: It is responsible for controlling LCD display and get user inputs from LCD
3. Motor Control Module: It controls operation of Motor that provides torque and controls rotation (measured by RPM)
4. Charge Control Module: It is used to control electric vehicle charger. It has its own communication module to communicate with charging station control module
5. Master Control Module: it controls and coordinates activities including configuration of other modules. It also collects data and passes to network for transmission to cloud.

2.1.2 Communication Module

It is responsible for communicating parameters to other subsystems. This module also sends data to the cloud for data analytics and get its response (only from main controller).

It may be divided into following

1. Bus Communication Module: it is used to transmit data over the bus e.g. CAN bus to send data or serial bus.
2. Network Communication Module: it is used to send data to cloud for analytics. It may use LPWAN or any other network stack. It is present only in the master control module.

2.2 The Charging Station Firmware

The Charging station firmware can also be vertically divided into.

1. Control Module
2. Communication Module.

2.2.1 Charging Station Control Module

It is responsible for getting data related to charging station. For charging station firmware, parameters are charging current, charging voltage, charging indicator, time of charging and charger temperature. It may optionally report maximum voltage and maximum current. If more than one vehicle can be charged, each charging point has different firmware.

Charging Station Communication Module

It is responsible for communicating parameters to other subsystems. This module also sends data to the cloud for data analytics and get its response (only from main controller). It also communicates with charger control module in electric vehicle if required.

The communication uses

1. Serial Peripheral Interface and/or UART to get data from charging station subsystems
2. Use a LPWAN to send data to cloud for analytics.

3 Addressing the Security Concern

Security needs to be addressed in design phase and not as afterthought. The security is proposed to be compliant with [7]. The security is added in the following manner

1. Boot up software security
2. Run time software security
3. Security in software download
4. Key Management.

3.1 Boot Up Software Security

A Crypto core is added to system for boot up along with secure RAM. The software and keys (non-fused) are stored in secure EEPROM in encrypted form. The crypto core boots starts and verifies the software (using SHA-4 hash, see [8]). Alternately Authentication Encryption (see [9, 10]) may be used. After successful verification, control is passed to two stage bootloader which finally boots up the kernel and application. The security for run time security is given below.

3.2 Run Time Software Security

This is ensured by the following

1. A Secure RAM is provided. This RAM is accessible only in specified operating mode. No debugger access is available to this RAM unless specified instructions are given.
2. Debugging through debugger is disabled in field. Debug can be done only by using prints and that too through kernel log or on console.
3. Software does not access secure EEPROM for anything than keys, if required, during runtime.
4. Even the new software is in this RAM before being flashed to secure EEPROM.
5. The downloaded software is always checked against the Secure HASH transmitted signed by the sender. The signature match and hash) are checked using crypto core (or authenticated encryption check). This is discussed in next section.

3.3 Software Download

The software needs to be updated in field for maintenance which includes feature change and bug fixes. The generic protocol is

1. The software to be downloaded is encrypted with AES 128 bit key (see [11]). Authenticated encryption is proposed to be used (see [9, 10]). Secure Hash Standard (SHS) [8] can also be used but two different algorithms may need further consideration in terms of crypto core processing and memory.
2. The upgrader sends the software and its hash signed with key.
3. The software is downloaded into RAM and also the key.
4. The crypto core is then requested to verify the software.
5. Upon verification, software is written to secure EEPROM.
6. The new software runs on next boot.

3.4 Key Management

Key Management is proposed to be conformant to [12]. The keys required and processing is given below.

3.4.1 The Keys

The following keys are proposed to be present and used

1. Signature key which is used to compute secure hash of software (it is used for software verification).
2. Software OTA decryption key: used to decrypt software sent over the air.
3. Software encryption/decryption (SE/D) key: The software is stored encrypted in secure ROM using this. This is a symmetric key i.e. same key used for encryption and decryption.
4. Encrypter Key: used to encrypt above keys. This is symmetric key.

3.4.2 Key Update

The basic process for update is always same and is as follows:

1. The sender sends request to update key and request is encrypted using previous key and signed by sender.
2. The request is validated using crypto core.
3. The recipient, on successful verification, sends the acknowledgment.
4. The sender sends new key encrypted and signed.
5. The keys is decrypted and verified.
6. The key is stored in secure EEPROM upon successful verification and acknowledgement sent to sender.

Each of key requires more steps after basic process

1. For Signature key: the new hash of software must be sent and hash verified with new key before key is stored.
2. For SE/D key: It must be possible to decrypt the existing software in secure EEPROM using this key else following steps must be done.
    a. The full software is copied from secure EEPROM and copied to RAM
    b. The software is encrypted using new key and stored in secure EEPROM.
3. For Encrypter key: the other two keys must be stored afresh in RAM after encrypting with new key.

4 Conclusion and Further Work

A full software system architecture and high level design is given for electric vehicle charging infrastructure is given. This is an initial architecture. More refinements and partitioning will be present in the architecture. Further, the software also has to comply with ISO 26262 [13], Autosar [14], Coding guidelines such as MISRA [15] also need to be followed to implement a full working system. The communication modules also need a full design thought like choosing which communication protocol to be followed. Usually, Controller Area Network (CAN) [16] is used for communicating for automotive parts so it is expected to be used in communication between motor, BMS and master controller. However, there is a flexibility that other protocols may be used. It is left to implementer to work with actual protocols, hardware components, etc.

References

1. Xing Y, Ma EWM, Tsui KL, Pecht M (2011) Battery management systems in electric and hybrid vehicles. Energies 4:1840–1857. https://doi.org/10.3390/en4111840
2. Renesas (2018) Battery management system tutorial. Retrieved from www.renesas.com/doc/whitepapers/battery-management/battery-management-system-tutorial.pdf on 25th Dec 2018
3. Hu R (2011) Battery management system for electric vehicle applications. Electronic Theses and Dissertations. Paper 5007. Retrieved through https://scholar.uwindsor.ca/etd
4. Kettles D (2015) Electric vehicle charging technology analysis and standards. FSEC Report Number: FSEC-CR-1996-15, U.S. Department of Transportation's University Transportation Centers Program
5. Autosar (2018) General specification of basic software modules AUTOSAR CP Release 4.4.0, 31st Oct 2018. Retrieved from www.autosar.org
6. Meissner E, Richter G (2003) Battery monitoring and electrical energy management precondition for future vehicle electric power systems. J Power Sources 116:79–98
7. Information Technology Laboratory National Institute of Standards and Technology, FIPS PUB 140-2 Security requirements for cryptographic modules, federal information processing standards publication, 25 May 2001
8. Information Technology Laboratory National Institute of Standards and Technology, FIPS PUB 180-4, Secure hash standard (SHS), federal information processing standards publication, Aug 2015. https://doi.org/10.6028/NIST.FIPS.180-4
9. Dworkin M (2007) NIST Special Publication 800-38D, recommendation for block cipher modes of operation: galois/counter mode (GCM) and GMAC
10. Dworkin M (2001) NIST Special Publication 800-38A, recommendation for block cipher modes of operation methods and techniques
11. Daemen J, Rijmen V (2001) Federal information processing standards publication 197. Announcing the Advanced Encryption Standard (AES), 26 Nov 2001
12. Barker E, Roginsky A NIST special publication 800-133 recommendation for cryptographic key generation. https://doi.org/10.6028/NIST.SP.800-133
13. International Standards Organisation, ISO 26262 Road vehicles—functional safety—all parts, latest versions. www.iso.org
14. Autosar. www.autosar.org
15. The Motor Industry Software Reliability Association (2012) MISRA-C: 2012 guidelines for the use of the C language in critical systems
16. International Standards Organisation, ISO 11898-Road vehicles—controller area network (CAN)—all parts, latest versions, www.iso.org

Automating Battery Management System and Billing Using Machine Learning

Moreshwar Salpekar

Abstract Battery Management System (BMS) is one of core parts of Electric Vehicles. The charging station/point and electric vehicles must both accurately track the charging and the billing. The billing will also help the upstream electricity generators estimate the power used by a single charging as well as total power used by a station or a group of stations in a time period. Machine Learning can automate BMS tracking and billing and also provide alerts if supplemented by communication systems. It can also help predict power needed in a given period and also number of charging points required in a given area. The paper proposes a BMS that uses machine learning. It starts with Battery Management System (BMS) conceptual diagram and its explanation. It then gives the parameters to be measured. This is followed by proposed system design and it will work along with the ML algorithms that can be used. This is expected to aid the design of practical BMS system using machine learning.

Keywords Battery management system · Electric vehicles · Supervised machine learning · Online machine learning · Data analytics

1 Introduction

Electric Vehicles consume electricity. However, electricity is not a free resource and there is a cost involved in production and distribution of electricity. This cost is recovered from consumers by raising invoices for electricity consumed. It is also imperative to keep a track of demand and supply of electricity at all times. Calculating and predicting the demand is not an easy task as demand varies in same day and also varies from season to season. Calculating supply involves keeping track of all areas where electricity is supplied and then totalling the supplied electricity. The gap needs computation and is different for times in day as well as seasons.

Different electricity consumption systems need different computations and demand and supply categorisation. One such upcoming system is electric vehicles. Government of India is promoting electric vehicles with curbing pollution and

M. Salpekar (✉)
Boson Motors Pvt. Ltd., Pune, India

R. K. Pillai et al. (eds.), *ISUW 2019*, Lecture Notes in Electrical Engineering 764,
https://doi.org/10.1007/978-981-16-1299-2_13

reducing dependence on fossil fuels as two of goals. Already policies, regulations and standards have been put into place [1, 2]. Electric vehicles will require also require billing and demand and supply computation systems. As the vehicles grow in number, setting up of infrastructure including electricity supply and charging stations will be needed. This will involve more monitoring on top of current monitoring systems. A system design is proposed in system that can help in automating monitoring and billing accurately and depends on communication infrastructure (IoT) and machine learning.

Before proceeding to design of such a system it is important to introduce the Battery Management System.

2 Battery Management System

The electricity storage and consumption in an electric vehicle is managed by a Battery Management System (BMS). It is responsible for monitoring the following parameters with respect to batteries in electric vehicles [3]:

a. State of Charge
b. State of Health
c. Charging and Output Current
d. Voltage of each cell in the battery
e. Battery Temperature
f. Automatic Cell Balancing.

The power consumption can be calculated from above parameters. It should be noted that the state of health of battery determines how much charge can the vehicle battery can hold and how frequently the battery needs to be recharged.

The block diagram is given below.

As seen in Fig. 1, the BMS consists of a master controller called Battery Controller which controls and coordinates the operation of the estimators. Each estimator uses a specific algorithm for its operation e.g. a State of Charge estimator may use Coloumb estimator along with Kalman filter to determine state of charge. The Battery controller along with its estimators is called the Battery Control Unit (BCU). The readings of estimators are then transferred to Electronic Control Unit (ECU). The ECU also receives data from motor controller for motor control system. The motor controller along with its estimators is called the MCU. The ECU coordinates and manages the MCU and BCU and the coordination amongst themselves.

3 The Proposed System

The diagram of proposed Charging Network System is given in Fig. 2.

The components of the proposed system are given in sub-sections below.

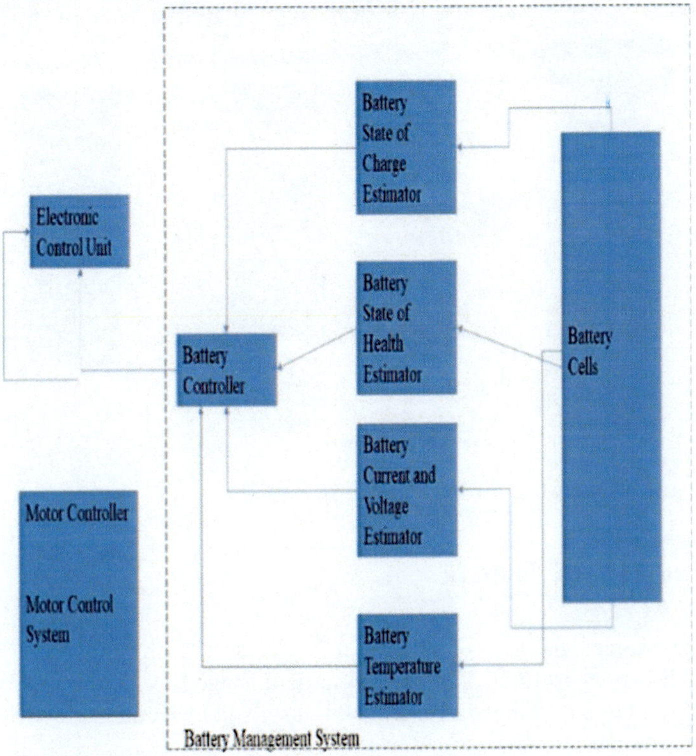

Fig. 1 Battery management system (derived from [4, 6, 8])

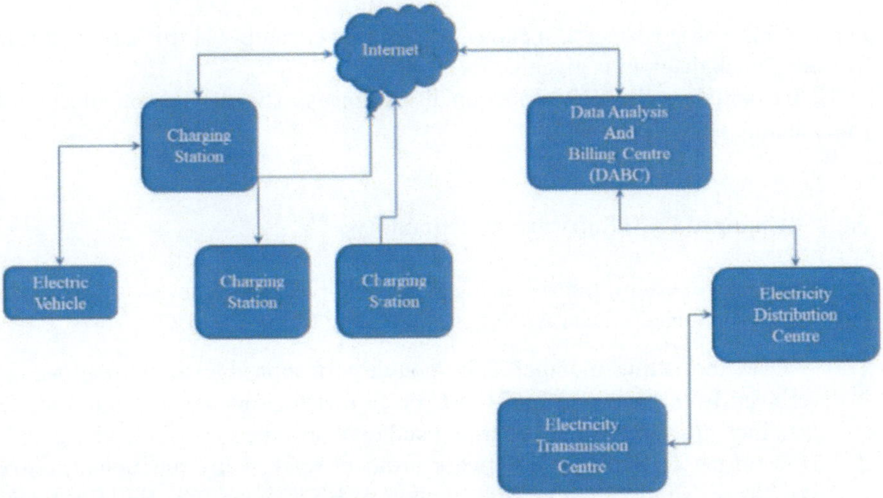

Fig. 2 The proposed system (derived from [4])

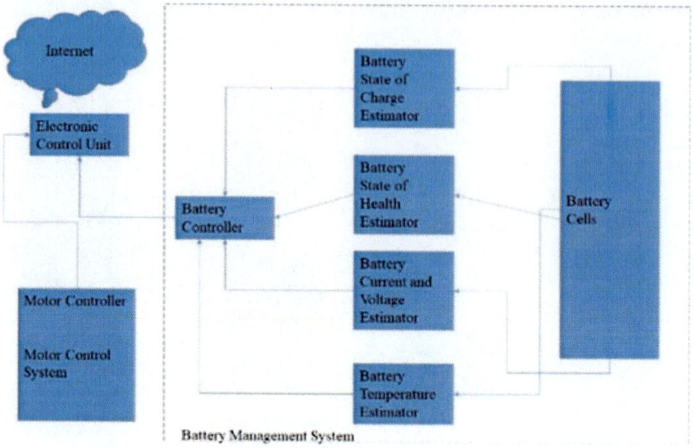

Fig. 3 System with network (derived from [4–8])

3.1 The Electric Vehicle

The electric vehicle has a set of batteries to be charged. Each battery has one or more cells. Each battery is 12 V. The battery set is managed by BMS. The ECU inside BMS will undergo a change in functionality as given in section below:

3.1.1 The Change in ECU Functionality

For ECU, the proposed system adds a networking capability at the hardware level. The new block diagram is given in Fig. 3.

At the software the ECU functionality changes. The description of changed functionality is given below.

3.1.2 Changed ECU Software Functionality

In the proposed system, the following functionality is added to ECU software to enable it to send data to Data Analytics and Billing Centre (DABC).

1. A Data formatting module: This module is responsible for formatting data received from BCU and MCU in CSV format for forward transmission. The data includes time spent in charging and total power consumed in charging.
2. Networking capability and network protocol stack. Fully functional network stack is added to ECU to enable support of HTTP over TCP. HTTP is chosen as all network systems allow HTTP protocol packets through firewalls. Other protocols may be stopped at firewalls.

3. An optional display for displaying statistics, analysis and billing from DABC. The DABC may also send this information to user defined email, phone or any other option specified. The display may be scrolling or page based.
4. An optional keypad connector to connect keypad for data entry and display control. The ECU controls operation of keypad and restricts data that can be entered through keypad (data read from BCU and ECU cannot be altered) and display control.
5. An optional support for printer to print analytics and bill if user requests so.
6. Each Unit comprising of ECU, BCU and MCU is programmed with a unique id that distinguishes it from other units. The unique id is assigned at factory time and cannot be changed at all. Any change in it results in unit invalidation and non-availability of BMS and MCU, i.e. vehicle cannot be used as electric vehicle.
7. Support for securing (with encryption and authentication) of data sent to DABC and response/data received DABC. This includes secure key management also.

3.2 The Charging Station

The Charging function functionality remains same. It gets vehicle registration number or unique id of BMS in vehicle and computes power/electricity consumed by electric vehicle and sends it over to the DABC. Optionally it may send other data if configured to do so. Other data may include photo of driver, location of charging station, etc. It also sends daily and weekly update for power consumed to the DABC.

The charging station may receive alerts and messages from DABC. It shall take action and send response to DABC acknowledging actions.

Additionally, a charging station may detect faults or have information for distribution centre e.g. loss of power, wiring faults etc. It shall send these messages to DABC for further transmission to distribution centre.

It is required by design to act on any messages received from DABC or the vehicle (optional and depends on vehicle) and acknowledge the messages with action taken.

3.3 Data Analytics and Billing Centre (DABC)

This receives data from charging stations. It is responsible for following

1. Computing bill for vehicle charged and send it to vehicle owner/operator as configured.
2. Predicting power requirement for vehicle.
3. Predict when vehicle owner/operator should change batteries using machine learning algorithms and pass on analysis to vehicle owner/operator. For this, the DABC uses BMS id, State of Health, State of Charge, Total battery capacity and

recharge period. This information can be used to predict battery requirements for zone of operation of DABC.
4. Store billing information in server of billing database along with unique id of BMS
5. Use the power consumption data received from all charging stations in its zone to compute total power and store total power in the database.
6. Send the total to Distribution centre for its analysis and storage.
7. Predict power requirement for its zone using machine learning algorithm (online learning for example) and pass on its prediction to Distribution centre.

A Cluster of charging stations is placed under a DABC. The number of charging stations under a DABC is chosen using following factors

1. The data transmission from charging station to DABC and versa is not long.
2. The DABC can process data and send response to charging station without any noticeable lag. The load on DABC is kept small enough to operate it efficiently and give quick responses.
3. The data transmission from DABC to distribution centre and vice versa is not long.

It is required by design to act on any messages received from distribution centre or charging station and acknowledge the messages with action taken.

3.4 Distribution Centre

The Distribution centre receives data from DABC. The data contains power consumption and predicted requirement for its zone. It may optionally use local analytics unit (a machine learning server) to analyse the data it receives to compute the total power consumed and predict the total power required in its zone. The computed power and required power are then sent to the Electricity Transmission Centre which may be a zonal centre or national centre.

The distribution centre can send alerts and messages to DABC if required. The alerts may include loss of electric supply or scheduled outage, emergency messages for charging station. The DABC passes these messages to charging station.

It is required by design to act on any messages received from transmission centre or DABC and acknowledge the messages with action taken.

3.5 Electricity Transmission Centre

It receives the data from distribution systems under its control and computes the demand and supply gap. It can also compute total transmission and distribution (T&D) losses if the network is lined up with sensors to track the transmission parameters like current, voltage (and resultant power), wire temperatures at various points.

The transmission centre can send alerts and messages to distribution if required.

It is required by design to act on any messages received from distribution centre and acknowledge the messages with action taken.

The Machine Learning Training ([9–11]).

The training for the system shall proceed in the following phases

1. Supervised Learning Period: In this period machine learning system shall be trained using pre-generated actual data. This data comprises training, validation and test set. The confusion matrix can be used along with least mean square and deviation to improve system
2. Online Training: The system above shall be trained using actual data and improvise its algorithm.

After above steps, the system shall be deployed and the data analysed for anomalies and deviations and the system improvised based on the analysis. This may result in re-training requirement steps if formula change is required.

4 Benefits from Proposed System

The following benefits are identified in the proposed system

1. Use of Machine Learning Systems will help automate power consumption calculation and prediction, and billing. It can also flag anomalies in system and even in power consumption patterns. This will help in freeing up persons involved in these task for important tasks such as system planning, operation, maintenance and help them by providing the required data. For example, if a charging station reports lower power requirements and other requires power, the distribution centre may readjust power supply.
2. The clustering of DABC and charging stations can give faster response time to both charging stations (and vehicles) and the distribution centre. The users can also get billing information details very fast.

5 Future Extensions

Additional features can be added to system easily if required, for example, user data drive may also be stored on user request. This information can be passed to municipal and traffic police who can then use it to identify difficult stretches. This data, with extensions to system, can be used by vehicle users to request assistance.

The system is easily extensible to allow battery swapping and this can also help calculating battery types and numbers in a zone in a given period.

References

1. Department of Heavy Industries Ministry of Heavy Industries and Public Enterprises, Government of India (2012) National Electric Mobility Mission Plan 2020
2. Department of Heavy Industries Ministry of Heavy Industries and Public Enterprises, Government of India (2018) Standardization of protocol for EV charging infrastructure Bharat public EV charger specifications
3. Xing Y, Ma EWM, Tsui KL, Pecht M (2011) Battery management systems in electric and hybrid vehicles. Energies 4:1840–1857. https://doi.org/10.3390/en4111840
4. Ministry of Law and Justice (Legislative Department): Electricity Act 2003, amended up to 2007
5. Meissner E, Richter G (2003) Battery monitoring and electrical energy management precondition for future vehicle electric power systems. J Power Sources 116:79–98
6. Renesas (2018) Battery management system tutorial. Retrieved from www.renesas.com/doc/whitepapers/battery-management/battery-management-system-tutorial.pdf on 25th Dec 2018
7. Hu R (2011) Battery management system for electric vehicle applications. Electronic Theses and Dissertations. Paper 5007. Retrieved through https://scholar.uwindsor.ca/etd
8. Kettles D (2015) Electric vehicle charging technology analysis and standards. FSEC Report Number: FSEC-CR-1996-15, U.S. Department of Transportation's University Transportation Centers Program
9. Chiu Y-W. (David Chiu) (2015) Machine learning with R Cookbook. Packt Publishing Ltd.
10. Raschka S (2015) Python machine learning. Packt Publishing Ltd.
11. Mitchell TM (1997) Machine learning. Mc Graw Hill

Consumer Perspectives on Electric Vehicle Infrastructure in India: Survey Results

Jyoti Maheshwari, Srinivas Cherla, and Amit Garg

Abstract Transport sector accounts for about 20% of global energy use with around 25–30% emissions resulting from vehicles alone. Electric vehicles (EVs) are considered to be non-polluting and environment friendly substitutes to conventional fuel vehicles because they have a zero tail pipe emissions and electric traction is more efficient than regular engines. A primary survey was conducted across 10 cities (7 tier-1 cities and 3 tier-2 cities) in India and around 6000 sample surveys were administered in order to understand consumer perceptions for EVs. The purpose of the study was to understand current vehicle ownership and driving patterns of consumers, their awareness on EVs, purchase criteria and expectations for EVs, their views on public charging infrastructure and incentives offered by government. It would enable us to understand the drivers for improving acceptance of EVs and estimate their market potential. The survey analysis indicated that over 90% of the consumers travelled within 60 km per day and travel time was less than 2 hours. Also, consumers prioritized basic EV infrastructure as the most important criteria among the others impacting their purchase decision, followed by cost of EV and its performance. More than 50% of consumers favored quick charging option over normal charging option and were willing to pay double price for charging at quick charging stations. Lack of charging infrastructure, high upfront cost, low driving range etc. were found to be main barriers for large scale adoption of electric vehicles.

Keywords Electric vehicle · Charging infrastructure · Consumer perspective · Electric mobility

J. Maheshwari (✉) · S. Cherla · A. Garg
Indian Institute of Management Ahmedabad, Ahmedabad, India
e-mail: jyotim@iima.ac.in

S. Cherla
e-mail: x14srinivas@iima.ac.in

A. Garg
e-mail: amitgarg@iima.ac.in

R. K. Pillai et al. (eds.), *ISUW 2019*, Lecture Notes in Electrical Engineering 764,
https://doi.org/10.1007/978-981-16-1299-2_14

1 Introduction

Transport sector accounts for about 20% of global energy use. Around 25–30% of emissions were resulting from vehicles alone [1]. Electric vehicles (EVs) play an important role for strategic developments as pollution free, noiseless and environment friendly substitutes to conventional fuel vehicles because of their high efficiency and low carbon footprint [2, 3]. They also enhance environment quality and sustainability without significantly affecting convenience [2].

National Electric Mobility Mission Plan (NEMMP) 2020, launched in 2013 by Government of India (GoI), aimed to accomplish fuel security by promoting hybrid and electric vehicles in the country. It targets to sale 6–7 million units of hybrid and electric vehicles such as electric cars, buses, light commercial vehicles, two-wheelers and three-wheelers, etc. every year starting from 2020 onwards by providing financial incentives. GoI has also launched "Faster Adoption and Manufacturing of Hybrid and Electric Vehicles (FAME)" scheme in 2015 to facilitate the major push for early adoption and market creation for hybrid and electric vehicles [4, 5].

Few studies indicated that lack of charging infrastructure has been one of the critical obstacles encountered by EVs than conventional vehicles. Also, the integration of new mobility services into existing infrastructure systems may create problems of acceptance, cooperation and compatibility. This would make EV charging infrastructure deployment a key issue for their use [3, 6, 7]. The current state of deployment of electric vehicles in Brazil and China was studied to promote their usage and their charging infrastructure in both private and public sectors [8]. The results of charge pricing model study for EV charging infrastructure based on public–private partnership projects in China indicated that the operating costs, electricity price, and charge volume were the main factors responsible for reasonable charge price [9]. Another study explored the potential of public charging infrastructure to encourage the sales of battery electric vehicle (BEV), enhancing electrified mileage and lowering greenhouse gas (GHG) emissions in US [10]. A study conducted by Davidov and Pantos, used an optimization model for planning EV charging stations which aimed at reducing the overall cost by enabling charging reliability and service quality anticipated by EV owners [11].

1.1 Abbreviations and Acronyms

2W	Two Wheeler
4W	Four Wheeler
BEV	Battery Electric Vehicle
EV	Electric Vehicles
FAME	Faster Adoption and Manufacturing of Hybrid and Electric Vehicles
GHG	Green House Gas
GoI	Government of India

HEV Hybrid Electric Vehicles
ICEV Internal Combustion Engine Vehicle
NEMMP National Electric Mobility Mission Plan

2 Methodology

The main objective of current research is to understand the consumers' perspectives for EV and drivers that improve acceptance of EVs by reviewing the following aspects:

- Current vehicle ownerships and driving patterns
- Awareness on EVs
- Purchase criteria and expectations for EVs
- Expectations for public charging infrastructure.

The methodology for current research includes following main steps.

1. Shortlisting cities to conduct EV surveys
2. Conducting comprehensive survey in order to understand consumer perceptions on EVs and its infrastructure
3. Review the current state of infrastructure and barriers affecting the adoption of EVs.

Considering social, economic, technical and political aspects (refer Appendix 1) to conduct EV surveys, 10 cities (7 tier-I and 3 tier-II cities) were shortlisted. Total 6000 consumers were surveyed by conducting primary survey in selected cities (600 surveys in each city) to understand the consumers' perceptions for EVs. The cities selected for surveys include New Delhi, Mumbai, Kolkata, Chennai, Bangalore, Ahmedabad, Pune, Nagpur, Jaipur and Ludhiana.

3 Results

Figure 1 presents the share of surveyed consumers by age group and gender-wise while Fig. 2 presents gender wise vehicle ownership. About 61–68% of survey consumers were of young age group i.e. below 35 years of age (see Fig. 1). Both male and female consumers have high preference for 2W, followed by public transport (or taxi) in females while hatchback type of 4-wheeler (4W) in males (see Fig. 2).

Figure 3 compares vehicle ownership with mode of travel preferred by consumer for daily commute. It is obvious that higher ownership of 4Ws would enhance their preference of travelling by 4Ws. Also, the inclination for 2Ws drops and increases for 4Ws by 33, 62 and 66% with increase in the ownerships of 4Ws.

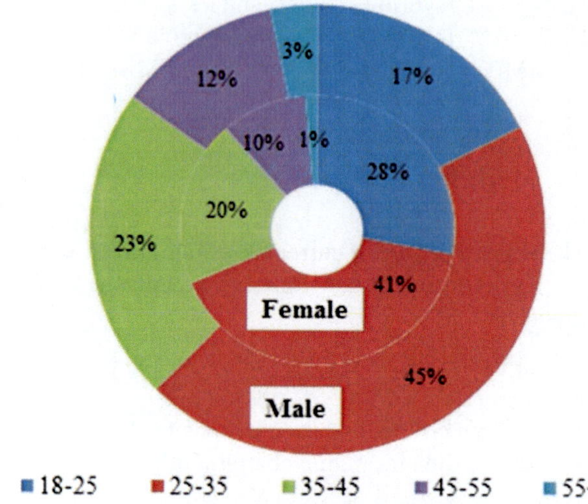

Fig. 1 Age group and gender-wise distribution of survey consumers

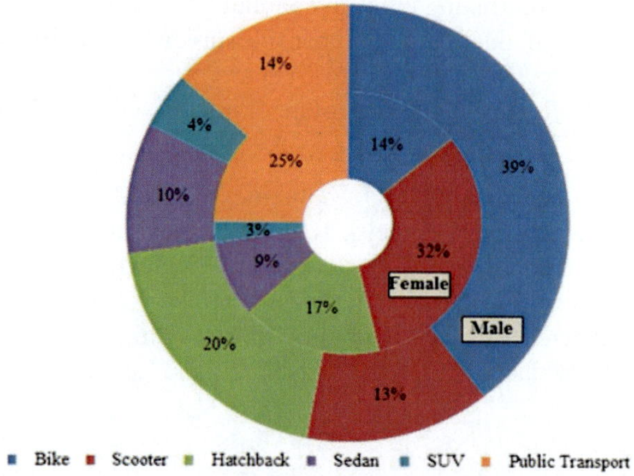

Fig. 2 Gender-wise vehicle ownership

Figure 4 shows city-wise the current and likely ownership of battery electric vehicles (BEV) or hybrid electric vehicles (HEV). Around 18–41% of internal combustion engine vehicle (ICEV) owners were interested to buy EVs in future. Also, their intention to own hybrid EV is more than battery EV (see Fig. 4).

Figure 5 shows average travel time of surveyed consumers with preferred modes of transport. As can be seen that travel time was within an hour (<60 minutes) for majority of consumers using 2Ws, while it was more than an hour for significant number of consumers using 4Ws.

Figure 6 shows multiple statements describing various aspects of EV to estimate the level of awareness on EV among the survey consumers. On an average, around 39% of consumers don't know enough about EV to agree or disagree on

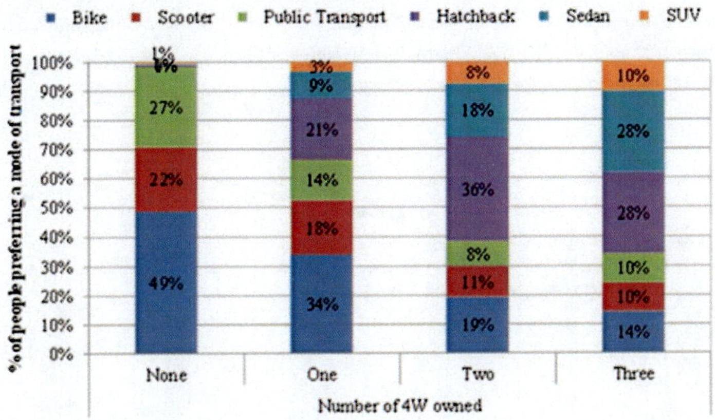

Fig. 3 Vehicle ownership versus preferred travel mode

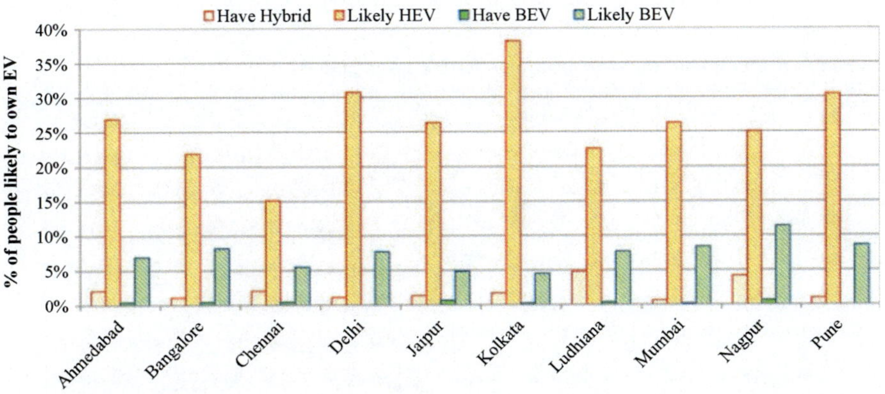

Fig. 4 City-wise current and likely ownership of BEV or HEV

a particular statement. The statements on which consumers agreed incorrectly or disagreed incorrectly both were highlighted with red colour.

Figure 7 presents the consumers' view on EV ownership and battery recharging. About 71% of consumers agreed for battery swapping instead of recharging and 45% agreed on leasing EV instead of purchasing. Also, significant number of consumers preferred the charging of EVs at home because of non-availability of public charging stations at present.

Figure 8 presents various attributes of EV that can be considered for making purchase decision by estimating their significance and uncertainty. Analysis indicated that consumers rated functional attributes such as time to recharge, range and charging infrastructure higher than personal attributes such as price, availability of variants, service and running cost etc. It does not mean that price, service or running cost won't

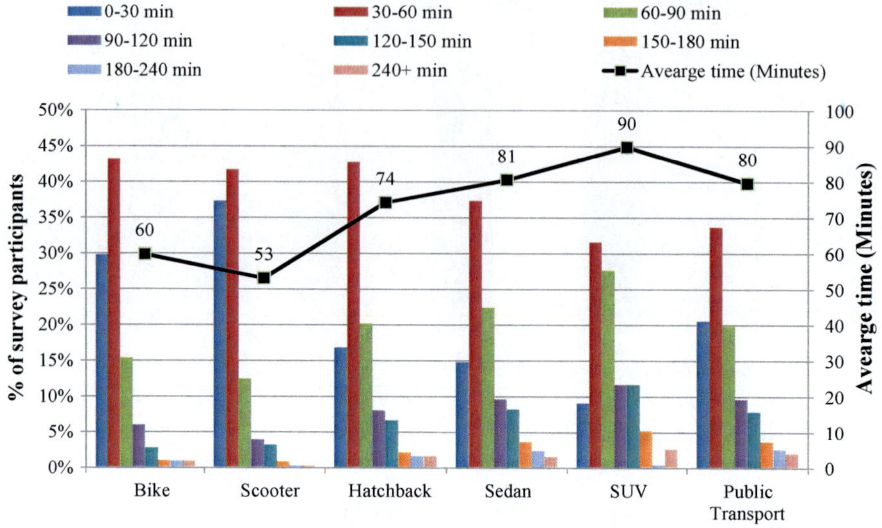

Fig. 5 Distribution of travel time with preferred travel mode

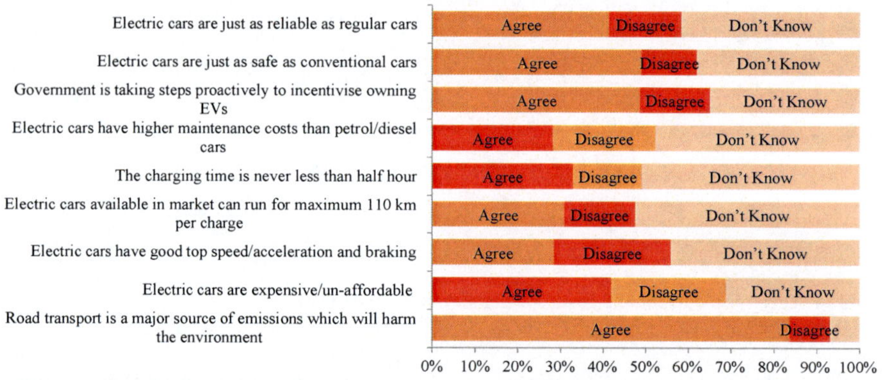

Fig. 6 Awareness level on various aspects of EVs among the consumers

affect purchase decision but consumers look for best value for money for EV purchase if the externalities such as charging cost, charging time etc. can be addressed.

The current study also attempted to estimate for how much time the consumers were willing to wait at public charging station with a booster charge with range extension of 30, 60 and 90 km (see Fig. 9). Results indicated that about 50% of consumers were time sensitive and were willing to pay double for quick charging services than regular power charges if quick charging services would be available. Around 20% of the consumers were cost constrained and preferred normal charging even though the waiting time was 8 times more for normal charging.

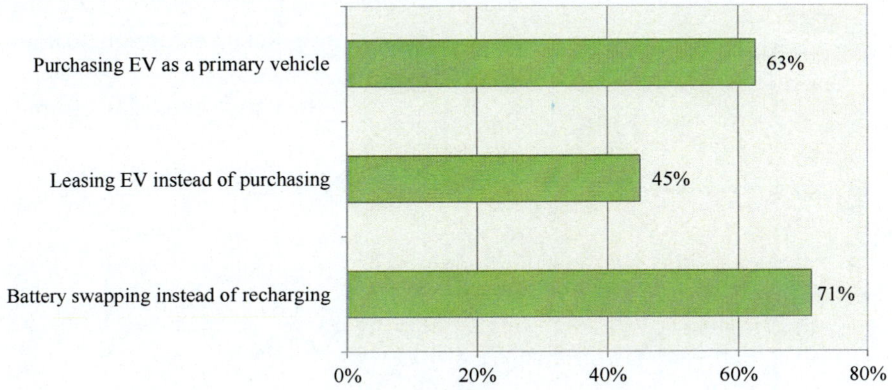

Fig. 7 Consumers' view for EV ownership and recharging

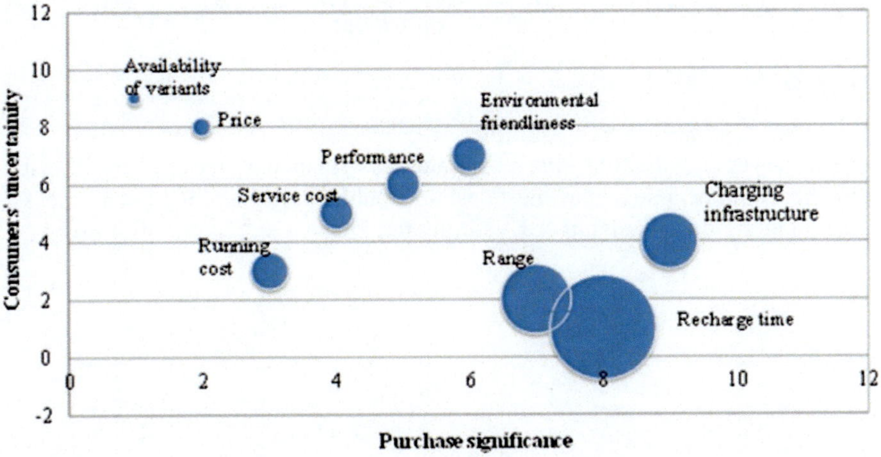

Fig. 8 Consumers' perspectives on various attributes for EV purchase versus uncertainty

4 Conclusions

The survey analysis indicated around 90% of the consumers travelled under 60 km per day and their travel time was less than 2 hours. The consumers ranked basic EV infrastructure (charging time and charging stations) higher than its range, performance, price, running cost, etc. More than 50% of consumers favoured quick charging option over normal charging and showed willingness to pay double price for charging at quick charging stations. However, lack of basic charging infrastructure, high upfront cost, low driving range etc. were estimated as main barriers for large scale adoption

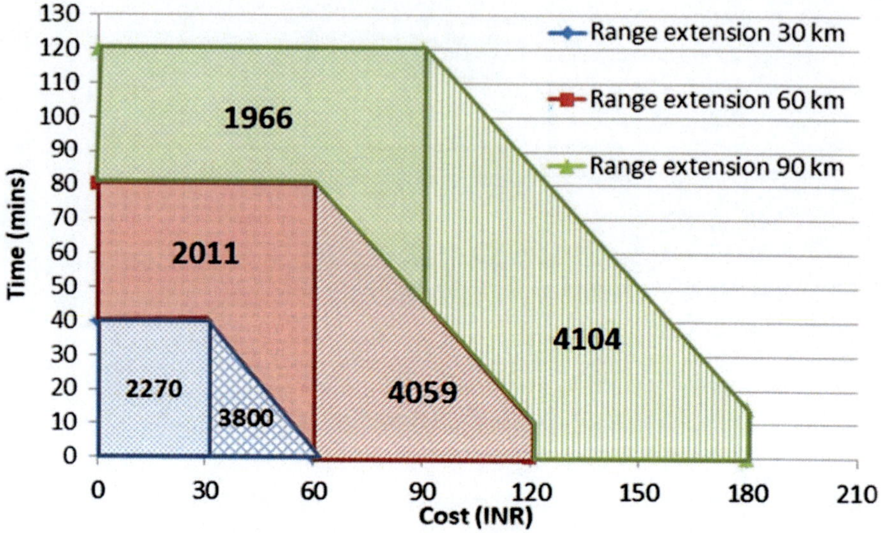

Fig. 9 Public charging preferences

of electric vehicles. Enabling tax rebates, low cost financing, free charging infrastructure etc. could be assessed as most desirable public policy and regulatory measures for promoting the ownership of EVs in future.

Appendix

Criteria Used for Shortlisting Cities to Conduct EV Survey

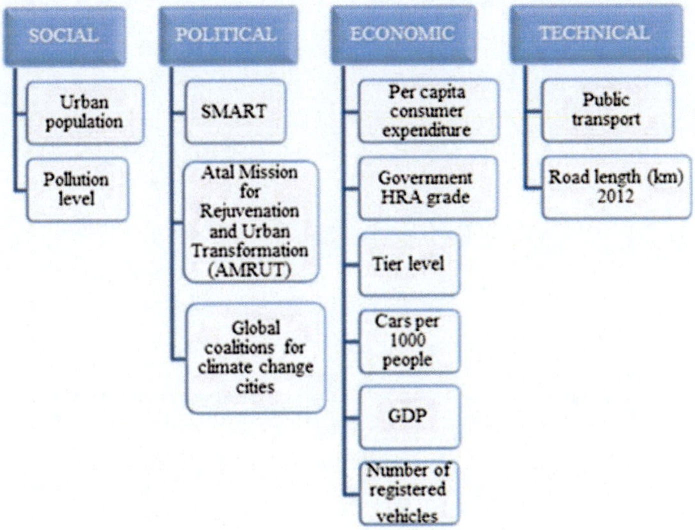

References

1. Pillai RK, Ahuja A (2015) Electric vehicles: a sustainable solution to air pollution in Delhi. ISGF white paper- 2015/00012 v.1.0, India Smart Grid Forum (ISGF), Government of India
2. Liu J, Wei Q (2018) Risk evaluation of electric vehicle charging infrastructure public private partnership projects in China using fuzzy TOPSIS. J Clean Prod 189:211–222
3. Sun Z, Gao W, Li B, Wang L (2018) Locating charging stations for electric vehicles. Transp Policy (Article in Press) https://doi.org/10.1016/j.tranpol.2018.07.009
4. National Electric Mobility Mission Plan, Department of Heavy Industry, Ministry of Heavy Industries and Public Enterprises, Government of India. https://dhi.nic.in/UserView?mid=1347
5. Kumar AG, Anmol M, Akhil VS (2015) A strategy to enhance electric vehicle penetration level in India. Procedia Technol 21:552–559
6. Gnann T, Funke S, Jakobsson N, Plötz P, Sprei F, Bennehag A (2018) Fast charging infrastructure for electric vehicles: today's situation and future needs. Transp Res Part D 62:314–329
7. Soylu T, Anderson JE, Böttcher N, Weiß N, Chlond B, Kuhnimho T (2016) Building up demand-oriented charging infrastructure for electric vehicles in Germany. Transp Res Procedia 19:187–198
8. Li Y (2016) Infrastructure to facilitate usage of electric vehicles and its impact. Transp Res Procedia 14:2537–2543

9. Zhang L, Zhao Z, Xin H, Chai J, Wang G (2018) Charge pricing model for electric vehicle charging infrastructure public-private partnership projects in China: a system dynamics analysis. J Clean Prod 199:321–333
10. Levinson RS, West TH (2018) Impact of public electric vehicle charging infrastructure. Transp Res Part D 64:158–177
11. Davidov S, Pantos M (2017) Planning of electric vehicle infrastructure based on charging reliability and quality of service. Energy 118:1156–1167

Pumped Hydro Storage Technology as Energy Storage and Grid Management Element for Renewable Energy Integration in Karnataka

Y. Nagarjun

Abstract The increased penetration of wind and solar into existing grid poses more challenges, which brings the need for energy storage schemes and grid management assets to ensure power system stability. For which Pumped storage plants can be used as both energy storage and grid management element instead of energy generation source alone. Before this pumped storage power generation was not of in much interest to many states although its contribution towards grid frequency stabilization and load control was well proven. A recent trend of power consumption pattern in Karnataka predicts the need for 'Pumped Storage Technology'. With availability of about 5GW of wind and solar power, Karnataka almost meets its 60% needs. So, taking into consideration the growth of renewable energy in the state, Government of Karnataka intends to set up pumped storage plants for grid management and energy storage. The idea of pump storage is to use the excess energy and balance the grid. A pre-feasibility study carried out on the construction of 2000 MW pumped storage plant in Sharavathi valley project, Shivamogga district has been detailed in this paper. This will be a first-of-its-kind project in Karnataka and would perhaps be one of the biggest Pumped storage Schemes in the range of 2000 MW in India. The study shows that the proposed project is techno economically viable and is planned as an additional structure utilizing the existing Sharavathi hydro project consisting of Liganamakhi, Talakalale and Gerusoppa Dam.

Keywords Pumped hydro · Energy storage · Grid management · Renewable integration · Smart grid

1 Development Need of Pumped Storage Plants

In order to fulfill the electricity demand during peak hours and for managing the imbalance in thermal: hydel mix, pumped storage schemes were developed in the country during 1960s, Now in recent times the increasing imbalance of thermal: renewable mix (mainly wind and solar) is again bringing need for developing pumped

Y. Nagarjun (✉)
Karnataka Power Corporation Limited, Shakthinagar, Karnataka, India

R. K. Pillai et al. (eds.), *ISUW 2019*, Lecture Notes in Electrical Engineering 764,
https://doi.org/10.1007/978-981-16-1299-2_15

storage schemes. Figure 1 shows the Karnataka state demand-wind-solar profile on a particular day. During a day wind generation does not take place during peak hours. Solar generation also occurs partly during off peak hours and partly during peak hours. Wind and solar plants are greatly subject to natural variations [1]. They fail to meet the electricity demand at required amount and at required time. So, excess renewable energy has to be stored so has to make it available when needed.

Integrating a large amount of MW scale wind and solar facilities into the electric grid system requires grid balancing and storage techniques. Though different forms of energy storage techniques have been tried and proven globally, pumped hydro storage plants are still playing an important role in meeting peak demand and helping maintain grid stability in many of the developed countries. Pumped hydro technologies can be thought of the only long term solution which can be technically effective, economical, more efficient and operationally flexible in large scale for energy storage use at a short notice. It is only pumped storage hydropower that can meet many of the grid-scale energy storage needs as no other storage system currently available can meet all grid demands. Pumped storage plants (PSP) has added benefits to reduce the effects of greenhouse gases on the environment. Developing pumped hydro plants particularly near sites with large scale wind and solar power generation, can improve grid reliability. The additional benefits of pumped storage schemes is the availability of spinning reserve to regulate the system frequency during sudden load changes and providing power factor and voltage correction when acting as synchronous condenser with no additional investment costs.

The first pumped hydro storage scheme was built in India at Nagarjunasagar Dam in 1970 with 700 MW installed capacity. Since then, several such pumped storage schemes were successfully built in India, which includes Bhira (150 MW), Kadana

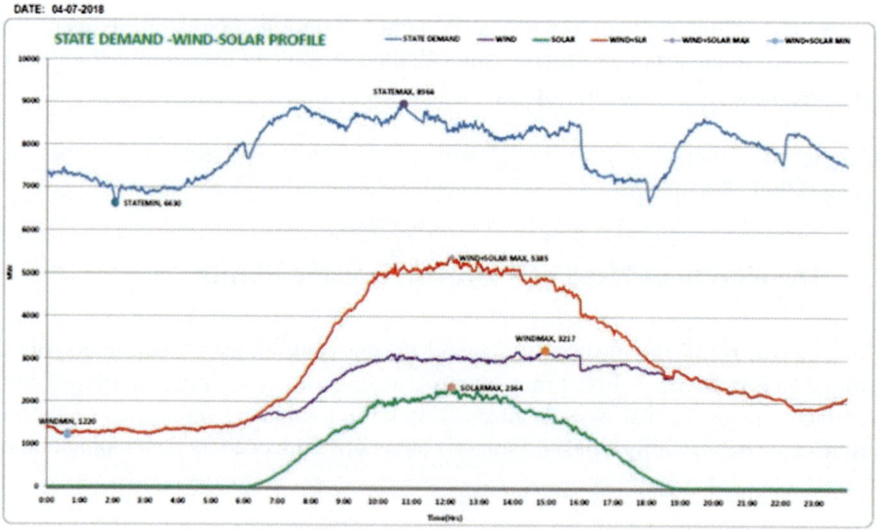

Fig. 1 State demand -wind-solar profile

(240 MW), Ghatghar (250 MW), Kadampari (400 MW), Srisailam (900 MW), Purulia (900 MW) storage projects [2, 3]. Currently nine such schemes are in operational with a total installed capacity of 4785.6 MW. Off these, only five schemes of total 2600 MW are being operated in pumping mode due to various techno-economic reasons. Now the support to develop grid scale energy storage for effective integration of new generation is bringing the need for development of pumped storage hydro power. Therefore Sardar-Sarovar Pumped Storage Scheme at 1200 MW and Tehri Pumped Storage Scheme at 1000 MW capacities are already under construction and 63 other such schemes identified across different parts of the country with an total installed capacity of 96,524 MW are planned for future.

2 Feasibility of Pumped Storage Schemes in the State of Karnataka

It is well known information that Government has devised an ambitious program for developing 175 GW of renewable energy generation by year 2022, comprising majorly Solar (100 GW) and Wind (60 GW). Whereas the total installed renewables in India reached around 75 GW at the end of 2018, representing 21% of the total installed capacity and making a record high of 11.9% of the total electricity generated in the September 2018 quarter. This clearly shows that solar and wind power generation has increased during past few years and contributes for a significant proportion of the total generation in the grid. This significant portion of renewable generation is mainly concentrated in few states of Tamil Nadu, Karnataka, Andhra Pradesh, Telangana, Gujarat, Maharashtra and Rajasthan. These states contribute more than 80% of total renewable generation.

Karnataka is one of the state which is blessed abundantly with all forms of renewable energy resources consisting of Biomass, Small Hydro, Solar, Wind, Cogeneration, Waste-to-Energy, and Tidal. As on November 2018 Karnataka has a renewable energy installed capacity of 12,662.67 MW of which wind (4736.76 MW) and solar (5265.26 MW) playing the major role. Considering the significance of this, Karnataka has long been a pioneer in the area of developing renewable energy projects in the country. As a part of its role to fulfill nation's renewable energy obligations and to reduce the increasing imbalance of thermal (7680 MW): renewable mix (12662 MW), the state is planning for developing pumped storage schemes. Karnataka has a good opportunity for adopting a sustainable model for development of Pumped Storage Schemes for integration of large renewable energy projects and to be a model for other states of the country. In the case of Karnataka state, it is observed from previous years data that most of hydropower projects were able to run at their full potential only during monsoon months. For remaining period, due to low reservoir levels the power generation varies accordingly and are not able to cater for peak hour demands. To overcome these challenges, the state's topography provides an opportunity for developing Pumped Storage Schemes. Compared to conventional hydro projects Pumped

Storage Schemes doesn't require large dams to be built and potential sites can be identified where a reasonably small reservoir can be built at different altitude to serve the purpose. As Pumped Storage Schemes require small storage to generate electricity for duration of up to 6–8 h during peak hours the water used can be pumped back to upper reservoir during off peak hours. Also, these projects will not have much of rehabilitation and resettlement issues, which is a big and problematic issue in conventional hydropower projects. Considering the tariff competition trend in recent times, the Pumped Storage Schemes are proven to be techno-economically feasible during peak hours. Having realized the above advantages the state utility has identified two such potential sites for developing pumped storage schemes on existing projects namely Sharavathy valley (2000 MW) and Varahi valley (700 MW). Karnataka Power Corporation Limited (KPCL) has submitted a draft pre-feasibility report (PFR) to Ministry of Environment and Forests (MoEF) for clearance. Topographical survey in Varahi Valley is done on the downstream of tailrace and PFR is under preparation for Varahi Pumped Project. A PFR study [4] submitted on the benefits of developing of Pumped Storage Scheme in Sharavathy valley has been detailed in this paper. The study shows that the Project is techno-economically viable.

3 Sharavathy Pumped Storage Project

The proposed Sharavathy pumped storage installation is planned within the existing Sharavathy Valley Hydro project. Figure 2 shows the flowchart of Sharavathy Valley Hydro project.

The Sharavathy Pumped Storage Scheme with an installed capacity of 2000 MW is proposed on the existing Talakalale and Gerusoppa reservoir which are situated at downstream of Liganamakhi reservoir on Sharavathy river. The reservoir formed by the Liganamakhi dam across the river Sharavathy is the key to the optimum development of water resources of the river comprising regulating dams, diversion structures and associated 4 power stations having an aggregate installation of 1469 MW. The Water resources of the Sharavathy River and adjacent streams have been optimally utilized for power generation in Sharavathy Basin. Five reservoirs plays a major role in regulating surplus waters of the Sharavathy river and adjacent streams during monsoon. The present scheme is a very attractive scheme both in terms of technical feasibility and from economical consideration. The proposed scheme contemplate for the utilization of the potential of the Sharavathy River released from Liganamakhi dam through dam toe Power house by a hydel channel in to Talakalale reservoir, which is a balancing reservoir for existing Sharavathy H.E. Project of 1035 MW. The proposed pumped scheme envisages for power generation during hours on a Pumped storage type model, utilizing a head of about 460+ m between Talakalale as upper reservoir and Gerusoppa as Lower reservoir. Since the Upper and Lower reservoirs of Sharavathy Pumped Storage Project (Sharavathy PSP) has effective storage capacity equivalent to four to five hours of generation daily at full rated output, it is not possible for Sharavathy PSP to operate on weekly or seasonal basis. Therefore,

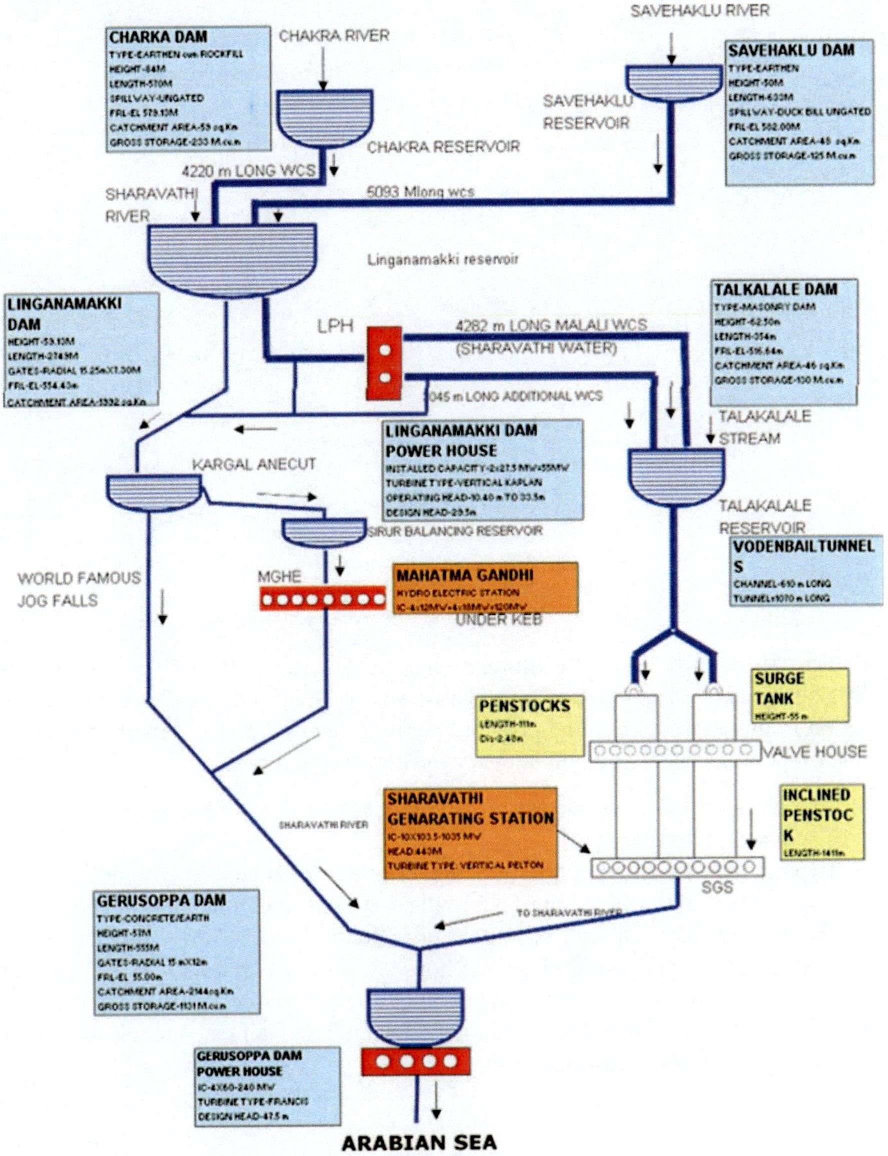

Fig. 2 Flowchart of Sharavathy valley hydro project

the Project is deemed to be operational on Daily basis. Figure 3 shows the proposed Integrated Sharavathy Basin Development Plan.

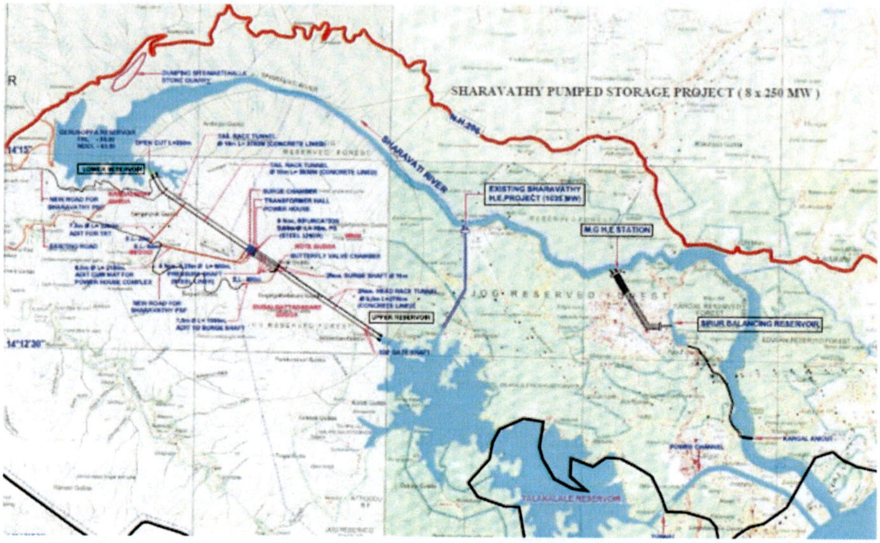

Fig. 3 Integrated Sharavathy basin development plan

The Sharavathy Pumped Storage project envisages to utilize the existing Talakalale dam as upper dam and Gerusoppa as lower dam without any modification in these structures. The present operating levels are also remain unchanged. The proposed scheme contemplates for the construction of:

- Two numbers of intake structure having necessary trash rack arrangement with mechanical raking system.
- Two numbers of 2.72 km long 9 m dia, concrete lined headrace tunnels.
- Two number of 0.82 km long 5.25 m dia, inclined steel lined pressure shafts.
- Two number of 52 m high 16 m dia Surge Shafts.
- A state of art underground power house with eight numbers of Francis type reversible pump-turbine driven generators of 250 MW capacity each.
- Two number of 3.78 and 3.83 km long concrete lined tail race tunnels to carry the releases to lower Gerusoppa reservoir.

3.1 Installed Capacity and Power Generation

For any given pumped storage scheme main influencing factor for installed capacity at a site are the requirement of daily operational peaking hours; operating head, live pondage in the dams and their capacity characteristics with regard to area. The technical details are summarized below in Table 1.

Table 1 Capacity and generation data

Installed capacity (MW)	2000
No of units	8
Unit size (MW)	250
Head (max)—generating	478 m
Head (Min)—generating	476 m
Hours of daily peaking operation	6
Energy generation (MWh)	12,000
Pumping energy (MWh)	14,833
Cycle efficiency	80.90%

3.2 *Power Evacuation Arrangement*

The power generated from the proposed scheme will be at 18 kV which is further stepped up to 400 kV. This power shall be further evacuated from Pothead yard area by following two 400 kV D/C transmission lines from Sharavathy PSP to 400 kV Sub-station at Talaguppa which is about 60 km.

3.3 *Environmental Aspects*

Based on the preliminary assessment of environmental issues considered in the present study, it can be concluded that the project is in proximity to ecologically sensitive area Sharavathy Wildlife Sanctuary (about 3 km). It is proposed to conduct, a detailed Comprehensive Environmental Impact Assessment (EIA) study with an objective to assess various impacts likely to arise as a result of construction and operation of the proposed project on various aspects of Environment. Appropriate management measures too shall be delineated as a part of Environmental Management Plan (EMP), which will be covered as a part of the Comprehensive EIA study. The water conductor system from Talakalale Reservoir is about 3.4 km from Sharavathy wildlife sanctuary. Hence, the project would require clearance from National Board of Wildlife.

3.4 *Project Cost Estimates*

At 2017 price levels the proposed project is estimated to cost about 5000 Crores. The guidelines of CEA/CWC was referred to prepare the preliminary cost estimate of the project. The cost estimate break-down is given in Table 2.

Table 2 Project estimate

Item estimated cost	(Rs. Lacs)
Civil works	273,980.38
Electro-mechanical works	227,764.24
Total	501,744.62

3.5 Financial Aspects

As indicated above, the Sharavathy Pumped Storage extension project, with an estimated cost (Generation only) of INR 5017.44 Crores and design peak energy generation of 4380 GWh is devised to be completed in a short period of five years. The generation tariff has been worked out considering a debt-equity ratio of 70:30, and annual rate of interest on loan at 12.50%. The tariff for initial year and levellised tariff has been worked out at INR 5.73/unit and INR 5.33/unit respectively in a 90% dependable year.

4 Conclusion and Recommendation

The project case detailed here has stressed for the need of utilizing existing hydropower projects to develop pumped storage schemes with an example of Sharavathy Hydro Project in Karnataka to cater for peak hours demands using pump storage technology. The proposed scheme has been planned such that it does not affect the original purpose of the existing dams for generating electricity. Based on the data reported following conclusion can be drawn.

The project is outside of the Sharavathy Wildlife Sanctuary. However, due to in proximity to the Wildlife Sanctuary the project layout has been prepared in such a way that it is completely underground. Dams are existing and the powerhouse complex is underground. Sharavathy Pumped Storage scheme requires minimum civil works and can be completed in 5 years. The project will be capable of generating a design energy generation of 4380 GWh in a 90% dependable year. The installation cost per MW works out at INR 2.50 Crores. The PFR encourages the scheme merits and for taking up for further detailed survey and investigation and preparation of DPR. The levellised tariff for the proposed 2000 MW scheme is of about INR 5.33/per unit which appears to be bit high. But per-capita energy consumption becoming bigger day by day, the construction of 8 × 250 MW will bring more benefits, as peak hours tariff can be very high.

A 2013 study report prepared for KPCL, involved load forecasting study, generation planning and prospective generation plan for the period from 2012–13 to 2021–22, it also had forecasted peak demands, demand met, deficiencies as well wind capacity credit (MW) for the years between 2014 and 2017. As per this study, the maximum monthly peak demand has progressively increased from 10,695, 11,468 and 12,302 MW in these three years. The report has concluded that improving wind capacity credit will depend on the flexibility of operation of future conventional generations like storage based hydro/pumped storage power plants and short term wind forecasting tools. Subsequent to this report, the Power System Operation Corporation Limited (POSOCO) and National Institute of Advance Studies (NIAS) have conducted study in March 2015 on "Wind and Solar Energy for meeting Karnataka's future electricity demand". The association developed a method of estimating hourly unrestricted demand and the likelihood of meeting bulk of it from conventional sources available then and the remaining from renewable sources to reduce the extent of deficit in Karnataka, which is substantial leading to economic losses and inconveniences. Analysis for 2017 and 2022 scenarios was done. The analysis for 2022 was based on the assumption of having a 1000 MW, 10 h pumped storage scheme. The report recommended a total of 4800 MW of wind power, 2500 MW of solar power and addition with energy storage facility (for 2022). It has concluded that combined, these are expected to reduce defict by 69% with no deficit for 7 months of the year and generate about 22% excess as expressed with deficts as a base. Thus, 2000 MW Sharavathy Pumped Storage Scheme will go a long way in meeting Karnataka's future electricity demand.

References

1. Ramachandra TV et al (2014) Potential assessment and decentralized applications of wind energy in Uttara Kannada, Karnataka. Int J Renew Energy Res 4(1)
2. Varma CVJ, Rao ARG (2007) Pumped storage schemes in India. CBIP Publication, New Delhi
3. Chalisgaonkar R, Mohan M (2015) Prospects of developing pumped storage projects utilising the reservoir of existing hyropower project in the State of Uttarakhand. ICHPSD-2015
4. KPCL-WAPCOS Limited, Pre Feasibility Report on 2000 MW Sharavathy Pumped Storage Project (2017)

Role of Power Utility in the Future of E Mobility in India

B. Nushreen Ahmed and Swati Mamidi

Abstract Urban air quality issues, coupled with a rising awareness of the problems associated with the world's appetite for oil, have created interest in Electric Vehicles (EV). Tata Power has an installed gross generation capacity of 10,857 MW and a presence in all the segments of the value chain in the power sector viz. Fuel Security and Logistics, Generation (thermal, hydro, solar and wind), Transmission, Distribution and Trading. Tata Power is navigating the digital transformation of utilities to integrated solutions by looking at new business growth in EV charging. Tata Power, being a responsible utility, has been in the forefront in setting up Infrastructure in India for Electric Vehicles. This paper explains how the company has been covering areas where EV users wish to have charging stations and thereby making India truly ready to usher in the EV wave.

Keywords Electric vehicle · EV charging infrastructure · Tata Power EV charging station · Business models for EV charging stations in India

1 Introduction

The auto industry has pointed to the dearth of EV charging infrastructure in public locations as one reason for relatively few EVs on road. With the Government of India efforts towards shifting to electric vehicles by 2030 has been relentless, Tata Power aims to build an integrated network of EV charging stations to make it convenient for people to adopt EVs and support the growth of infrastructure requirements.

In addition to promoting EVs and educating customers on the benefits of electric drive, Tata Power is well positioned to adopt EV and unlock the numerous advantages of EVs.

B. Nushreen Ahmed (✉)
Corporate Digitalization Group, Tata Power Company Limited, Mumbai, India
e-mail: nushreen.ahmed@tatapower.com

S. Mamidi
Business Collaboration, Tata Power Company Limited, Mumbai, India
e-mail: swati.mamid@tatapower.com

R. K. Pillai et al. (eds.), *ISUW 2019*, Lecture Notes in Electrical Engineering 764,
https://doi.org/10.1007/978-981-16-1299-2_16

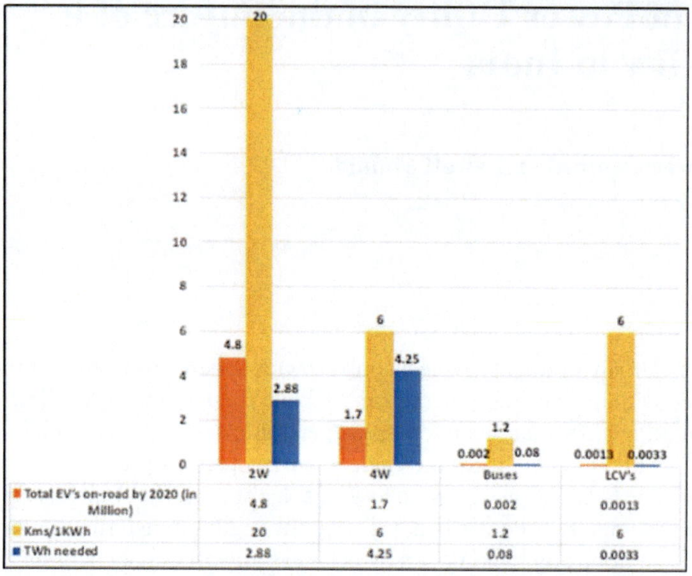

Fig. 1 Total electricity generation needed to power EVS in India in year 2020

2 Overview

There are programs initiated by the government such as National Electric Mobility Mission Plan (NEMMP), Faster Adoption and Manufacturing of Hybrid and Electric Vehicles (FAME) to help the Indian EV market to gain momentum.

It is estimated that a total number of 1.7 million 4-wheeler (4W) EVs will be on road in India by 2020. The average distance travelled by a 4W is 15,000 km/year. Hence, total electric km for 4Ws in India is 25,500 million electric km/year. Typically, 1 kWh of electric energy can drive a 4W for around 6 km. Hence, 4250 GWh of energy is required by 2020 to power 4W EVs on road. Energy required for 2W EVs, Buses and LCVs is also deduced with the same methodology. Total energy required for EVs (4W, 2W, Buses and LCVs) is estimated to be 7208.25 GWh by 2020 (Figs. 1 and 2).

3 International Standards

Indian Standards for EVs in India are still under preparation and in draft stage. Indian car manufacturers are evaluating several charging infrastructure technology. However, the Central Electricity Authority (CEA) recommends the European Standard because it is well accepted globally (Table 1).

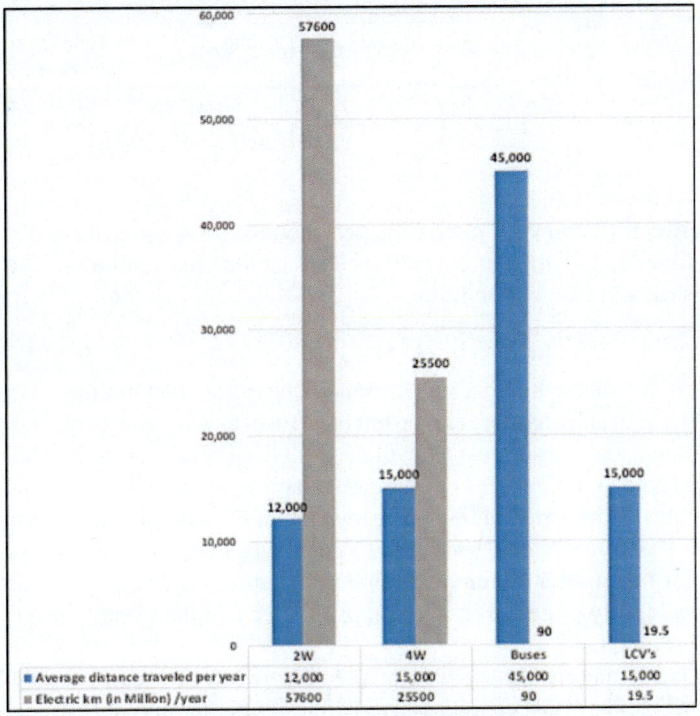

Fig. 2 Average distance travelled depicted in electric km/year

Table 1 Different standards for EV charging

Organisation	Place	Standard
International Electrotechnical Commission (IEC)	Europe	IEC 61,851
Society of Automotive Engineers (SAE)	USA	SAE J1772
GuoBiao (GB)	China	GB/T 20,234
CHAdeMO	Japan	CHAdeMO

4 Case Study

A project is being run by an aggregator in collaboration with an automobile company and electric cars were launched as a part of cab fleet in a city of India.

Key Parameters of the Case Study

The average distance travelled by car with 100% charged battery is approx. 90–100 km (with air conditioner ON) (Table 2).

The cars are charged at least 3 times a day. After completing a few rides (6–7 rides equivalent to 100 km), the car is charged at the nearest charging station. The

Table 2 Types of charging station

Type of charging station	Rating	Time required for full charge (h)
Fast	DC, 10–15 kW	2–3
Slow	AC, 2–4 kW	6–7

hours between 12 noon to 2 pm are the peak hours during which the drivers even face waiting time (up to 5 h). The car is charged a second time at around 6 pm depending on the usage and battery condition.

Key Learning from the Case Study

- Long wait times at the Charging station—Due to a smaller number of fast chargers, the drivers usually have no choice left but to use the slow chargers. The wait time at a charging station is almost 5 h which could have been productive. Slow charging is reasonable only in cases of overnight charging or charging required for top up.
- Location of Charging Station—The location of EV station is also of utmost importance. The station should be set up at a place such that it is easily accessible in their route of travel with minimum diversion. Another deciding factor for location is the availability of nearest power supply source and the existing transformer load at the source.
- High infrastructure Costs—Overall the costs involved to build a fast EV charging station is high. A fast charger is nearly 5 times higher in cost than a slow charger. Besides charger, infrastructure also includes land, power supply, branding and back end management system.

5 Pilot Project by Tata Power to Provide EV Charging Stations

The Planning Process

Step 1 Identifying Locations for EV Charging Points

A survey is conducted to find strategic locations for setting up EV charging infrastructure. The locations are shortlisted based on data like the usual route of the EV cars, the start and end point of their route, their preferred location of charging and availability of power distribution network of Tata Power (Fig. 3).

Step 2 Conduct Technical Feasibility Study of identified EV Station Locations

Site visits are conducted to assess the possibility of the setting up EV charging infrastructure. The following factors are considered for the study.

- the number of charging stations needed
- availability of electrical grid capacity

Fig. 3 Mapping of survey data points to find strategic locations for charging stations. work location, charging location, preferred charging location, Tata Power HT/LT network

- availability of space at the location
- type of charging station needed at each location
- rules and regulations at relevant locations (Fig. 4).

Step 3 Install, Operate and Manage the EV Charging Stations

As per the rollout plan, the EV chargers are installed and selected personnel at the station are given basic training on the safety and operation standards (Fig. 5).

Step 4 Sustaining the EV Network

Continuous study is being carried out along with the Strategy team to develop a sustainable business model for Tata Power. This is being developed taking into consideration the evolving business scenario, the policy initiatives, role definitions, tariff structure, customer segment etc. Two such business models will be explained

Fig. 4 Site Visit to identify possible locations to set EV charging stations

Fig. 5 Installation activity being carried out at one of the sites

in the next section. Timely feedbacks from end users and communication are being maintained and actions for improvement are being taken (Fig. 6).

6 Business Models

Option 1 Business-to-business (B2B): Charging Network owned, operated and managed by Power Utility (Fig. 7)

Fig. 6 Tata power launches India's first ever public charging station for EVs (response from EV pioneers in the city)

Fig. 7 EV charging station at Vikhroli, Mumbai (owned, operated and managed by Tata Power)

Business Model Characteristics

- Increase power sale by integrating an EV charging station into existing power network
- Utility can leverage the existing distribution customer base to increase scale
- Utility shall provide space and infrastructure.

Target Segment of Customers

- Fleet operators for destination charging
- Frequent users of National Highway

Fig. 8 EV charging station at one of the reputed malls, Palladium at lower Parel, Mumbai (operated and managed by Tata Power, Owned by Palladium)

- Workplace charging, if the charger point is in the vicinity.

Strategic Partners

- Establish alliance with technology platform developers who shall aggregate all stakeholders in the ecosystem

Option 2 Business-to-consumer (B2C): Power Utility to operate and manage the network while alliance owns the charging station (Fig. 8).

Business Model Characteristics

- Build a network spanning the value chain and customer segments to mitigate risk and maximize revenue
- Utility can charge the alliance a subscription fee form managing the network
- Alliance shall provide space and infrastructure.

Target Segment of Customers

- Parking places in business hubs like Malls, Cinema, Supermarkets, Airport, Parks, Gym etc.
- Oil/Gas Stations.

Partners

- Establish alliance with technology platform developers who shall aggregate all stakeholders in the ecosystem
- Establish alliance with Commercial business owners like Estate Builders, Oil & Gas Companies etc.

Fig. 9 Tata Power creative minds brainstorm in idea crucible on EV charging station business model to come up with path breaking ideas such as 'charge on the GO' and 'enabling power entrepreneurs'

7 Challenges for Power Utilities in India

- India does not have Lithium ion reserves to support a large domestic market for electric vehicles
- Lack of finalization of Indian Standards on charging stations in India
- Lack of clear policies for supporting the growth of demand, manufacturing and recycling of batteries
- Lack of policies for EV chargers
- Safety concerns/perceptions around electric vehicles (Fig. 9).

8 Role of Tata Power in EV Charging Infrastructure

After gaining insight from numerous studies, trials, pilots and simulations Tata Power is in a position to adept to the evolving EV market growth in India. As a power utility, the forte of Tata Power in building the EV network lies on the key points as given below.

Grid Stability

As the adoption of EV increases, the stress on electric power system will increase and consequently there will be an increase in maintenance costs. Presently, 25% of Tata Power's distribution transformers are loaded. However, with proper planning and preparation, transportation electrification can result in more efficient and less costly operation of the grid, provide value add services, lower charging rates.

Reliable and Safe EV Charging Network

Tata Power can provide a perfect "partner" for independent providers of charging stations. Tata Power's focus is to provide safe, reliable electricity to its customers.

There will be a number of operational issues associated with running a successful and reliable charging infrastructure, including the ability to provide 24 × 7 service support and security cameras to monitor for vandalism.

Real Time Tarif Structure

With increased customer involvement and penetration of more complex systems, such as dynamic pricing will require more advanced and compact systems to offer even greater benefits to consumers and utilities. Smart systems with end to end functionality can ensure long term stakeholder and customer engagement.

With the increasing price volatility and higher costs for peak hours, dynamic pricing options such as time of use (TOU), critical peak pricing (CPP), critical peak rebate (CPR), real time pricing (RTP), and variable peak pricing (VPP) will be used in India to encourage peak load management. Implementation of such programs will require smart meters and increased customer involvement. Utilities can entice EV owners to charge their cars during off peak hours and avail the rebates.

Smart Charging

Smart charging can unlock the full benefits of electric vehicles. Smart charging will allow utilities to effectively use this storage capacity to stabilize the grid and lower net costs, creating savings that can be passed on to electric vehicle owners directly and to all consumers through lower rates.

9 Conclusion

Tata Power in partnership with software platform providers can leverage their capabilities to develop a smart infrastructure. Tata Power can select the locations where they want the charging stations to be installed. Tata Power can pay some portion of the capital expenditure required to install the infrastructure and in return earn a portion of the indirect revenue from the service provider. The benefits from this approach is manifold like: charging stations are deployed to maximize societal benefit, competition is promoted by having multiple infrastructure providers, Infrastructure providers can finally make the business case work as the utility realize a new source of revenue and customers benefit by getting access to easy to use, safe, reliable charging stations. Broadly, utilities can play a key role in promoting and facilitating the transformation of our transportation sector away from petroleum-based fuels.

References

1. Jennings S, Parkin R, Del Maestro A (2018) Powering ahead! Making sense of business models in electric vehicle charging
2. Bansal S (2018) EV charging infrastructure: the evolving business models

3. Hall D, Lutsey N (2017) Literature review on power utility best practices regarding electric vehicles
4. National mission on electric mobility
5. Tata Power Corporate website

Design and Development of PV/FC Based Integrated Multilevel Inverter for Smart Grid

P. S. Suvetha and R. Seyezhai

Abstract The rapid growth of industrialization and globalization leads to the exploitation of non-renewable fossil fuels and these fuels cause carbon emission, global warming, climate changes, and atmospheric pollution. In the current scenario, the demand for electric energy of growing population is a major concern. This led to the development of new power generation systems at the distribution level by using various non conventional and renewable energy sources such as solar energy, wind power, fuel cells and bio-gas. It gives an opportunity to utilize renewable energy sources for green and clean environment. Apart from this, a hybrid generation scheme extends the attainable flexibility and scalability for the outstanding capability of energy management. Owing to clean, safe, eco-friendly conditions, the photovoltaic cell and fuel cell are imperatively used as main power sources. This work focuses on the unique hybrid power generation, employing a novel single-switch non-isolated DC–DC converter integrated to the grid using nine-level inverter. The proposed circuit is designed by integrating a boost with quadratic boost converters which provides lower product cost and improves circuit efficiency. A nine-level inverter employing only one input source with less number of switching devices is proposed smart grid applications. The nine level output is achieved by the series and parallel combinations of one voltage source and two capacitors. In order to achieve more reliability and enhance the performance of the system, it is necessary to operate and control the smart grid in an appropriate way. The main objective of the control technique is to provide equal load sharing among the DERs in per unit and also to maintain the terminal voltage constant. Simulation studies of the proposed integrated MLI comprising of high gain DC–DC converters along with nine-level output is carried out in MATLAB/SIMULINK. A suitable controller will be designed to control the output power. A prototype of the proposed power electronic circuit will be developed to validate the simulation results.

P. S. Suvetha (✉) · R. Seyezhai
Department of Electrical and Electronics Engineering, SSN College of Engineering, Chennai, India

R. Seyezhai
e-mail: seyezhair@ssn.edu.in

R. K. Pillai et al. (eds.), *ISUW 2019*, Lecture Notes in Electrical Engineering 764,
https://doi.org/10.1007/978-981-16-1299-2_17

Keywords Multilevel inverter · Smart grid · PV · Fuel cells · DC–DC converter

1 Introduction

In recent years, numerous applications such as dynamic voltage restorer, industrial sectors, electric vehicle and flexible ac transmission systems require high power and medium voltage. A power semiconductor switch which endures a medium voltage is not discovered yet and it is also difficult to produce minimum voltage level using one switching device [1]. Multilevel inverters (MLIs) can be used as the key component to work with medium voltage and high power applications. The MLI is nothing but a combination of power semiconductor switches and voltage sources to generate a stepped output voltage waveform. They have various advantages such as high quality waveform, reduced stress across the switches due to lower frequency, better electromagnetic interference and the THD of the output voltage and currents is also low when compared to a conventional two level inverter [2].

However, the major drawback in MLI is that it requires more number of switching devices which leads to a complex circuit configuration. Thus, the cost of the MLI has become high than the conventional inverter topologies. Many researchers have been worked on this aspect and proposed several topologies in order to overcome the disadvantages [3, 4].

This work proposes a nine level MLI with less number of power semiconductor switches and a single DC source. The input for the inverter configuration is fed from a DC–DC converter. This high gain DC–DC converter utilizes a hybrid source which is a combination of solar and fuel cell for its operation. This low level in boosted to a higher level with a single switch converter topology. Here, the renewable energy sources such as solar and fuel cell is preferred due to its advantages such as pollution free and clean form of energy. The analysis of the topology is done and is validated using a simulation.

2 High Gain DC–DC Converter

The proposed high gain DC–DC converter is an integration of conventional boost converter with a quadratic boost converter. The converter is supplied with a hybrid source which is a combination of solar and fuel cell [5]. The circuit diagram for the proposed high gain DC–DC converter is shown in Fig. 1,

The operation of the converter is explained in two modes.

Mode 1: In this mode, the switch gets turned-on and the diodes D_1 and D_4 are also turned-on simultaneously. The current through the capacitors are equal to that of the opposite inductors. The difference in the voltage levels i.e., ($V_{c1} - V_{c3}$) and ($V_{c2} - V_O$) leads to the turning off of the diodes D_2 and D_5. Then diode D_3 is reverse bias by the voltage ($-V_{c2}$).

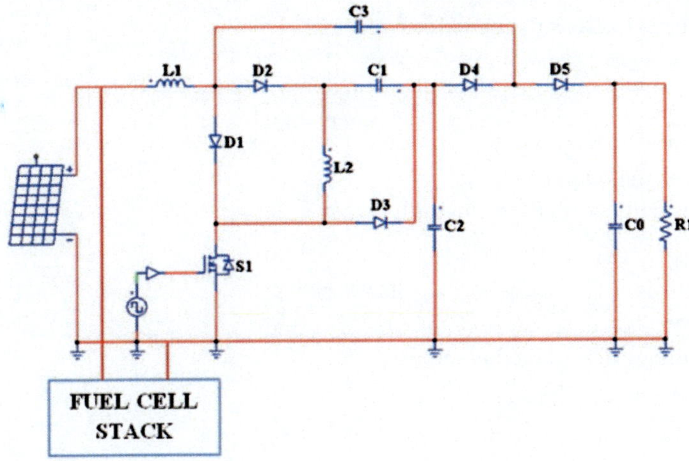

Fig. 1 High gain DC–DC converter

The voltage across the inductor L_1 becomes equal to the input voltage V_g and the voltage across L_2 equal to the difference between the voltage in C_1 and C_2. The current through C_1 capacitors is equal to inductor L_2 current. The current flowing through the capacitor C_2 is equal to that of the current flowing through inductor L_1. At this condition the charging of capacitor C_2 takes place.

Mode 2: In this mode, the switch gets turned off, and the diodes D_2, D_3 and D_5 get turned-on. Now the voltage across capacitors becomes equal and the current through the inductors starts decreasing linearly. The diode D_1 gets reverse biased by the voltage across the capacitor C_1 and the diode D_4 is turned off by the negative voltage $(V_{c2} - V_{c0})$ the voltage across capacitor C_2 and C_3 are equal. The decrease in inductor currents i_{L1} and i_{L2} is proportional to the voltage $(V_g + V_{c3} - V_O)$ and $(-V_{c1})$. Capacitor C_2 and C_3 are being charged by the currents across inductor L_1. The high gain DC–DC converter is designed using the following equations,

$$\text{The value of inductor, } L_1 = \frac{D.Vin}{2fs\Delta IL1} \tag{1}$$

$$\text{The value of inductor, } L_2 = \frac{D.Vin}{2fs\Delta IL2} \tag{2}$$

where,

V_{in} is the input voltage
f_s is the switching frequency
D is the duty ratio
I_{L1} is the current flowing through inductor L_1
I_{L2} is the current flowing through inductor L_2.

The current across the inductor L_1 is,

$$I_{L1} = \frac{I_o}{(1-D)^2} \quad (3)$$

where, I_o is the output current.

The current across the inductor L_2 is,

$$I_{L2} = \frac{Io}{1-D} \quad (4)$$

The voltage across the inductor is,

$$V_{L1} = V_{in} \quad (5)$$

where, V_{in} is the input voltage.

The voltage across the inductor,

$$V_{L2} = V_{c1} - V_{c2} \quad (6)$$

where, V_{c1} and V_{c2} are the voltages across capacitors C_1 and C_2 respectively.

The value of the capacitor C_1 is,

$$C_1 = \frac{I_{0D}}{(1-D)\Delta Vc1fs} \quad (7)$$

The value of the capacitor C_2 and C_3 is,

$$= \frac{I_{0D}}{\Delta Vc2fs} \quad (8)$$

where,

$$V_{C1} = V_{C2} = \frac{Vs}{1-D} \quad (9)$$

The voltage across the capacitors is,

$$V_{c2} = V_{c3} \quad (10)$$

$$V_{c1} = 2V_{c2} - V_o \quad (11)$$

$$V_{c2} = [1/(2-D)]V_o \quad (12)$$

Table 1 Specifications of high gain boost converter

Parameters	Range
Input voltage (Solar + Fuel cell), V_{in}	61.8 V (solar panel-37.8 V Fuel cell-24 V)
Inductance, L_1	5 mH
Inductance, L_2	25 μH
Capacitance,C_1	50 μF
Capacitance, C_2 and C_3	100 μF
Output capacitance, C_0	1000 μF
Resistance, R_1	250 Ω
Duty ratio	0.4
Switching frequency, fs	90 kHz

$$V_{in} = D'(V_o - V_{c3}) \tag{13}$$

where,

$$D' = (1 - D)$$

The voltage gain of this topology is given by,

$$\text{Voltage gain, } M = (V_0/V_{in}) = \frac{2 - D}{(1 - D)2} \tag{14}$$

where,

D is the duty ratio
V_{in} is the input voltage
V_o is the output voltage.

Using the above equations, the simulation parameters are designed for the high gain converter and it is shown in Table 1.

3 Multilevel Inverter

The proposed hybrid MLI consists of two circuits, one is a developed SC circuit (DSCC) at the frontend of the inverter and another circuit is a conventional H-bridge circuit. The H-bridge circuit connected at the backend of the inverter is used for producing the negative voltage levels of the output [6–8]. The hybrid multilevel inverter can be operated based on the following switching sequence and it can produce nine level output. The switching sequences are given in Table 2.

Table 2 Switching sequence

V_0	S_1	S_2	S_3	S_4	S_5	S_6	S_7	S_8	S_9
$2V_{dc}$	On	Off	Off	On	On	On	On	Off	Off
$3V_{dc}/2$	On	Off	Off	On	On	On	Off	Off	On
V_{dc}	On	Off	Off	On	On	Off	Off	On	Off
$V_{dc}/2$	On	Off	Off	On	Off	Off	Off	On	On
0	Off	On	Off	On	Off	Off	Off	On	On
$-V_{dc}/2$	Off	On	On	Off	Off	Off	Off	On	On
$-V_{dc}$	Off	On	On	Off	On	Off	Off	On	Off
$-3V_{dc}/2$	Off	On	On	Off	On	On	Off	Off	On
$2V_{dc}$	Off	On	On	Off	On	On	On	Off	Off

(1) To achieve the voltage level of $+2V_{dc}$ the switches S_1, S_4 of the H-bridge circuit and the switches S_5, S_6 and S_7 of the DSCC circuit is in ON state.
(2) To achieve the voltage level of $+3V_{dc}/2$ the switches S_1, S_4 of the H-bridge circuit and the switches S_5, S_6 and S_9 of the DSCC circuit is in ON state.
(3) To achieve the voltage level of $+V_{dc}$, the switches S_1 and S_4 of the H-bridge circuit and the switches S_5 and S_8 of the DSCC circuit is in ON state.
(4) To achieve the voltage level of $+V_{dc}/2$, the switches S_1 and S_4 of the H-bridge circuit and the switches S_8 and S_9 of the DSCC circuit is in ON state.
(5) To achieve the zero level voltage, the switches S_2 and S_4 of the H-bridge circuit and the switches S_8 and S_9 of the DSCC circuit is in ON state.
(6) To achieve the voltage level of $-V_{dc}/2$, the switches S_2 and S_3 of the H-bridge circuit and the switches S_8 and S_9 of the DSCC circuit is in ON state.
(7) To achieve the voltage level of $-V_{dc}$, the switches S_2 and S_3 of the H-bridge circuit and the switches S_5 and S_8 of the DSCC circuit is in ON state.
(8) To achieve the voltage level of $-3V_{dc}/2$, the switches S_2 and S_3 of the H-bridge circuit and the switches S_5, S_6 and S_9 of the DSCC circuit is in ON state.
(9) To achieve the voltage level of $-2V_{dc}$ the switches S_2 and S_3 of the H-bridge circuit and the switches S_5, S_6 and S_7 of the DSCC circuit is in ON state.

The switching pattern of the hybrid multilevel inverter is shown in Fig. 2,

4 Simulation Results

The PV modelling is done using MATLAB/SIMULINK and the results are shown in Fig. 3. The model has been designed for 250 W [9, 10], the PV and IV characteristics for the panel is shown in Figs. 4 and 5 respectively,

From Fig. 4., it is clear that the maximum power occurs at a value which is nearer to 250 W with respect to that of the open circuit voltage.

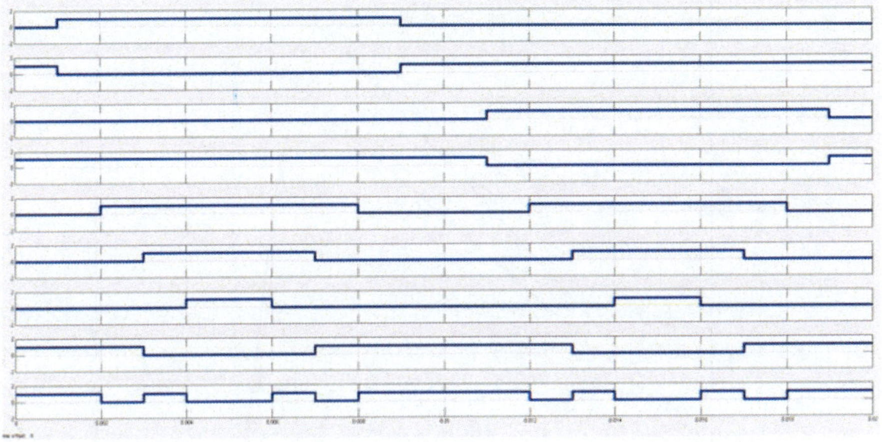

Fig. 2 Switching pattern of hybrid multilevel inverter

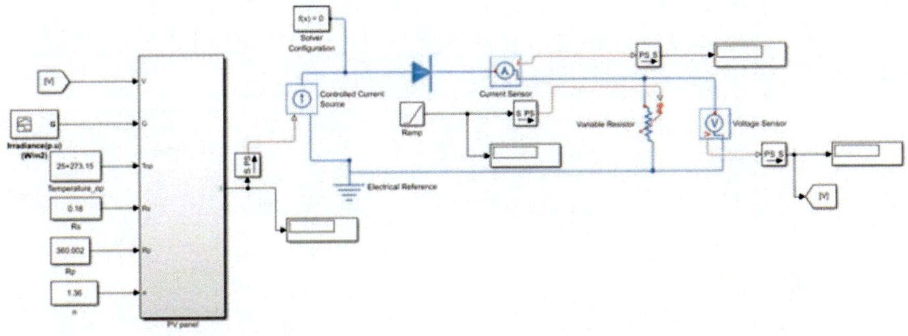

Fig. 3 Simulink model for PV panel

Figure 5, shows the waveform for short circuit current with respect to that of the open circuit voltage. The curve indicates that the rated value of current is obtained.

The simulation of high gain DC–DC converter is carried out using MATLAB/SIMULINK with a hybrid source. The simulink model is shown in Fig. 6,

From Fig. 7, it is clear that the output voltage of the DC–DC converter is in the range of 145.5 V.

The simulink model for the multilevel inverter topology is shown in Fig. 8. The output of the DC–DC converter is fed at the input terminal of the inverter.

Figure 9, shows the nine level output voltage of the inverter. At the level of 2Vdc an output voltage of 291 V is attained.

Figure 10, shows that the THD of output voltage is 21.98% without using any filter components, which is very low compared to conventional topologies. In this proposed inverter configuration the number of switches used to attain nine level output is less when compared to conventional topologies.

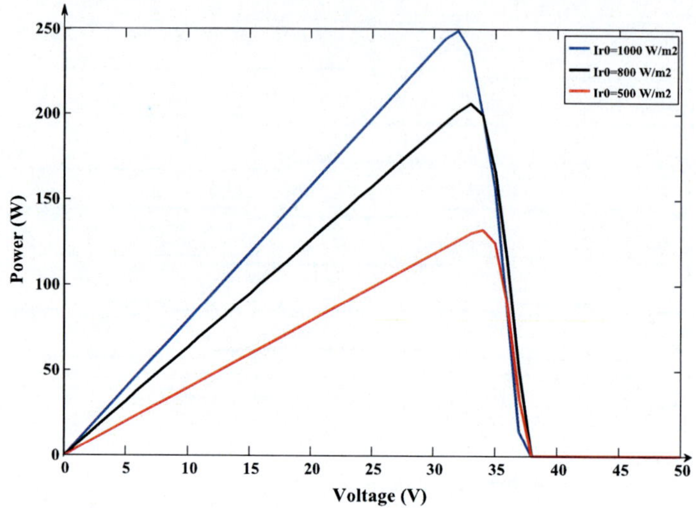

Fig. 4 P–V characteristics

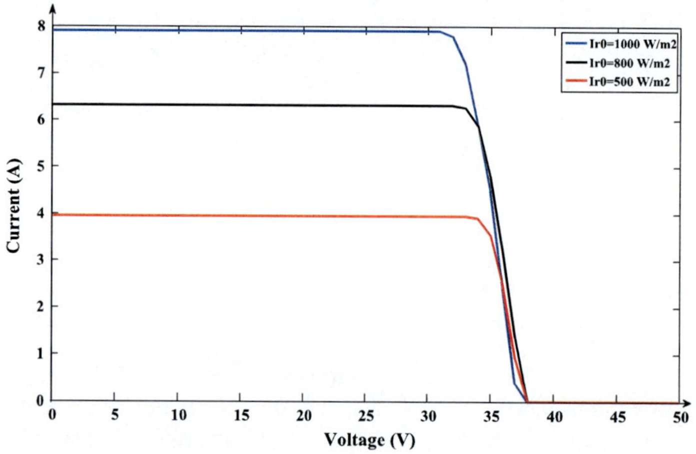

Fig. 5 I–V characteristics

5 Conclusion

In this paper, a nine level inverter configuration for smartgrid application is proposed. The inverter is supplied with a high voltage with the help of a DC–DC converter with PV-FC power generation scheme. The converter used in this scheme is designed with a single power electronic switch which leads to reduction of switching losses. The nine level output of the inverter is also attained using single dc source with less number of switching devices, The THD of the output voltage is also less when compared to

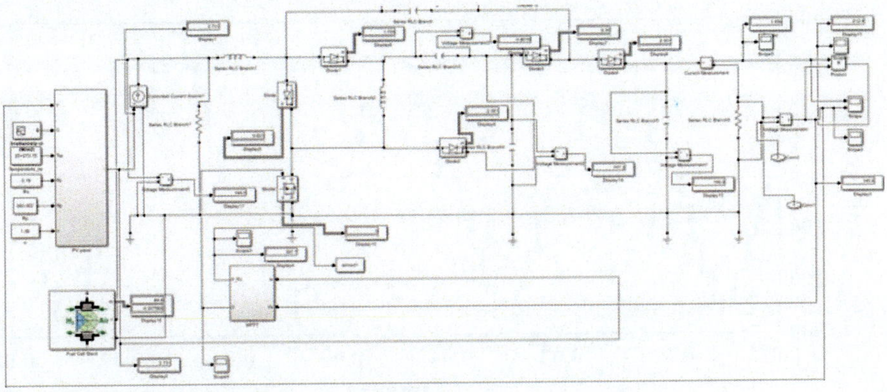

Fig. 6 MATLAB/SIMULINK model for high gain DC–DC converter

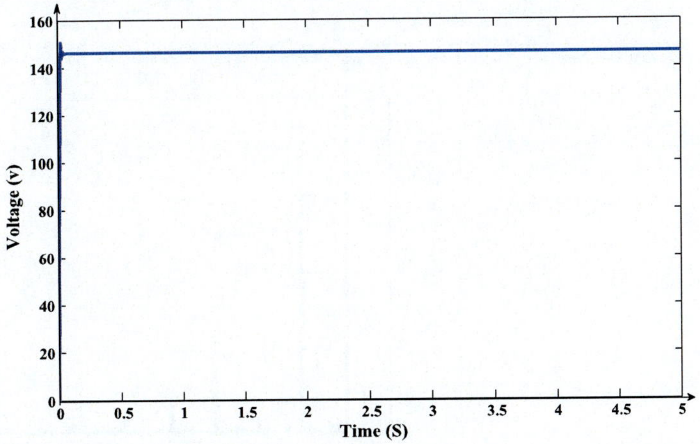

Fig. 7 Output Voltage waveform for high gain DC–DC converter

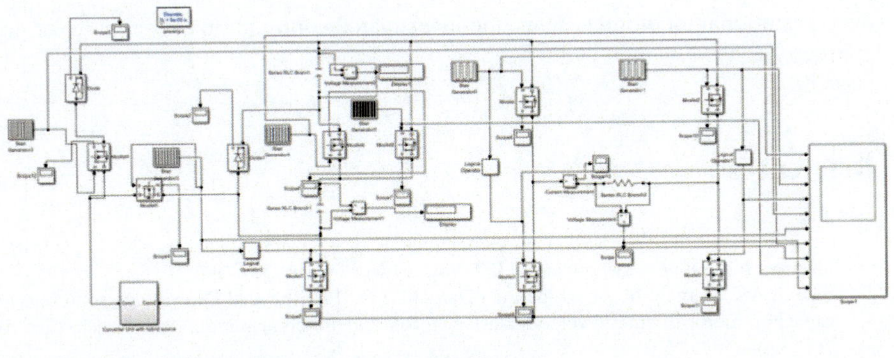

Fig. 8 MATLAB/SIMULINK model for multilevel inverter

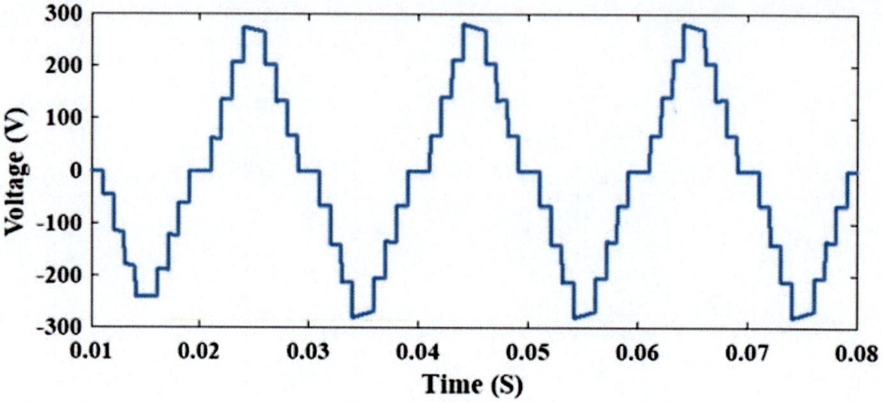

Fig. 9 Output voltage waveform for nine level inverter

Fig. 10 THD of 9-level output voltage

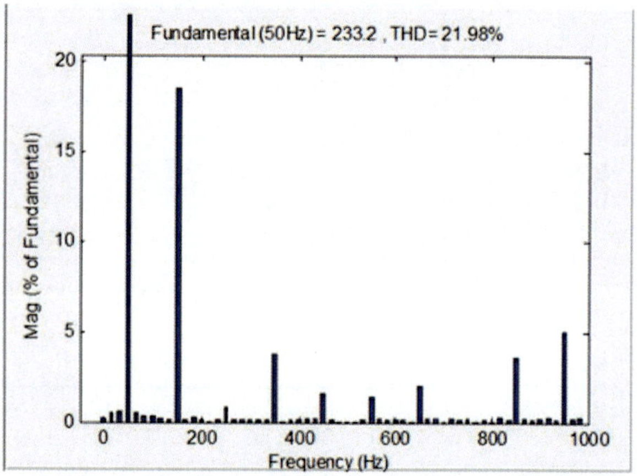

the conventional topologies. Thus, the investigated configuration is suitable for grid applications.

References

1. De S, Banerjee D, Siva Kumar K, Gopakumar K, Ramchand R, Patel C (2011) Multilevel inverters for low-power application. IET Power Electron 4(4):384–392
2. Sadigh AK, Dargahi V, Abarzadeh M, Dargahi S (2014) Reduced DC voltage source flying capacitor multicell multilevel inverter: analysis and implementation. IET Power Electron 7(2):439–450
3. Rodriguez J, Lai JS, Peng FZ (2002) Multilevel inverters: survey of topologies, controls, and applications. IEEE Trans Ind Appl 49–4:724–738

4. Nedumgatt JJ, Vijayakumar D, Kirubakaran A, Umashankar S, A multilevel inverter with reduced number of switches. Select VIT
5. Mohamed HE, Fardoun AA (2016) High gain DC–DC converter for PV applications. In: IEEE transactions 59th international midwest symposium on circuits and systems (MWSCAS), pp 591–596
6. Liu J, Cheng KWE, Ye Y (2014) A cascade multilevel inverter based on switched-capacitor for high-frequency AC power distribution system. IEEE Trans Power Electron 29(8):4219–4230
7. Babaei E, Gowgani SS (2014) Hybrid multilevel inverter using switched capacitor units. IEEE Trans Ind Electron 61(9):4614–4621
8. Hemachandu P, Veera Reddy VC, Jagan Mohan Reddy VC, Usha Reddy V (2015) A PV/FC co-generation based micro-grid system using compact integrated 7-level inverter. IEEE Trans 252–258
9. Selvan S, Nair P, Umayal (2016) A review on photo voltaic MPPT algorithms. Int J Electr Comput Eng (IJECE) 6(2):567–582
10. Villalva MG, Gazoli JR (2008) Comprehensive approach to modelling and simulation of photovoltaic arrays. IEEE Trans Power Electron 24(5):1198–1208

Energy Storage Technologies: Past, Present and Future

Pruthiraj Swain and Ashoka Shyamaprasad

Abstract Decentralization of the main grid into microgrid levels largely depends upon the energy storage penetration level. The limits of the energy storage duration have been pushed with the increase in the penetration of renewables, from intermittent to hours based upon the application requirement. Energy storage technologies are majorly categorized into mechanical, chemical, thermal, electromagnetic and its combination depending upon the application requirement. Energy storage helps in decoupling the energy production and demand, thereby reducing the effort of constant monitoring of the load demand. Storage offers economic benefits of reduction in generation station energy to meet the average demand rather than the peak demands. This also helps in the appropriate sizing of the transmission lines and balance of plant for the average power demands. The storage technologies are compiled and evaluated based upon project/market requirement parameters such as energy/power density, specific energy/power, efficiency, cycle life, capital energy/power costs, technical maturity and its environmental impact, keeping in view their capacity and its microgrid application. Although every storage technology has its own advantages and disadvantages, with focus on the incremental development of existing technology, certain storage technology has the potential to meet the requirement with increased reliability and longevity of the decentralized system.

Keywords Energy · Power · Storage · Density · Cycle life · Technical maturity · Cost

P. Swain (✉) · A. Shyamaprasad
Grid Automation, ABB India Limited, Bengaluru, India
e-mail: pruthiraj.swain@in.abb.com

A. Shyamaprasad
e-mail: ashoka.shyamaprasad@in.abb.com

R. K. Pillai et al. (eds.), *ISUW 2019*, Lecture Notes in Electrical Engineering 764,
https://doi.org/10.1007/978-981-16-1299-2_18

1 Introduction

The modern energy economy has undergone rapid growth change, focusing majorly on the renewable generation technologies due to dwindling fossil fuel resources, and their depletion projections [1]. Figure 1 shows an estimate increase of 32% growth worldwide by 2040 [2, 3]. Asia, North America and Europe has the highest share whereas Asia, Africa and Latin America has shown 1.9%, 2.7% and 1.5% respectively increase per year (2015–2040) [4].

The overall consumption of the electricity worldwide has gain momentum as shown in Fig. 2, where Asia and Africa have the highest YoY % increase of 3.4 and 4.5% respectively [2–4].

The global challenge is well-known, i.e. transforming power and transportation systems through the integration of reliable energy storage systems which can provide a solution to complete the energy transition towards renewables.

Energy storage (which is not only batteries) systems represent a set of technologies and methods that are used to store various forms of energy. Energy storage can be used to manage power supply, to create a resilient energy system and to bring cost savings to both prosumers and utilities.

Energy storage will play a major role in the future for residential, commercial and industrial sectors, and will lead to a transformation of both the power and the transportation sectors. Depending on the sector and the needs, energy storage applications will be a significant part of the future energy system. The goal for a 100%

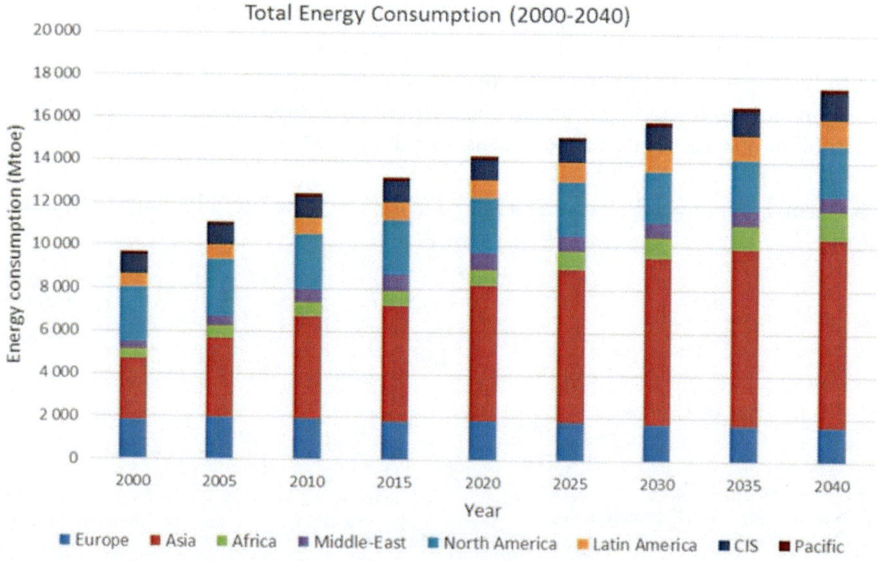

Fig. 1 Total energy consumption of world in Mtoe (2000–2040) [3]

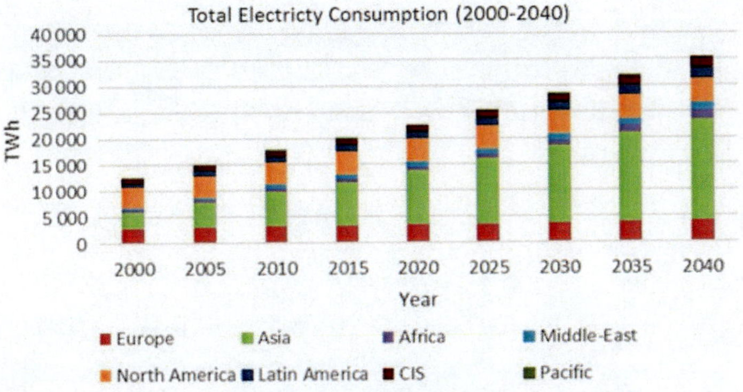

Fig. 2 Total electricity consumption of world in TWh (2000–2040) [2, 3]

renewable energy system could be achieved in the future, thanks to state-of-the-art batteries and development in the other forms of storage systems.

2 Energy Storage Technologies Overview

There are different forms of energy storage depending on two scales, power and time. Certain energy storage technologies are used to store power for different periods of time based upon the application requirement. In this context, understanding which energy storage technology is appropriate in each case is crucial.

As shown in Fig. 3, it is broadly classified into four categories; namely mechanical, electromagnetic, chemical and thermal storage. Out of this, currently pumped hydroelectric (of mechanical storage system) is dominant in terms of deployed forms of energy storage (nearly 99%) [5, 6].

Storage system has no immediate environmental or air quality impacts, helps in demand charge reduction, allows participation in demand response programs and in maximizing time-of-use rates. In remote areas, storage systems plays a vital role of resilience power supply as emergency backup. With the aforesaid commercial advantages, it also helps in the optimal sizing of the transmission lines and equipment followed by the mitigation of problems associated with the intermittency of the renewable energy generation.

In 2017, pumped storage accounts for 96.28% (153 GW) out of the Global utility scale energy storage capacity (by technology), followed by electro-mechanical (1.3 GW), electro-chemical (2.3 GW) and Thermal (2.3 GW) [3]. More than 75% of stationary grid-connected storage capacity was operating in only 10 countries as of 2017 [5].

Technologies such as flywheels and superconductors are having high parasitic losses, are useful for very short duration applications in power quality and regulation. Technologies with lower parasitic losses drop (e.g., pumped hydro), become

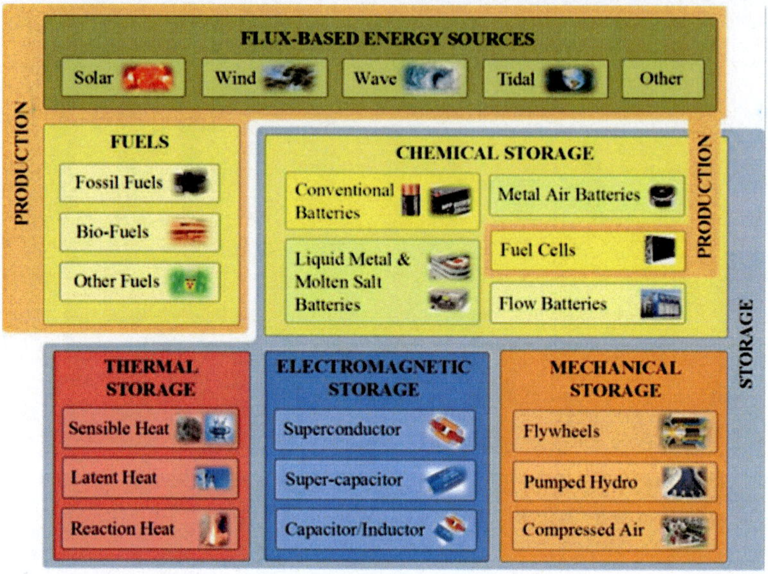

Fig. 3 Pictorial view of the energy storage systems and generation [3]

more relevant for longer term energy management. In the next subsection, positive and negative aspects of each class of technology are represented and concluded in identifying key issues and likely future trends in the energy storage landscape. Figure 4 represents the share of renewables and its trend (region-wise) over the period from 1997 to 2017. The drive for the renewables (due to paradigm shift in the application requirements) has pushed the limitation of the storage duration from seconds to months.

The energy storage system scales are categorized based upon their power rating as well as their application (storage duration) based upon the categories mainly power quality and regulation, bridging power and energy management as shown in Tables 1

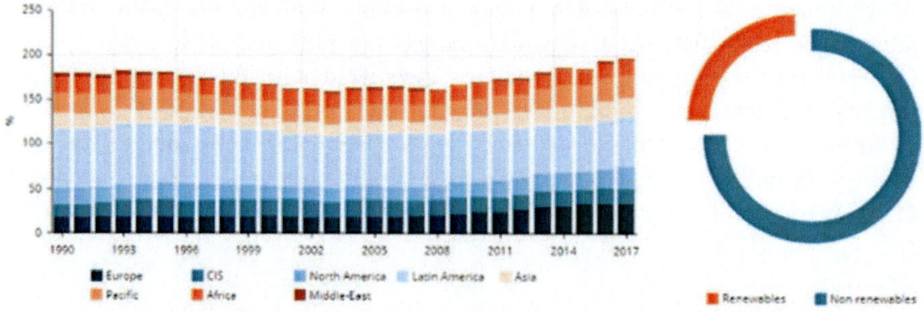

Fig. 4 Distribution trend of renewables (region-wise) and its overall share [4]

Table 1 Categorization of energy storage scales and their applications

Category	Applications	Power rating
Small scale	Mobile devices, electric vehicles, satellites, etc.	≤1 MW
Medium scale	Office buildings, remote communities, etc.	10–100 MW
Large scale	Power plants, etc.	≥300 MW

Table 2 Categorization of energy storage scales and their applications [7]

Category	Applications	Storage duration
Power quality and regulation	Fluctuation suppression/smoothing	≤1 min
	Dynamic power response	
	Low voltage ride through	
	Line fault ride through	
	Uninterruptable power supply	
	Voltage control support	
	Reactive power control	
	Oscillation damping	
	Transient stability	
Bridging power	Spinning/contingency reserves	1 min–1 h
	Ramping	
	Emergency backup	
	Load following	
	Wind power smoothing	
Energy management	Peak shaving/generation/time shifting	1–10 h
	Transmission curtailment	5–12 h
	Energy arbitrage	
	Transmission and distribution deferral	
	Line repair	
	Load cycling	
	Weather smoothing	
	Unit commitment	Hours–days
	Load leveling	
	Capacity firming	
	Renewable integration and backup	
	Seasonal storage	≥4 months
	Annual smoothing	

and 2. The energy storage technologies are classified based upon the application requirement with storage duration.

2.1 Mechanical Energy Storage

Mechanical energy storage has the highest share across all the energy storage technologies. It is comprised of systems such as, pumped hydro storage (PHS), flywheels (FES) and compressed air energy storage (CAES). These systems are widely used and are advantageous on large scale in various commercial, industrial, and residential uses (Table 3).

Mechanical energy storage systems have a huge potential to grow, pertaining to its various beneficial factors such as, technical maturity, regulation of power and frequency, relatively lower environmental impact, high energy/power densities and long duration [8–10]. However, lack of site availability, high capital costs, safety issues and disapproval of governments in initiating capital inducement projects, along with the development of recent modern technologies can hinder the mechanical energy storage market [11, 12].

Table 3 Comparison of different mechanical energy storage technologies based upon listed parameters [7]

Parametrics	PHS	CAES	FES
Scale/application	Large/energy management	Large/energy management	Medium/power quality
Technical maturity	Mature/fully commercialized	Proven/commercializing	Mature/commercializing
Environmental impact	High/medium	Medium/low	Low
Specific energy (Wh/kg)	0.3–1.5	3.2–60.0	5.0–200.0
Energy density (kWh/m^3)	0.5–1.5	0.4–20.0	0.25–424.00
Specific power (W/kg)	0.01–0.12	2.20–24.0	400–30,000
Power density (kW/m^3)	0.01–0.12	0.04–10	400–2000
Efficiency (%)	65–87	57–89	70–96
Cycle life (cycles)	10–60 k	8–30 k	10–100 k
Energy cost (USD/kWh)	1–291.20	1–140	200-150 k
Power cost (USD/kW)	300–5288	400–2250	30.28–700

2.2 Chemical Energy Storage

This type of energy storage has the highest diversity of research and energy storage products which are commercialized presently. This includes traditional batteries, molten salt/liquid metal batteries, metal air batteries, fuel cells and flow batteries.

(1) *Traditional batteries*

Tables 4 and 5 shows comparison between most known and generally used forms of chemical storage in traditional batteries category, namely Zinc Silver oxide (Zn–Ag), Alkaline Zinc Manganese Dioxide (Zn–Mn), lead Acid (Pb–Acid), Lithium-ion (Li–Ion), Nickel Metal Hydride (Ni–MH), Nickel Cadmium (Ni–Cd), Nickel–Iron (Ni–Fe) and Nickel Zinc (Ni–Zn) [13].

Batteries can help when the demand of energy is higher than the available energy at specific times and places. Batteries can help to compensate for different limitations of the power system [13–15].

Table 4 Comparison of different chemical energy storage technologies based upon listed parameters [7]

Parameter	Zn–Ag	Zn–Mn	Pb–Acid	Li–ion
Scale/application	Small/energy management	Small/energy management	Medium/energy management	Small, medium/energy management
Technology maturity	Fully commercialized	Fully commercialized	Fully commercialized	Fully commercialized
Environmental impact	Low	Medium	High	High/medium
Specific energy (Wh/kg)	81–276	80–175	10–50	30–300
Energy density (kWh/m^3)	4.2–957	360–400	25–90	94–500
Specific power (W/kg)	0.09–330	4.35–35	25–415	8–2000
Power density (kW/m^3)	0.36–610	12.35–101	10–400	56.80–800
Efficiency (%)	20–100	36–94	63–90	70–100
Cycle life (cycles)	1–1500	1–200	100–2000	250–10,000
Energy cost (USD/kWh)	3167–20,000	100–1000	50–1100	200–4000
Power cost (USD/kW)	741,935–7,140,620	1000–11,900	175–900	175–4000

Table 5 Comparison of different chemical energy storage technologies based upon listed parameters [7]

Parameters	Ni-MH	Ni–Cd	Ni–Fe	Ni-Zn
Scale/application	Small/energy management	Small, medium/energy management	Small, medium/energy management	Small/energy management
Technology maturity	Mature/fully commercialized	Mature/fully commercialized	Mature/limited development	Mature/limited development
Environmental impact	High	High	Low	Low
Specific energy (Wh/kg)	30–90	10–80	27–60	15–110
Energy density (kWh/m^3)	38.90–300	15–150	25–80	80–400
Specific power (W/kg)	6.02–1100	50–1000	20.57–110	50–900
Power density (kW/m^3)	7.80–588	37.66–141.05	12.68–35.18	121.38–608
Efficiency (%)	50–80	59–90	65–80	80–89
Cycle life (cycles)	300–3000	300–10,000	1000–8500	100–500
Energy cost (USD/kWh)	200–729	330–3500	444.27–1316	250–660
Power cost (USD/kW)	270–530	270–1500	8167–16,312	270–530

(2) *Molten Salt, Liquid metal and Metal Air batteries Traditional batteries*

Molten salt (Sodium Sulphur-NaS and Sodium Nickel Chloride-NaNiCl) and liquid metal are the typical class of high temperature chemical batteries (as in Table 6) which uses molten salts and liquid metals which acts as electrolyte as well as the electrodes [13, 16].

Metal air batteries (Zinc Air and Iron Air) are the cross-over of fuel cells and traditional chemical batteries, having one anode electrode and other as oxygen electrode (catalyzes the production of hydroxyl ions).

Although the typical nature of chemistries of electrolytes, electrodes and its stability, non-distributed charge–discharge patterns due to dynamic load variations, and sensitivity to change in environmental conditions, still traditional batteries are the mostly used storage devices till date in portable applications mostly due to high energy/power densities [17–19].

(3) *Fuel Cells*

Fuel cells (FC) are generally energy generation devices rather energy storage devices, which takes hydrogen and oxygen as input and produces electricity and water as output. The fuel is oxidized at anode and reduced at the cathode [20, 21].

Table 6 Comparison of different chemical energy storage technologies based upon listed parameters [7]

Parameters	NaS	NaNiCl	Zinc air	Iron air
Scale/application	Medium, large/energy management	Medium, large/energy management	Small/energy management	Small/energy management
Technology maturity	Proven/commercializing	Proven/commercializing	Mature/commercialized	Early stage research/developing
Environmental impact	Medium/low	Medium/low	Low	Low
Specific energy (Wh/kg)	100–240	85–140	10–470	8–109
Energy density (kWh/m^3)	150–345	108–190	22–1673	100–1000
Specific Power (W/kg)	14.29–260	10–260	60–225	18.86–146
Power density (kW/m^3)	1.33–50	54.20–300	10–208	250
Efficiency (%)	65–92	21–92.50	30–50	42–96
Cycle life (cycles)	1000–4500	2000–3000	1–500	100–5000
Energy cost (USD/kWh)	150–900	100–345	10–950	10–150
Power cost (USD/kW)	150–3300	150–10,000	100–4000	950

The hydrogen fuel can also be derived from the natural gas, methanol, ethanol, hydrocarbon gas and ammonia with the help of reformers.

Table 7 shows the comparson between four main types of fuel cell system namely, Proton Exchange Membrane (PEMFC), Solid Oxide (SOFC), Direct methanol (DMFC) and Molten Carbonate (MCFC).

Fuel cells technologies is cleaner, quieter and higher efficiency as compared to other technologies. They do have higher specific power/energy and energy densities. However, water and thermal management [22] along with the setup of Hydrogen ecosystem are the critical areas of concern along with its high capital intensive and lifespan, which restricts its economical viable uses to aerospace and large scale grid backup generation application [23–25].

(4) *Flow batteries*

Flow batteries are very much similar in operation to fuel cell systems. To generate electricity, the electrolytes containing dissolved active material flow through the fuel cell. This are classified into 2 types, namely Redox flow batteries and halide/metal batteries as compared in Table 8 [13].

Redox flow batteries usually have 2 tanks storing electrolytes (known as catholyte and anolyte), series/parallel connected bipolar cell stacks and pumping system. Metal/Halide batteries are those which utilizes deposition of metals as a means of storing energy [13, 26, 27].

Flow batteries are having good depth of discharge, simplicity in operation and uses non-toxic materials in their operation but due to lack of cost competitive ness, loses its race with other chemical storage technologies.

Metal/halide batteries has high efficiency and utilises inexpensive materials, resulting in potential of becoming cost competitive due to abundance availability of the electrolyte materials [27, 28].

2.3 Electromagnetic Storage

Electromagnetic storage generally covers storage in inductors (magnetic field) and capacitors (electric field) [29, 30]. With advancement in the technologies, this has been extended to super conductors and supercapacitors (Electrochemical double-layer capacitors) [31] for large scale applications as compared in Table 9.

Super conductors have long cycle life, high efficiency, fast response with very high discharge rates, because of which is used mostly in power quality and stability applications [29, 32]. The major hurdle faced by such technology is the high costs (to maintain the stringent operation parameters) and the mismatch between power capacity and its energy performance.

Supercapacitor usually have low capacitance resulting in making it not suitable for high power applications, however combining them other storage systems such as battery systems will help in increasing life span of battery system by absorbing the

Table 7 Comparison between types of fuel cell systems based upon listed parameters [7]

Parameters[a]	PEMFC	DMFC	MCFC	SOFC
Scale/application	Small, medium/energy management	Small/energy management	Medium/energy management	Medium/energy management
Technology maturity	Proven/commercializing	Proven/developing	Proven/developing	Proven/commercializing
Environmental impact	Low	Low	Medium/low	Medium/low
Specific energy (Wh/kg)	100–450	140.30–960	369–607	410–1520
Energy density (kWh/m^3)	112.2–770	29.90–274	25–40	172–462.09
Specific power (W/kg)	4–150	2.10–20	12–36.70	10–63.34
Power density (kW/m^3)	4.20–35	1–300	1.05–1.67	4.20–19.25
Efficiency (%)	22–85	10–40	45–80	50–65
Energy cost (USD/kWh)	70–13,000	30,670–3190	146–175	180–333
Power cost (USD/kW)	0–10,200	15,000–125,000	3500–4200	481–8000

[a]Cycle life is not applicable for fuel cell systems as source is external to the system

Table 8 Comparison among the types of flow batteries based upon listed parameters [7]

Parameters	Vanadium redox (VR)	Zinc bromine (ZB)	Polysulphide bromine (PB)
Scale/application	Medium, large/energy management	Large/energy management	Large/energy management
Technology maturity	Proven/commercializing	Proven/developing	Proven/developing
Environmental impact	Medium/low	Medium	Medium
Specific energy (Wh/kg)	10–50	11.10–90	10–50
Energy density (kWh/m^3)	10–33	5.17–70	10.80–60
Specific power (W/kg)	31.30–166	5.50–110	NA
Power density (kW/m^3)	2.50–33.42	2.58–8.50	1.35–4.16
Efficiency (%)	60–88	60–85	57–83
Cycle life (cycles)	800–16,000	800–5000	800–4000
Energy cost (USD/kWh)	100–2000	110–2000	110–2000
Power cost (USD/kW)	175–9444	175–4500	330–4500

Table 9 Comparison among the types of flow batteries based upon listed parameters [7]

Parameters	Superconducting magnetic (SMES) [30]	Supercapacitors (ELDC) [33]
Scale/application	Medium, large/power quality	Small, medium/power quality
Technology maturity	Proven/commercializing	Proven/commercializing
Environmental impact	Low	Low
Specific energy (Wh/kg)	0.27–75	0.07–85.60
Energy density (kWh/m^3)	0.20–13.80	1–35
Specific power (W/kg)	500–15,000	5.44–100,000
Power density (kW/m^3)	300–4000	15–4500
Efficiency (%)	80–99	65–99
Cycle life (cycles)	10,000–100,000	10,000–1,000,000
Energy cost (USD/kWh)	500–1,080,000	100–800
Power cost (USD/kW)	196–10,000	100–800

surges and spikes during the transient operations and also fast delivery of the stored energy during requirement of the load [30, 31, 33].

2.4 Thermal Energy Storage

Thermal storage systems (TES) are used in mainly thermal power plants (industry scale) [34, 35]. Since mechanical, chemical and electromagnetic storage technologies are focusing on electricity storage, however the thermal storage needs to be coupled to heat engines or some thermoelectric generators for electricity generation (useable form).

Thermal storage is further categorized into three types namely, sensible heat, latent heat and reaction heat whose comparson is given in Table 10. This system contains three main components: the containment system, the thermal material and heat exchanger.

Thermal storage has wide range of applications in existing power plants and potential for the solar power plants, where the heat loss can be utilized to increase

Table 10 Comparison of different types of thermal storage systems based upon listed parameters [7]

Parameters	Sensible heat (STES)	Latent heat (LTES)	Chemical reaction heat (CTES)
Scale/application	Medium/bridging power	Medium, large/energy management	Small, medium/energy management
Technology maturity	Mature/commercializing	Proven/commercializing	Proven/developing
Environmental impact	Low	Low/uncertain	Low/uncertain
Specific energy (Wh/kg)	10–120	150–250	250
Energy density (kWh/m^3)	25–120	100–370	300
Specific power (W/kg)	NA	10–30	NA
Power density (kW/m3)	NA	NA	NA
Efficiency (%)	7–90	75–90	75–100
Cycle life (cycles)	NA	NA	NA
Energy cost (USD/kWh)	0.04–50	3–88.3	10.90–137
Power cost (USD/kW)	2500–7900	200–300	NA

the overall system efficiencies (also with the use of phase change material, PCM for storing solar heat [36, 37]). Research is underway for the underground thermal storage technology, which can take advantage of the planet's inner heat.

3 Conclusion

The energy storage technologies are vast and out of which twenty-seven types of storage technologies are considered. The technologies are compared based on parameters such as technical maturity, specific energy/power, energy/power density, efficiency, cycle life, energy/power cost, environmental impact and its applications.

Ni–MH, Ni–Cd, ZnAg, ZnMn and Pb-Acid from chemical energy storage systems and PHS spearheaded the technology maturity stage, however recycling and disposal of the batteries are still critical to the environment. Due to low loss storage and high share among energy storage systems (nearly 99%), PHS is mostly used for the energy management applications.

FES (followed by SMES) and flow batteries has the lowest and lowest impact respectively on the environment among other storage technologies. Superconducting at room temperature condition would be a game changer for energy storage Flow batteries has the potential for cost competitive when compared with CAES along with the improvements in power/energy capacities,

Fuel cells (SOFC followed by MCFC), FES, Metal Air (Zn–Air followed by Fe–Air), Super capacitors, SMES, SMES, leads in the specific energy (Wh/kg), specific power (W/kg), energy density (kW/m^3), power density (W/m^3), efficiency (%), cycle life (cycles) parameters respectively.

Fuel Cells (SOFC followed by MCFC) is far more than PHS, CAES and flow batteries in terms of energy performances, which means huge reduction of land requirement/usage, whereas Metal Air (Zn–Air followed by Fe-Air) shows higher power performances among others. This helps in putting fuel cell systems for large scale applications in future.

Li-ion has significant potential in the small scale applications such as electric vehicle applications due to its high energy/power densities as against molten salt, NaS, NaNiCl, NiCd, NiZn and NiMH which are having shorter life cycles and requires complex thermal management.

Energy cost (USD/kWh) and power cost (USD/kW) is highest for SMES (followed by Supercapacitors) and Zn–Ag (followed by DMFC) respectively, which makes its commercialization restraint. CAES has the lowest energy cost, with high power/energy capacities making it a good contender for utility scale energy storage applications.

Thermal energy storage systems are the major focus areas for the already installed generation systems as well for the renewables energy systems (mainly PV solar) for efficiency improvements. Hence, these systems are going to be integrated part of other storage systems. Furthermore, with respect to grid power quality management and regulation, FES, supercapacitors and SMES systems are the three main storage systems.

Other storage systems which are yet to be explored are biomass energy storage [38], gravity-based storage [39], mechanical spring systems-based storage [40], rail energy storage (RES) [41], phase change material (PCM) storing solar heat energy [36, 37].

The advancements in the energy storage systems from small scale to large scale, with duration from seconds to months are largely driven by the application requirements as well as the policies, standards and regulations adapted by the countries to reduce the impact of fossil fuel consumption on the environment with rapid modernization of the industry and transportation sector.

For the development of the energy storage technologies, continual effort needs to be in place for the improvement of the existing technologies as well as disruption of new technologies. But due to decreased energy cost of the competitors and lower investment in the new technology, the consumer patterns are unchanged which focuses on lower cost and increased performance of technology.

4 Assumption

Discharge rate assumed is 1C-rate. Data has been collected from multiple online databases available online. Specific power and energy are calculated based upon the dry mass of the systems. In case of energy generation systems such as fuel cell, 24 h time duration is considered for evaluation. Specific system dimensions for concerned systems are considered based upon the online available brochure/technical datasheet for the volume-based densities computation. Maximum efficiency has been considered in case of non-availability of information. Cost comparisons are made based on the recent trends in the current market scenario assuming exchange rates are in the acceptable variable range.

References

1. Shafiee S, Topal E (2009) When will fossil fuel reserves be diminished. Energy Policy 37:181–189
2. https://yearbook.enerdata.net/total-energy/world-consumption-statistics.html
3. https://www.enerdata.net/publications/world-energy-statistics-supply-and-demand.html
4. https://eneroutlook.enerdata.net/forecast-renewables-in-electricity-generation-share.html
5. Key World Energy Statistics (2013) International Energy Agency (IEA). Paris, France

6. Renewables (2013) Global status report; Renewable energy policy network for the 21st century. Paris, France
7. Swain P, Kashyap K (2019) Energy storage technologies: A qualitative analysis. Int Conf Energ Storage. New Delhi, India
8. Bueno C, Carta JA (2006) Wind powered pumped hydro storage systems, a means of increasing the penetration of renewable energy in the canary islands. Renew Sustain Energy Rev 10:312–340
9. Deane JP, Gallacho BPO, McKeogh EJ (2010) Techno-economic review of existing and new pumped hydro energy storage plant. Renew Sustain Energy Rev 14:1293–1302
10. Droste-Franke B (2012) Assessing environmental impacts of storage technologies and competing options for balancing demand and supply in 2050. In: Proceedings of the international renewable energy storage conference (IRES). Berlin, Germany, 12 Nov 2012
11. Yang C-J, Jackson RB (2011) Opportunities and barriers to pumped-hydro energy storage in the United States. Renew Sustain Energy Rev 15:839–844
12. Steffen B (2012) Prospects for pumped-hydro storage in Germany. Energy Policy 45:420–429
13. Buchmann I, Battery University. Available online: https://batteryuniversity.com/
14. Tarascon J-M, Armand M (2001) Issues and challenges facing rechargeable lithium batteries. Nature 414:359–367
15. Linden D, Reddy TB (2002) Handbook of batteries. McGraw-Hill, New York, NY, USA
16. Dueber R (2008) Strategic overview of silver-zinc rechargeable batteries. Battery Power Prod Technol 12:1–2
17. Olivetti E, Gregory J, Kirchain R (2011) Life cycle impacts of alkaline batteries with a focus on end-of-life. Massachusetts Institute of Technology, Cambridge, MA, USA
18. Huang Y-J, Lin Y-L, Li W-S (2012) Manganese dioxide with high specific surface area for alkaline battery. Chem Res Chin Univ 28:874–877
19. Ghaemi M, Amrollahi R, Ataherian F, Kassaee MZ (2003) New advances on bipolar rechargeable alkaline manganese dioxide–zinc batteries. J Power Sources 117:233–241
20. Swain P, Jena D (2015) Modeling, simulation and optimal control of non-linear PEM fuel cell with disturbance input. In: 2015 international conference on innovations in information, embedded and communication systems (ICIIECS). Coimbatore, pp 1–7
21. Swain P, Jena D (2015) PID control design for the pressure regulation of PEM fuel cell. In: 2015 international conference on recent developments in control, automation and power engineering (RDCAPE), Noida, pp 286–291
22. Cheng X, Shi Z, Glass N, Zhang L, Zhang J, Song D, Liu Z-S, Wang H, Shen J (2007) A review of PEM hydrogen fuel cell contamination: impacts, mechanisms, and mitigation. J Power Sources 165:739–756
23. Ren X, Zelenay P, Thomas S, Davey J, Gottesfeld S (2000) Recent advances in direct methanol fuel cells at los Alamos national laboratory. J Power Sources 86:111–116
24. Cassir M, Ringuede A, Lair V (2013) Molten carbonates from fuel cells to new energy devices. In: Molten salt chemistry. Elsevier, Burlington, MA, USA
25. Zhao Y, Xia C, Jia L, Wang Z, Li H, Yu J, Li Y (2013) Recent progress on solid oxide fuel cell: lowering temperature and utilizing non-hydrogen fuels. Int J Hydrog Energy 38:16498–16517
26. Blanc C, Rufer A (2010) Understanding the vanadium redox flow batteries. In: Paths to sustainable energy. InTech, Glasgow, UK
27. Skyllas-Kazacos M, Menictas C (2013) Redox flow batteries for medium- to large-scale energy storage. In: Electricity transmission, distribution and storage systems. Woodhead Publishing Limited, Cambridge, MA, USA
28. Parasuramana A, Lima TM, Menictasc C, Skyllas-Kazacos M (2013) Review of material research and development for vanadium redox flow battery applications. Electrochim Acta 101:27–40
29. Superconductivity: present and future applications. Available online: https://www.ccas-web.org/pdf/ccas_brochure_web.pdf. Accessed on 25 Dec 2014
30. Buckles W, Hassenzahl WV (2000) Superconducting magnetic energy storage. Power Eng Rev, 16–20 May 2000

31. Sharma P, Bhatti TS (2010) A review on electrochemical double-layer capacitors. Energy Convers Manag 51:2901–2912
32. Ali MH, Wu B, Dougal RA (2010) An overview of SMES applications in power and energy systems. IEEE Trans Sustain Energy 1:38–47
33. Dai T, Tang Y, Shi J, Jiao F, Wang L (2010) Design of a 10 MJ HTS superconducting magnetic energy storage magnet. IEEE Trans Appl Supercond 20:1356–1359
34. Thermal Energy Storage (2013) Technology brief. International Renewable Energy Agency (IRENA). Abu Dhabi, UAE
35. Muthukumar P (2011) Thermal energy storage: methods and materials. Indian Institute of Technology Guwahati, Sevilla, Spain
36. Sharma A, Tyagi VV, Chen CR, Buddhi D (2009) Review on thermal energy storage with phase change materials and applications. Renew Sustain Energy Rev 13:318–345
37. Farid MM, Khudhair AM, Razack SAK, Al-Hallaj S (2004) A review on phase change energy storage: materials and applications. Energy Convers Manag 45:1597–1615
38. https://www.eia.gov/energyexplained/?page=biomass_home
39. https://techxplore.com/news/2018-11-brick-solution-topple-energy-storage.html
40. Rossi F, Castellani B, Nicolini A (2015) Benefits and challenges of mechanical spring systems for energy storage applications. Energy Proc 82:805–810. ISSN 1876-6102
41. https://www.aresnorthamerica.com/

Flexible Generation and Flexible Load for Large Integration of Renewable Generation into the Grid

Rishika Sharan

Abstract India is keen to attempt to work towards a low carbon emission pathway. As per goal set up for Intended Nationally Determined Contribution (INDC) by India, India has to reduce the emissions intensity of its GDP by 33–35% by 2030 from 2005 level and to create an additional carbon sink of 2.5–3 billion tones of CO_2 equivalent through additional forest and tree cover by 2030. Harnessing Renewable energy sources is one of the attempt to work towards a low carbon pathway. To accelerate development and deployment of renewable energy in the country, the Government is taking a number of initiatives like up-scaling of targets for renewable energy capacity addition from 30 GW by 2016–17 to 175 GW (out of total 479 GW Installed Capacity(IC)) by 2021–22 and further up to 275 GW (44%, out of total 619 GW IC) by 2026–27 (National Electricity Plan—vol. I: Generation (Notified vide Extra ordinary Gazette No. 1871,Sl. No. 121,under part-III, Section IV dated 28.03.2018)). Due to variable, intermittent and non-dispatchable generation from RE sources, the safe and reliable grid operation is the next step towards the readiness for integration of such huge capacity of RE into the grid. Presently the balancing of grid for variability and intermittency of RE generation is done by ancillary services, already in place and by increasing/decreasing of generation from conventional sources. In the year 2021–22, when the capacity of RE is expected to be 175 GW (37% of the total IC), other measures are also required to be identified for reliable, safe and economic operation of grid. These measures include flexible operation of existing coal-fired power plants, operation of some pre-identified units of coal, gas and hydro during morning peak and off peak hours and installation of Pump storage or battery storage system of appropriate capacity at suitable location. Next, at the end of the year 2026–27, when the installed capacity of RE generation in the grid is expected to further go up to 275 GW (44% of the IC), some additional measures are to be identified. The flexible load is one of them [Electrification and future of Electricity Market. IEEE Power and Energy Magazine, July/Aug 2018]. All the measures well planned and well placed into the system by that time, would certainly help safe, reliable and economic operation of the grid.

R. Sharan (✉)
Central Electricity Authority, New Delhi, India

R. K. Pillai et al. (eds.), *ISUW 2019*, Lecture Notes in Electrical Engineering 764,
https://doi.org/10.1007/978-981-16-1299-2_19

Keywords Renewable energy · Flexible generation · Flexible load

1 Introduction

India is keen to attempt to work towards a low carbon emission pathway, while simultaneously endeavoring to meet all the developmental challenges that the country faces today. As per goal set up for Intended Nationally Determined Contribution (INDC) by India, India has to reduce the emissions intensity of its GDP by 33–35% by 2030 from 2005 level and to create an additional carbon sink of 2.5–3 billion tonnes of CO_2 equivalent through additional forest and tree cover by 2030 [1]. Thrust on renewable energy, promotion of clean energy, enhancing energy efficiency, developing climate resilient urban centres and sustainable green transportation network are some of the measures for achieving this goal.

Harnessing Renewable energy sources will put India on the path to a cleaner environment, energy independence and, a stronger economy. Over the years India has successfully created a positive outlook necessary to promote investment in, demand for, and supply of, renewable energy. India's strategy on renewable energy is driven by the objectives of energy security, energy access and also reducing the carbon footprints of the national energy systems. It has evolved over the years through increasingly stronger commitment at government level.

To accelerate development and deployment of renewable energy in the country, the Government is taking a number of initiatives like up-scaling of targets for renewable energy capacity addition from 30GW by 2016–17 to 175 GW by 2021–22 and further upto 275 GW (out of total 619 GW) by 2026–27. As per 19th Electricity Power Survey report the peak electricity demand of the country will also increase to 226 GW in the year 2021–22 to 298 GW in the year 202,627 [1].

Due to large deployments of renewable energy generating plants, some significant changes in the physical supply–demand dynamics of the future Indian electricity system can be anticipated. This would be more prominent with the increased penetration of RE generation (30–40% of the total consumption) into the grid. With the help of improved sensing, and simple control functionality, we can reduce its effect on system including peak electricity demand and accommodate increased renewables penetration, creating a more sustainable electricity grid.

2 The Present Scenario (2017–18) of Electricity Supply–Demand Dynamics

At present most of the generation is dispatch able and load is almost passive. The status for the RE generation is a must run status. Balancing in the grid due to variability of RE generation is being fulfilled by conventional Generating Stations. The ancillary services are introduced to take care of ramp up and ramp down and balancing requirement of the grid.

3 The 2021–22 Scenario of Electricity Supply–Demand Dynamics

At the end of the year 2021–22, it is expected that, out of total 479 GW of All India installed capacity, 175 GW (37% of the total IC) would be from RE sources. The RE generation capacity will be non-dispatchable and the load is expected to be almost passive load. Due to large deployments of renewable energy generating plants, there are a number of additional challenges with large penetrations of renewable generation. Some significant changes in the future load/net load curve in the Indian electricity system can be anticipated.

The all India expected load curve for a typical day during the year 2021–22, is as shown in Fig. 1. The expected all India Net load curve (Duck Curve) obtained from the All India Load curve after deducting the load met from must-run plants i.e. wind, solar and nuclear generations, and generation from hydro plants. The net load curve in Fig. 1 would be met with coal generation. As the % RE generation into the grid would increase, the difference in maximum and minimum load (Duck belly peak demand and demand of the net load curve) to be met from coal generation during the day would also increase.

Due to increase in Duck belly peak demand and demand difference of net load curve, there would also be increase in ramp up and down capacity and its change

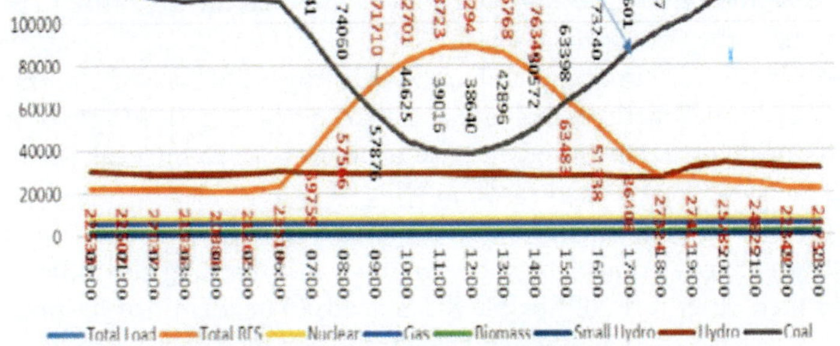

Fig. 1 Load and net load curve of a typical likely day in 2021–22

in rate, reserve capacity, residual capacity, spinning reserve and balancing capacity. The increase also lead to partial loading and two shifting i.e. cycling of fossil based power plants and hence low plant low factor, increase in requirement of ancillary services and hence increased system operation cost, Increased forced outage and O&M cost, equipment life time reduction, poor heat rate and high auxiliary power and requirement of enhanced transmission network and its under utilization.

Considering the high variability and intermittency of generation from renewable, efficient and economical grid operation becomes one of the critical challenges in India's Power system. The CERC has taken several initiatives to ensure integration of variable RE generation. The framework for forecasting, scheduling and deviation settlement for wind and solar has been put in place. To enable thermal generators to provide balancing support, necessary regulatory framework has been provided defining technical minimum for such plants and commensurate compensation for flexing such (thermal) generation up to technical minimum. The CERC has issued Suo Motu order delineating the road map for operationalizing reserves [3]. The SERCs are also taking necessary steps in this regard in their respective states.

In order to reduce the Duck belly peak demand and demand difference of net load curve, some measures like operation of some pre identified units of coal and gas during morning peak and off peak hours and also some capacity of hydro plants (to stop generation from morning 6:00 h to evening 18:00 h and shall run from evening peak,18:00 h to morning peak, 6:00 h) in the form of flexible generation may be made operational in the grid. The temporal shifting of Renewable energy generation with the help of energy storage like battery storage and pump storage (consume power during off peak and generating power at peak hours of net load curve) may also be helpful in decreasing the Duck belly peak demand and demand difference of net load curve [2].

4 Identification of Flexible Load

As the % increase in the capacity of RE plants and subsequently increase in % generation from RE, the duck belly Peak demand and demand difference of the net load curve would further increase and some additional measures are to be identified to reduce the same.

Flexible load has a large potential to solve energy imbalance due to RE generation. It can help in diurnal energy imbalance and also help in load Shifting by storing energy produced during off-peak period and use from on-peak to off-peak periods, thereby reducing the system peak load as well [2].

Flexible load in the form of flexible loads in buildings, Electric vehicle charging, and battery electric storage, may be effective in providing flexibility to net load curve. Further, other form of flexible loads needs to be identified. Demand-response technologies may also play a similar role by shifting load demand so that it coincides with RE generation or lowers the ramping requirements of the remaining generation fleet.

Some of the facts of the flexible loads are listed below [2]:

- There are many types of classification of Flexible loads. However, ultimately for each type of flexible load, the goal is to flatten net load curve.
- Flexible load can solve diurnal energy imbalance and reduce load at peak to avoid capital investments (i.e., substations and transmission lines). The other type of flexible load can be storage of excess energy due to over generation in one season and using it in energy deficits scenario in other times of year.
- Nonfirm load are those flexible load which can be curtailed and can be shifted. Those load which are with a high operating to-capital-cost ratio, can be chosen for non firm load.
- Flexible load may be of short duration or long duration.
- Interruptible flexible loads may be the small commercial and residential loads, for which applications can be developed for aggregating many small interruptible loads into a large one.
- Flexible loads benefits customers by reducing the utilitybill peak demand and improving flexible plant efficiency. The customers may be billed on Time-ofUse (TOU) rate schedule and may be incentivized also.
- In the scenario of surplus of energy (negative net load curve), the balancing solutions in the form of flexible load can be scheduled to absorb surplus energy.
- Under a high-renewable system, where supply is variable with near-zero marginal cost, and new technology enables unprecedented demand side flexibility, a case may also be anticipated where large loads may bid a demand curve in which different amounts of energy consumption would be offered at different prices, in a way that is completely analogous to how generators bid an energy supply curve today.

5 Way Forward

Exciting yet challenging times lie ahead. The future electricity systems, with the large capacity of RE generation, would be radically different from today's systems, which is dominated by dispatchable thermal capacity and inflexible loads. The future electricity systems may be dominated by non dispatchable & variable generation and technology enabled flexible load. The expansion capacity of renewable generation may also increase compared with no load flexibility incorporation [4].

Though the future electricity systems seems technically feasible, however, it requires identification of appropriate capacity for flexible operation of coal, gas, hydro and storage (battery and pump storage) generating plants at suitable location. Further, the flexible loads in the form of flexible loads in buildings, Electric vehicle charging, and battery electric storage etc. are also required to be identified into the system. Flexible load potential analysis is also needed to be carried out as a pilot project for future implementation.

The flatten the net load curve or lesser the Duck belly peak demand and demand difference of net load curve, the higher the capacity factors of residual capacity and lower the investment. Therefore, the generators and loads providing flexibility may be considered for incentivizing also.

Smart grid and smart metering is the essential part of efficient flexible load/demand side management and may be implemented in a phased manner.

Regulators may have to strive to understand which changes would be best accomplished through markets and which through regulations.

All the measures well planned and well placed into the system, would certainly help safe, reliable, economic and flexible operation of the grid with different types of energy resources.

References

1. National Electricity Plan—vol. I: Generation (Notified vide Extra ordinary Gazette No. 1871,Sl. No. 121,under part-III, Section IV dated 28.03.2018)
2. Electrification and future of Electricity Market. IEEE Power and Energy Magazine, July/Aug 2018
3. Discussion Paper on Re-designing Ancillary Services Mechanism in India, CERC, Sept 2018
4. Dang C, Wang X, Wang X, Li F, Zhou B (2018) DG planning incorporating demand flexibility to promote renewable integration. IET Gener Transmission Distrib 12(20):4419–4425

3D Zinc Sponge Battery Technology in Mobile and Stationary Energy Storage Applications—It's Advantages When Compared with Lead Acid and Lithium-Ion Batteries

Sauman Das Gupta and Buddha Burman

Abstract The main objective of the paper is to delve into understanding the potential of another advanced battery technology—3D-Zinc Anode sponge-based battery technology—Zinc battery technology has not been given proper importance by many, major concentration was towards development of Lithium Ion battery technologies. 3D Zinc Anode Sponge battery technology has made significant developments. Zinc batteries were thought of as a primary (non-rechargeable) battery technology, developments over the decade has made it possible to make it secondary (rechargeable). 3DZinc Anode-based battery technology has properties which make it cheaper then Lead Acid batteries while it is equivalent to most existing Lithium Ion battery technologies. Thus, it provides a very attractive solution for all mobile and stationary sustainable energy storage use cases. The paper will look primarily at 3D Zinc Anode sponge battery Technology developments and advantages when compared with lead acid and lithium-ion batteries.

Keywords 3D zinc anode · Batteries · Lithium ion

Present rechargeable battery energy-storage solutions are mainly dominated by lithium-ion batteries and Lead Acid with Nickel Hydride taking a minor share. Lithium-ion batteries are preferred mainly because of their cycle life, energy content, considerable improvement, resourced research, development and deployment programs. There have been reported numerous safety incidents, recycle ability issues, toxicity along with many other impediments, which include transportation restrictions, limited resource supply of both lithium and cobalt, cost -of-plant equipments, manpower skills availability and cost to set operations /supply chains.. In spite of these disadvantages, Lithium-ion batteries are widely used.

Electronic supplementary material The online version of this chapter (https://doi.org/10.1007/978-981-16-1299-2_20) contains supplementary material, which is available to authorized users.

S. Das Gupta (✉) · B. Burman
Indian School of Innovation in Sustainability (INSIST), New Delhi, India

R. K. Pillai et al. (eds.), *ISUW 2019*, Lecture Notes in Electrical Engineering 764,
https://doi.org/10.1007/978-981-16-1299-2_20

Thus to displace Lithium-Ion batteries the characteristics and it's attributes must replaced by alternative battery technology to be able to compete for market share.

At the research and development stage, many different technologies such as Sodium-ion, Lithium- Sulphur, Lithium Oxides as well solid Lithium-Air are being studied and tested. Most of these still use non-aqueous electrolytes that is prone to safety issues, recycling and environmental concerns.

Hence the next generation of high-performance batteries must provide alternative advanced chemistries which are safe. Thus, providing options that are safer and cheaper to operate than non-aqueous lithium ion technologies and toxic lead acid batteries.

3D-Zinc Sponge Anode technology with aqueous electrolytes- batteries can answer that challenge:

A high performance rechargeable zinc based battery has been of interest to battery developers since the time of Thomas Edison. He patented a zinc battery in 1901.

The series of zinc-based batteries—Nickel–Zinc, Silver–Zinc, Zinc–Magnesium, and Zinc–Air—does offer very competitive alternatives. They have practical specific energies ranging from 80 to 475 Wh kg^{-1}, with specific power up to a continuous 800 W kg^{-1}, with operational validation for decades in providing safety advantages using aqueous alkaline electrolytes.

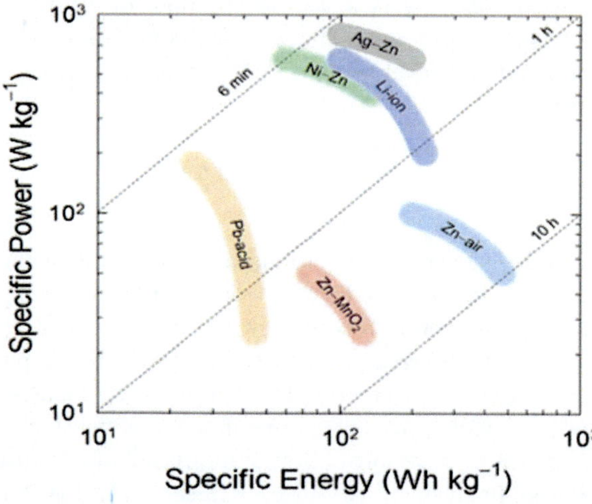

Ragone Plot of Zinc-Based Batteries using aqueous electrolytes (Zinc–MnO_2, Zinc–air, Nickel–Zinc, and Silver–Zn). Lead- acid and Lithium-ion systems are shown to provide a comparison.

What eluded Mr. Edison is why his zinc batteries did not last very long—very limited charge/discharge cycle.

Hence, the zinc battery was relegated to a primary or disposable battery. Many people have tried to make zinc commercially rechargeable and the two solutions used are either a flow battery (like a fuel cell) where the "bad" zinc is simply replaced. Or by using dendrite suppressing additives added to the electrolyte to slow down but not eliminate dendrite growth. Both of these limited the specific energy of the battery.

The historic limitation which has prevented using Zinc in rechargeable batteries is its poor recharge ability due to the formation of dendrites and eventual pre-mature failure.

This problem has been for increase cycling durability by redesigning the Zinc electrode into a monolithic structure, which is porous, has an aperiodic architecture in which there is a persistence of electron-conductive metallic Zinc in its inner core when subjected to deep levels of discharge. This is has been coined as 3D Sponge Zinc Anode.

The solution is elegant and simple. It is achieved by converting Zinc to 3D Sponge Zinc form making it a continuously wired conductive structure which enables current flow to be uninterrupted during charge and discharge. Disruption of current flow is the main reason of failure when conventional Zinc Slurry is used due to dendrite formation.

Now problematic hot-spots cannot form, dendrites growth is eliminated and anode cracking does not occur.

The use of 3D Zinc Anode-based alkaline batteries is the first that offers a structural solution for use of zinc electrode. Creating a battery that offers the specific energy of Lithium-ion batteries at cost more like lead-acidbattery and safer (to use as well as for the planet) than both. It also has a simpler supply chain and is not dependent on resource constrained components like Lithium or Cobalt (Fig. 1a)

The complete series of 3D Zinc Anode-based alkaline batteries (Using 3D Zinc anode vs silver oxide, nickel oxyhydroxide, or air cathode) is anticipated to emerge as a serious competition to replace Lithium-ion, lead acid and nickel metal hydride batteries. Zinc is Global available and inexpensive. Zinc has low polarizability, two redox electrons and delivers high specific capacity and power.

3D Zinc sponge anodes in nickel–zinc alkaline cells can be cycled hundreds to thousands of times without undergoing passivation or macroscale dendrite formation has been established.

3D Zinc sponge anodes with Nickel Cathodes can be used to produce secondary batteries which can be cycled thousands of times that allow for deep discharge is based on the results and performance cells in three fields of tests which confirms this:

(i) **>90% theoretical depth of discharge in primary cells,**
(ii) **>100 high-rate cycles at 40% Depth of Discharge at lithium-ion-commensurate specific energy, and**
(iii) **Tens of thousands of power-demanding duty cycles that is required for start-stop micro-hybrid vehicles.**

In primary 3D Zinc-air cells, 3D Zinc discharges >90% of the Zinc, which translating to an 50% improvement over existing technologies. When cycling 3D Zn anodes at the demanding current densities greater than 10 mA cm^{-2}, that otherwise induce dendrite formation in aqueous alkaline electrolyte. We observe that the 3D

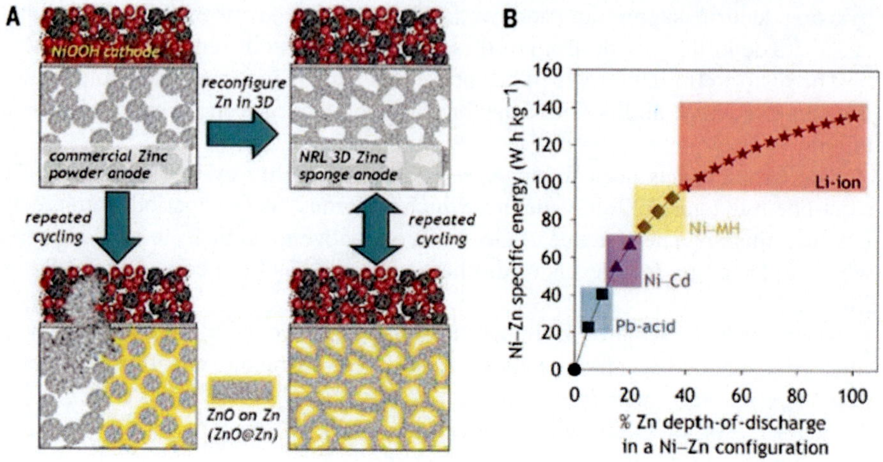

Fig. 1 Possibilities with rechargeable Ni-Zn. **a** Schematic of the effect of recharging Ni-Zn (conventional powder zinc anodes) versus Ni-3D Zn in which the anode is redesigned as a monolithic aperiodic sponge ensuring persistent 3D wiring of the metallic Zn core. Dendrites that form at powder-composite Zn anodes can reach hundreds of micrometers in length (30, 31). **b** The calculated specific energy of a fully packaged Ni-Zn cell as a function of increasing Zn depth of discharge versus a capacity-matched NiOOH electrode. The shaded areas highlight the specific energy range of common battery chemistries. For example at ≥40% DOD_{Zn} (percentage of theoretical utilization). Ni-Zn becomes competitive with Li-ion at the single-cell level

Zinc Anode restructures uniformly without the generation of dendrites that cause separator piercing and pre mature failure.

The performance 3D Zinc Anode used to explore the secondary 3D Nickel–Zinc system in which rechargeable cathode (NiOOH) is used is further development of the existing cathode of rechargeable Zinc-Nickel, and is much more economically feasible than Silver-zinc combination.

Nickel–3D Zinc batteries discharge via the oxidation of Zinc metal along with the reduction of nickel oxy-hydroxide according to reactions at anode (Eqs. 1–3; Zn) and cathode (Eq. 4; NiOOH).

$$\text{Electrooxidation: } Zn(s) \rightarrow Zn^{2+}(aq) + 2e^- \tag{1}$$

$$\text{Complexation: } Zn^{2+}(aq) + 4OH^-(aq) \rightarrow Zn(OH)_4^{2-}(aq) \tag{2}$$

$$\text{Dehydration/Precipitation: } Zn(OH)_4^{2-}(aq) \rightarrow ZnO(s) + H_2O(l) + 2OH^-(aq) \tag{3}$$

$$\text{Electroreduction: } 2\beta\text{ - }NiOOH(s) + 2H_2O(l) + 2e^- \rightarrow 2\beta Ni(OH)_2(s) + 2OH^-(aq) \tag{4}$$

The theoretical specific energy for Nickel–Zinc is 372 Wh kg^{-1}, at a practical level Ni-3DZinc battery delivers up to 135 Wh kg^{-1} (~ 300 Wh L^{-1} on a volumetric basis) which depends on battery design details and Zinc depth of discharge. Comparing the specific energy for a fully packaged Nickel–3D Zinc cell as a function of increasing depth of discharge of Zinc versus that for lead-acid, nickel–cadmium, and nickel–metal hydride shows that the performance of Nickel–3D Zinc is comparable or superior (Fig. 1a), even at modest utilization of the Zinc (10–20% Depth Of Discharge of Zinc). Deeper depths of discharge of Zinc are required (≥40%) to bring Nickel–3D Zinc to a specific energy that becomes competitive with common Lithium-ion batteries at the single-cell level.

The conservative calculations assumed that the Zinc and Nickel electrodes are present at 39% of the total packaged weight. The actual production level packaging weight (casing) is expected to decrease when scaling Nickel–3D Zinc cells for applications is vehicles and other real life usecases.

With regards to electrolyte formulations and electrode additives that minimize shape change of 3D Zinc Anode electrodes cycled 20 times to 20% Depth Of Discharge of Zinc in a Nickel–3D Zinc configuration.Proper electrolyte formulation included additives that force dehydration of soluble Zinc-ate [$Zn(OH)_4$, 2— (aqueous) to ZnO(solid)], (Eqs. 2 and 3) at lower concentrations is unadulterated 6 M KOH.

In the case of deep discharge and long term cycling conditions electrolyte formulation of 6 M KOH + 1MLiOH in conjunction with a $Ca(OH)_2$-infused 3D Zinc electrode was used.

This combination of additives provided superior round-trip cycling efficiency.

I. **Lithium + augments NiOOH rechargeability by suppressing O_2evolution;**
II. **Zinc-ate super-saturation is induced by $Ca(OH)_2$;**
III. **Use of pre-doped Zinc with 300 ppm of In and 300 ppm of Bi suppresses Hydrogen evolution**.

The higher cell voltage of Nickel–Zinc over single-use alkaline batteries (Magnesium O_2–Zinc) is big advantage if it can be coupled to use 3D Zinc anode. This holds a huge potential in the case of wearables, IOT, consumer electronics and other micro applications.

The ability of 3D Zinc anodes to discharge to high Zinc capacity and be recharged without inducing dendrite shorts was probed by exhaustively discharging Nickel–3D Zinc cells (Fig. 2a) at a current density of 10 mA cm^{-2} (C/9—the total capacity of the battery discharged in 9 h) subsequently recharging at the same rate. These cells reached an average 91% Depth Of Discharge of Zinc (743 mAhg $Zinc^{-1}$; 1202 Wh kg $Zinc^{-1}$) and were able to be recharged to >95% capacity from these extreme depths (Fig. 2b). Similar Zinc depths of discharge were obtained in previous 3D Zinc–air studies, but could not probe capacity recovery in this configuration because of the lack of a mature rechargeable air cathode. Using 3D ZincAnodes through the emulsion route provides great flexibility in application for different sizes and form factors, the anode size and shape being defined by the moulds.

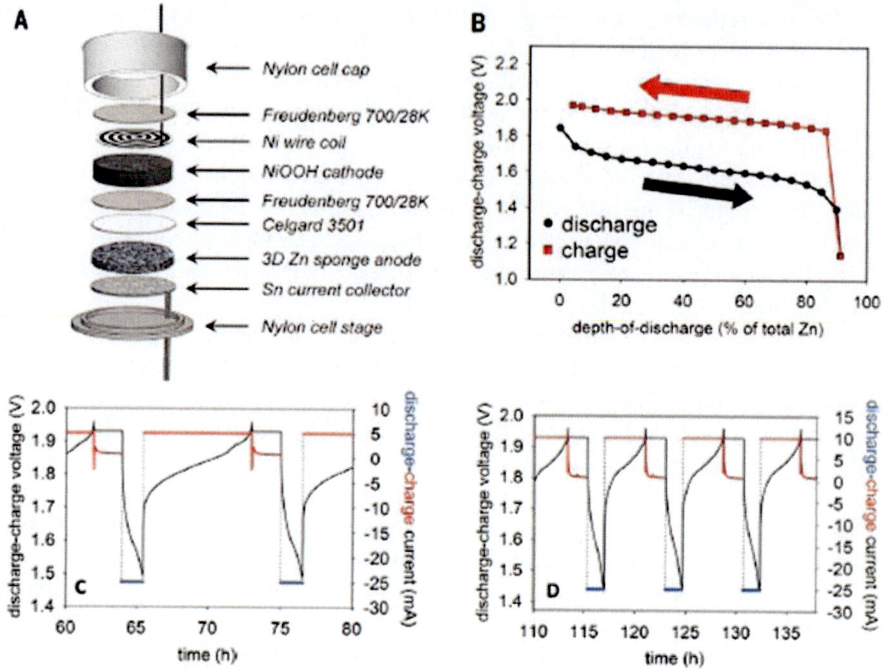

Fig. 2 Cycling performance of nickel–3D zinc cells. **a** Schematic design of the nickel–3D zinc coin cell used in this study. **b** Nickel–3D zinc cells tap >90% of the theoretical Zn capacity upon discharge (black circles, at 10 mA cm^{-2}) and > 5% of that discharged capacity can be recovered upon subsequent recharge (red squares, at 10 mA cm^{-2}) with a half-cycle voltage hysteresis of <300 mV. **c, d** The voltage–time curves for cells discharged at 25 mA cm^{-2} to 40% DOD_{Zn}, and recharged at either (**C**) 5 mA cm^{-2} or (**D**)10 mA cm^{-2}. The constant voltage at at 193 V indicates the potentiostatic region of the charge profile

To establish the feasibility of Nickel–3D Zinc battery in applications of use that demand multi-cell stacks, high cycle life, and power performance, Ni–3D Zincn cells were cycled to a Depth Of Discharge of Zince (40%) which translates to a specific energy that is competitive with Lithium-ion battery (Fig. 1b). The experiments were conducted for long term using current at 5-mA cm^{-2} breakin cycle consisting of a 50 mAh discharge (50% Depth Of Discharge of Zinc) with a recharge of 40 mAh. This capacity first cycle mismatch was chosen to ensure saturation of the electrolyte with Zinc-ate as well to introduce a buffering amount of Zinc-Oxide and $Ni(OH)_2$ into the respective electrodes for minimization of gas evolution upon charging. In following cycles, the cells were discharged at 25 mA cm^{-2} (with C/1.5 rate with respect to a nominal capacity current of 328 mAhg $Zinc^{-1}$) and recharged at either 5 mA cm^{-2} or 10 mA cm^{-2} Fig. 2c, d respectively.

In order to ensure exhaustive oxidation of the NiOOH electrode for avoiding O_2 evolution a 3 mAh potentiostatic hold at 1.93 V added at the end of each charge. For

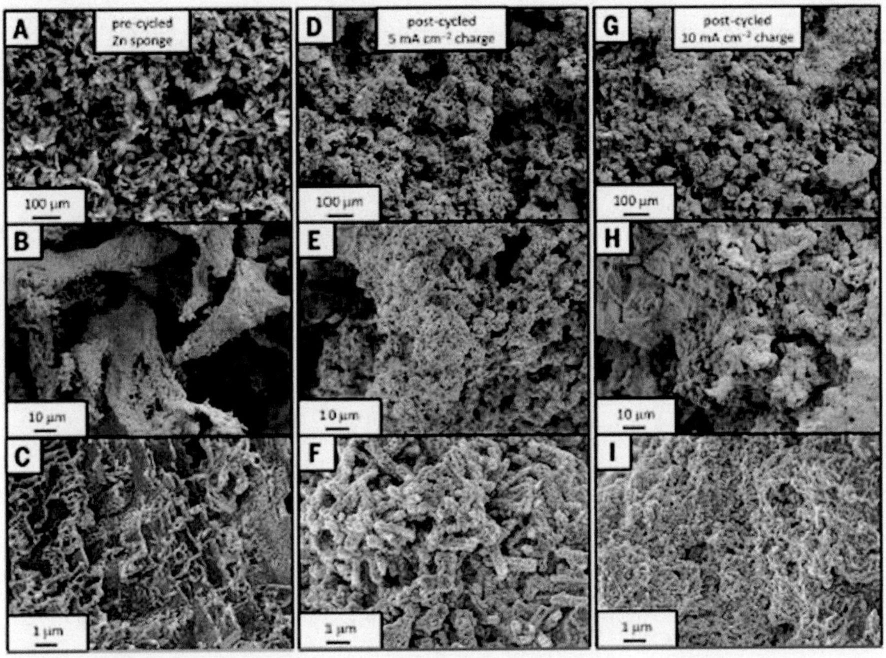

Fig. 3 Postcycling microstructural analysis of 3D Zn sponges. Scanning electron micrographic analysis of **a–c** precycled and **d–i** postcycled Zn Sponges after >100 cycles, verifying that minimal shape change occurs and no dendrites are formed when the Ni–3D Zn cell is discharged at 25 mA cm^{-2} to 40% DODzn and recharged at either **d–f** 5 mA cm^{-2} or **g–i** 10 mA cm^{-2}

charging cases of 5 and 10 mA cm^{-2} the cells ran for 111 and 141 cycles respectively, prior to falling below 50% of nominal cycling capacity.

Upon reaching >20% capacity fade, occuring at >80 cycles, injection into the cathode compartment of water or electrolyte revives the non hermetically sealed plastic cells back to nominal capacity, which demonstrates that the fade arises from dehydration and not irreversible passivation of either the cycled 3D Zinc Anode or Nickel cathode.

These Nickel–3D Zinc cells for 85 and 65 cycles maintained 100% of the required discharge capacity having an average energy efficiency of 84% before capacity fading (which compares to 85% energy efficiency found in Lithium-ion batteries).

Though some densification was reported compared to the precycled microstructure, but it showed achievement of cyclic stability of the electrolyte which was limited of the surfaces. It showed absence of dendrite formation, porosity of the 3D Anode and inter connectivity of the monolithic structure (Fig. 3a–c).

Ni–3D Zinc could compete in a another common use case—replacing lead-acid batteries within micro-hybrid vehicles.

The "startstop" for a normal operation duty cycles involve pulses for engine start and restart. In additional it has auxiliary constant-use loads such as air conditioning

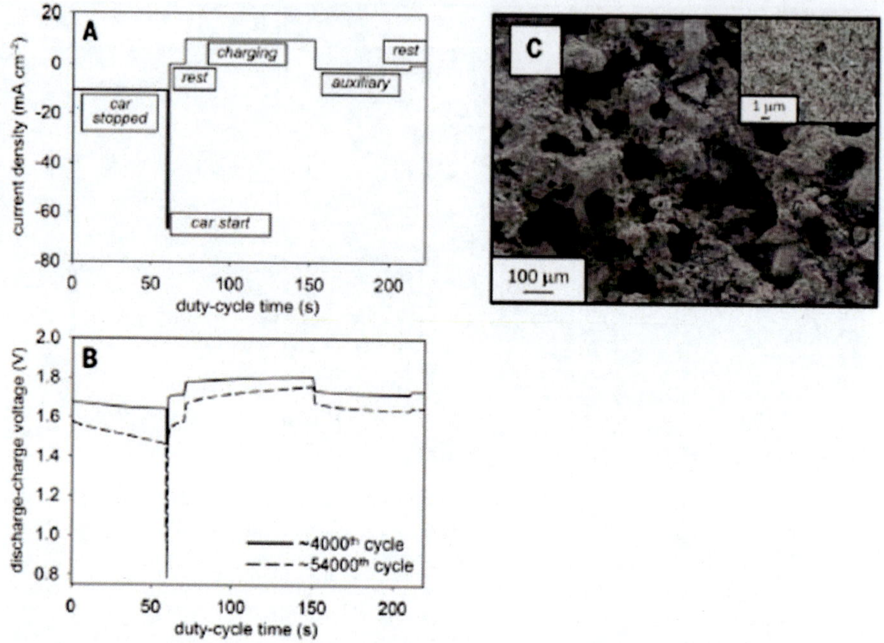

Fig. 4 Long-term performance of Ni-3D Zn single cell as cycled under start-stop conditions. **a** The current–time duly cycle modeled from a BMW AGM start-stop drive cycle (28) scaled to our l-cm^2 Ni-3D Zn coin cells. **b** The measured current–time curves tor Ni-3D Zn coin cells at early (solid line, 4000 cycles) and late (dashed line, 54,000 Cycles) points in the 4.5-month-long, nonstop cycling. **c** Micrographic analysis of a postcyded Zn sponge after—54,000 cycles, which verities that minimal shape change occurs and no dendrites are formed

and information-entertainment management systems. Currently lead–acid battery is used by micro-hybrid for start-stop applications and uses absorbed glass mat (AGM) technology.

AGM lead acid battery has the advantage of low cost and excellent shelf life in the charged state. Main disadvantages is low specific energy, low volumetric energy, low life-cycle, environmental concerns as the active materials are toxic (Pb and PbO_2), electrolyte instability at low charged state, and prices compared with standard starting-lighting-ignition (SLI) lead acid batteries.

To validate the use of Nickel–3D Zinc applicability as a SLI battery, by approximating the current versus- time duty cycle of the BMW microhybrid battery as scaled to a typical single- cell Dimensions was taken (Fig. 4a).

The following assumptions were used:

i. the specific power of individual Nickel–3D Zinc cells will match that of individual Lead acid cells within the AGM (SLI) battery commonly used in BMW's micro-hybrid systems;

ii. a scaled-up Nickel–3D Zinc battery will require eight cells to achieve the necessary voltage (i.e. 12 V) and would therefore deliver 33% more power than its six-cell Lead-acid counterpart;
iii. Zinc will occupy 19% of the packaged weight.

The inter-connected void structure of the 3D Zinc anode serves to improves transport limitations under high-rate demands (Fig. S2), similar to that required during the acceleration phase of a start-stop duty cycle.

It is customary to keep percentage of capacity used intentionally low for start-stop batteries to achieve >104 cycles. Per 4-min duty cycle, the capacity tapped of the Nickel–3D Zinc coin cells was kept to <1% Depth of Discharge of Zinc. More than 50,000 cycles (Fig. 4b) were achieved. The cycling stopped only when the high load pulse (65 mA cm^{-2}) reached a preset voltage limit of 0.8 V.

Thus for a nominal 20 start-stop cycles in a round-trip commute, Nickel–3D Zinc would provide in excess of 2500 days of start-stop performance (approx >6.8 years of daily use),closing towards the average 11.4-year age of U.S. cars. The 54,000 cycles cumulative discharge capacity is 3 times that achieved in the 40% Depth of Discharge of Zinc at 100+ cycles discussed above.

The non hermetically sealed months long cycled cells were subject to postmortem analysis. They revealed a dry cell concomitant with an increased cell resistance. The post cycled 3D Zinc anode remained visibly monolithic; scanning electron microscopy reveals that the pore solid architecture of the 3D Zinc anode is retained with no anomalous macro-scale dendrites being generated (Fig. 4c). The effect of 3D Zinc anode based battery on the energy-storage requirements of various electric vehicles applications where assessed. The quantitative assessment fixed the energy capacity for each application using the current state of the art for.

i. an electric bicycle (using standard lead acid),
ii. a start stop micro-hybrid (using lead acid AGM),
iii. an all-electric battery vehicle (using Lithium-ion).

For all three applications by using Ni–3D Zinc weight and volume savings is indicated (Table 1).

A projected Nickel–3D Zinc battery pegged to the specific capacity of the Nissan Leaf (using a battery capacity of 24 kWh) saves 100 kg of weight. Much of the weight and potential cost savings with Nickel–3D Zinc over Lithium based EV batteries is due to the the reduction or elimination of subsystems that are required for Lithium-ion battery packs, which are include thermal management, sophisticated electronic controls, and structural protection to manage any catastrophic events.

3D Zinc Anode–based batteries will not require comparably complex subsystems

Table 1 Summary of the projected effect of the nickel–3D zinc-based battery on various weight and normalized capacity metrics of relevance to electric vehicles (EVs)

	E-bike		Start-stop microhybrid		Battery electric vehicle	
	SLA	Ni-3D Zn[a]	AGM	Ni-3D Zn[a]	Li-ion[b]	Ni-3D Zn[a]
Energy capacity (Wh)	540	540	1720	1720	24.000	24.000
Weight (kg)	12.2	5.9	45.0	21.7	339	220
Specific energy ($Wh\ kg^{-1}$)	44.3	91.8	38.2	79.2	71	109
Energy density ($Wh\ L^{-1}$)	140	225	126	164	96	216

[a]Calculations for the Zn–3D Zn battery (scaled to match the capacity of a specific application) were made on the basis of a fully packaged battery system. The sensitivity of the energy density of the battery to variations in capacity of the zinc anode and nickel cathode is 20% (Fig. S3)
[b]Metrics for the Li-ion stack in the Nissan Leaf were used for comparison
SLA sealed lead-acid; *AGM* absorbed glass mat

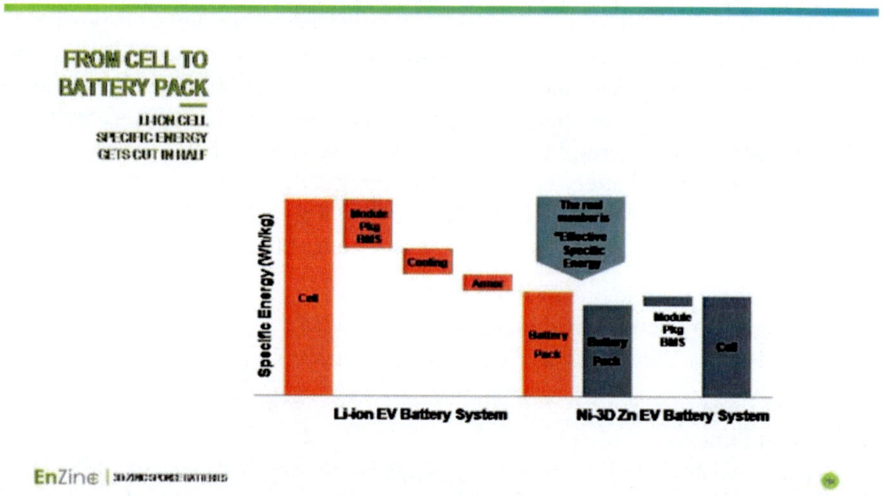

Lithium Ion Battery costs depends on size

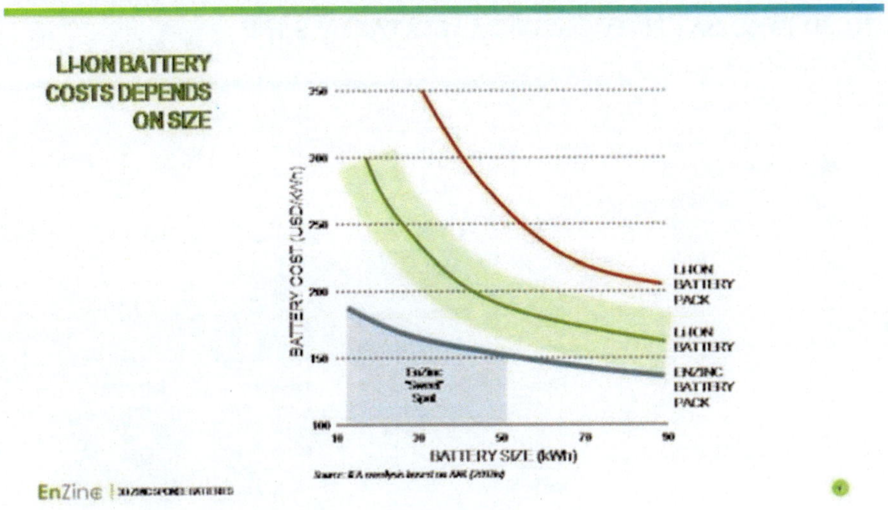

NI-3D ZINC BATTERIES ARE COST EFFECTIVE:

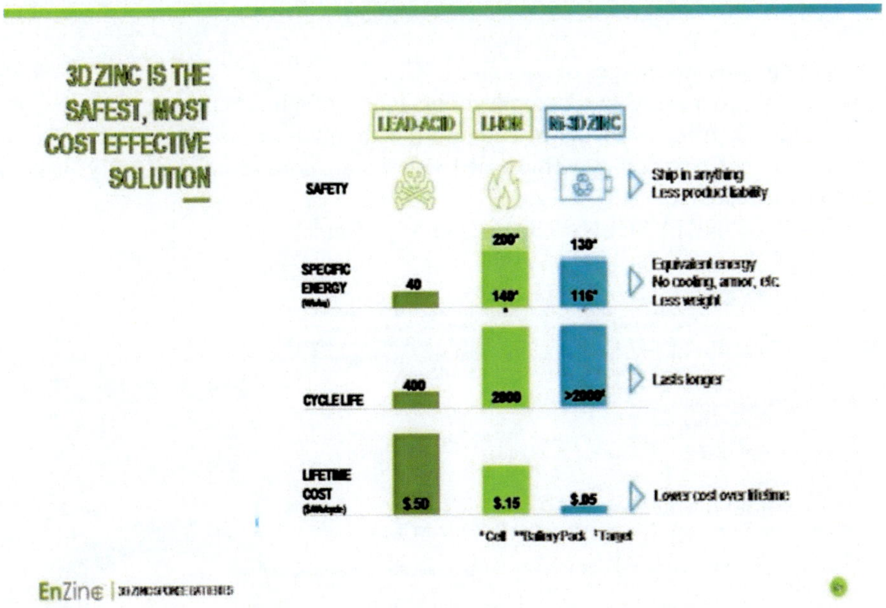

- **3Times Energy of lead acid with 2 to 3 times Cycle life**
- **1/3rd The Size,Weight and Volume of Lead Acid**
- **Equivalent cost to lead acid to manufacture but 1/10 the per cycle cost to operate**
- **Half cost of Li-ion with one-third the cost to operate.**

NI-3D ZINC BATTERY IS FUNDAMENTALLY SAFE

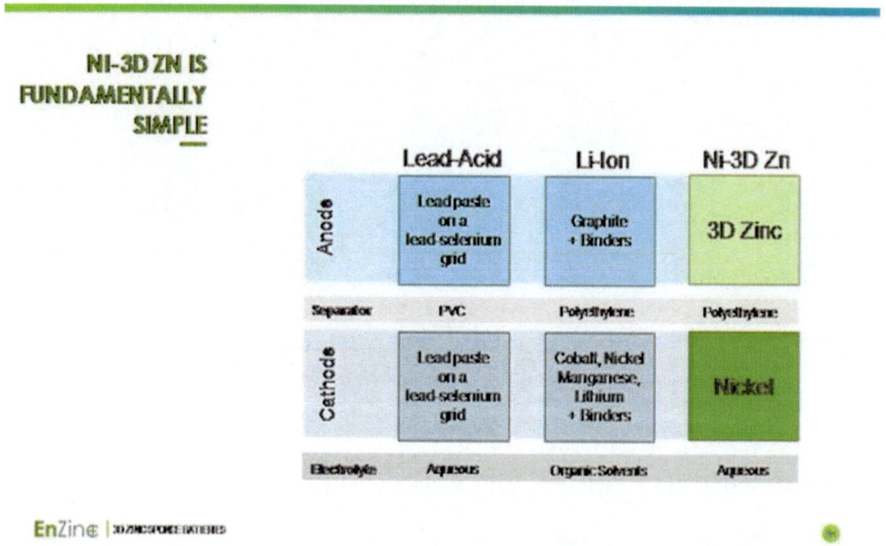

The obvious advantages of Nickel–3D Zinc–based batteries is the projected range and cost improvements in EV applications. It also eliminates the dangers associated with fire risk from incidents of Lithium-ion thermal runaway incidents. The raw materials used in Nickel–3D Zinc based batteries are nonstrategic, globally available, recyclable resource.

Possible "INDIA SCENARIO using 3D Ni-Zinc:

Globally India is ranked as the fourth largest producer of Zinc Ore with abundant reserves. The first documented use and production of pure zinc dates back to 2500 years ago in Jawar in the State of Rajastan. It is expected that India will produce 6 MT of Zinc by 2020 and is capable of producing much more. With the huge increase in EV and electric mobility (two, three wheelers, passenger cars and electric) uptake supported by friendly electric mobility programs that are being facilitated. Aligned with India's focus to reinforce grid capacity, stability and to fast track renewable energy goals is bound to increase the requirements of energy storage systems and subsystems at various levels. 3D Zinc Anode based batteries provides a very cost effective minimum carbon footprint solution for self sufficiency to cater to existing and future needs of the country.

Ackowledgements The authors wish to acknowledge the individuals of Enzinc Inc,the US Naval Research Laboratory, the Advanced Research Projects.With a special vote of thanks to Michael Burz of Enzinc for his support.

References

1. Parker JF, Chervin CN, Pala IR, Machler M, Burz MF, Long JW, Rolison DR (2017) Rechargeable nickel–3D zinc batteries: an energy-dense, safer alternative to lithium-ion. Science 356(6336):415–418. https://doi.org/10.1126/science.aak999
2. Ko JS, Geltmacher AB, Hopkins BJ, Rolison DR, Long JW, Parker JF (2018) Robust 3D Zn sponges enable high-power. Energy-Dense Alkaline Batteries. https://doi.org/10.1021/acsaem.8b01946
3. Three-dimensional zinc sponge electrodes for primary and secondary zinc-containing batteries. https://patents.google.com/patent/US20170338479A1/en
4. Zinc sponge anode makes Li-ion alternative rechargeable. https://www.electronicsweekly.com/news/research-news/zinc-sponge-anode-makes-li-ion-alternative-rechargeable-2017-04/
5. Li-Ion Batteries-a fire hazard. https://www.if-insurance.com/large-enterprises/insight/risk-consulting-magazine/risk-consulting-2019-1/li-ion-batteries-a-fire-hazard
6. Events with smoke, fire, extreme heat or explosion involving lithium batteries. https://www.faa.gov/hazmat/resources/lithium_batteries/media/Battery_incident_chart.pdf
7. A new dimension in energy: https://www.enzinc.com/index.html

Supplementary Material

8. https://www.sciencemag.org/content/356/6336/415/suppl/DC1

Planning of Charging Infrastructure for Urban Public Electric Bus Fleets in India

Shyamasis Das and Anirudh Ray

Abstract The present paper proposes a detailed and effective framework for selecting charging technology(ies) for urban public electric bus (e-bus) fleets, embedded in India's context. The study reviews the principal features of public bus transport networks in India's Tier-I and Tier-II cities, to identify the route-specific charging requirements of an e-bus fleet. It is followed by examining various charging technologies available in India, to elucidate the feasible charging options for e-buses. Subsequently, the study attempts to zero in on the "*most suitable*" technology(ies) for satisfying the charging requirements of an urban public e-bus fleet, based on a detailed and effective Multi-Criteria Decision Analysis tool. The study finds that an urban public e-bus fleet can be charged at a depot during operating hours (of public transport) as well as overnight. For longer routes, en-route charging of buses may be necessary to avoid being stranded. The investigation concludes that cable-connected DC Fast Chargers are most suitable for charging an e-bus fleet at depot as well as en-route during operating hours. In addition, AC-II charging systems are most suitable for overnight e-bus charging. This investigation, and the accompanying framework can potentially facilitate logical and data-driven planning of charging infrastructure for urban public e-bus fleets in Indian cities. State Road Transport Undertakings, interested investors, urban local bodies, power utilities, charging infrastructure and automotive manufacturers, policymakers and other stakeholders are the targeted beneficiaries of this study.

Keywords Electric bus · EVSE · Charging · Decision matrix

S. Das (✉)
Alliance for an Energy Efficient Economy (AEEE), New Delhi, India

A. Ray
School of Planning and Architecture, New Delhi, India

R. K. Pillai et al. (eds.), *ISUW 2019*, Lecture Notes in Electrical Engineering 764,
https://doi.org/10.1007/978-981-16-1299-2_21

1 Introduction

Despite the ambitious outlook of the Government of India towards vehicular electrification, large-scale adoption of electric vehicles (EVs) in the country is yet to garner momentum. Various reasons attributed to EVs, such as their high costs, low ranges, long charging hours, and lack of charging stations, pose significant barriers for scaled-up deployment of EVs. The seeming complexities and related costs of EV-adoption deter users to shift away from a tried-and-tested, age-old, and reliable drivetrain technology, i.e., of the internal combustion engine (ICE)-based vehicles, to a new format which has had limited precedence in the country and across the globe till now. However, urban public bus fleets are considered to be potential early movers toward an electric drivetrain technology. Also, it is comparatively easy for the government to encourage and incentivize the deployment of electric buses (e-buses) for urban public bus transport systems, since most urban public bus fleets are owned and operated by public agencies, such as state-owned bus operating companies, municipal bodies, and State Road Transport Undertakings (SRTUs). From the perspectives of reducing fossil fuel consumption, eliminating local air pollution, and meeting climate change mitigation targets, transitioning to e-buses appears to be a necessary action for the government, majorly as public bus fleets contribute to nearly 14% of the entire transport-related diesel consumption [1].

However, diligence must be exercised in such technological transitions, as any consequential disruptions in urban public bus services are undesirable—and could have grave implications—for the service providers and the users. In a country like India, where majority of the urban public transport demand is met through bus transport, avoiding service disruptions are even more critical. Methodical planning of charging infrastructure is key for successful operation of e-buses for urban public transport, and selecting the right technology(ies) for charging e-buses is an important cog in that planning exercise.

2 Objective and Scope of Work

One of the major barriers for public bus agencies to switch to an e-bus fleet, presently, is the lack of knowledge about charging infrastructure. The stake is high for these agencies, especially because the price of an e-bus is much higher than the price of a conventional diesel bus, in addition to the investments required for setting up adequate charging stations to support e-bus operations. In order to shift seamlessly to e-bus transport, it is imperative that the public bus agencies plan for the required charging infrastructure in advance . In this context, this paper reviews and comparatively assesses the possible charging requirements of a typical urban public e-bus fleet, and evaluates the suitability of various charging technologies for e-buses. This exercise is embedded in the context of specifications of charging technologies, fully electric bus models and Li-ion batteries presently available in India, and the main features of an urban public bus transport network in an Indian (Tier-I or Tier-II) city. The intended primary beneficiaries of this study are the public bus agencies, potential investors in charging infrastructure, charging infrastructure and automotive manufacturers, power utilities, urban local bodies, policymakers and other stakeholders.

3 Methodology

The study entails three main steps:

- Understanding the main features of an urban public bus transport network in an Indian (Tier-I or Tier-II) city to identify the route-specific charging requirements for an urban public e-bus fleet
- Examining charging technologies/practices prevalent in India to identify suitable options for charging e-buses
- Applying a Multi-Criteria Decision Analysis (MCDA) tool for selecting "*most suitable*" charging technology(ies) for meeting the various charging requirements of an urban public e-bus fleet.

3.1 Understanding the Main Features of an Urban Public Bus Transport Network in an Indian City

An essential principle of the study is that the carrying capacity and service frequency of a route, for an urban public e-bus fleet, should remain at par with the baseline. In other words, the service parameters of the route, post- and pre-electrification, must remain the same to avoid service quality implications from a drivetrain transition. Replacement of an existing [diesel or compressed natural gas (CNG)] bus by a lower capacity e-bus would require a higher frequency of service to maintain the route's carrying capacity, thereby necessitating a higher number of infrastructural units for the route, and increasing the volume of vehicular traffic on the route. On the other hand, bus service on the route must remain unaffected by the charging requirement. An increase in the bus service headway on a route (i.e. decrease in service frequency) would negatively affect the bus-service on that route.

To plan charging infrastructure, it is vital to consider:

- Where should the e-buses be charged, i.e., at the depots/terminals or en-route
- When must the e-buses be charged, i.e., overnight or during operating hours
- How must the e-buses be charged, i.e., the feasible and most suitable charging technologies.

Taking into account the current operations of an urban public bus transport network, and the complications associated with setting up charging stations en-route, the investigation understands that bus depots/terminals (also called transit nodes) are the most suitable locations for charging public e-bus fleets in Indian cities. At depots/terminals, e-buses can be charged after completion of trips during operating hours of public transport, or overnight, when public transport services are not in operation. Overnight charging is intended to reduce the fleet's charging requirement during operating hours, thereby ensuring that the availability of e-buses and service quality (frequencies) of the entire network are not affected in the critical hours of operation. To support e-bus operations on a route longer than the vehicle's range, it may be necessary to establish an en-route charging station at an intermediate halting point on the route. Such charging stations would provide "*top-up*" charging to e-buses on a route/multiple (overlapping) routes.

Hence, this study identifies and inspects the following methods for charging an urban public e-bus fleet:

- Charging at depots/terminals post completion of trips
- Charging at depots/terminals overnight (post operating hours)
- Charging at an en-route charging station, located between the origin and destination points of a route.

Identifying the aforementioned charging methods sets the stage for selecting the "*most suitable*" charging technology for each method, from amongst the different options available in the Indian market. This is irrespective of whether the e-bus is equipped with HVAC system or not.

3.2 Examining the Charging Technologies/Practices Available in the Indian Market

Recognising that it might not be possible to deploy every available charging technology for charging all typologies of EVs (such as buses, four-wheeler passenger cars, two- and three-wheelers, light commercial vehicles, etc.), the study undertakes a detailed comparative assessment of the charging options available in India, using a set of critical techno-economic parameters. The objective of such an assessment is to identify the charging technology(ies) which can practically cater to the charging requirements of e-bus fleets in urban areas (refer to Table 1). The identified plausible technologies are then further investigated to select the "*most suitable*" one(s) to meet the requirements of the aforementioned charging methods for public e-bus fleets.

Table 1 Comparative assessment of various EV charging technologies, in reference with urban public e-bus fleets

Parameters	AC-I	AC-II	DCFC	Inductive charging	Battery swapping
Input voltage from power distribution network (V)	108–120	415 or above	415 or above	415 or above	Depends on charging point used for charging; additional voltage required per swapping arm: >415 V
Output power of charging technology (kW)	1.3–2.3[a]	20–80[a]	50-250 for plug-in, up to 650 for pantograph[a]	50–250	Depends on charging point used for battery charging
Charging/swapping time for e-buses (rated battery capacity = 200 kWh)[b]	40–70 h[c]	5–13 h[c]	Approximately 45 min for plug-in and 20 min for pantograph[c]	1–3.4 h[c]	3 min for interchange one module of battery [2] (1 module = 50 kWh; an e-bus requires 4 modules)
Type of power connection required [High Tension (HT)/Low Tension (LT)]	Required grid voltage is not present in India	HT	HT	HT	HT (as battery swapping arms are required)
Additional infrastructure requirement	No additional infrastructure required (simple plug-and-play) [3]	DT[d]; SCADA[e] standard wire gauge; breakers and switches; operating system [3]	DT; liquid-cooled cables; breakers; operating system; SCADA; wire gauge; regulator; switches [3]	AC-AC transformer; SCADA; DT; wire gauge; alignment indicator; breakers; switches; operating system [3]	DT; SCADA; storage depot for batteries; battery movement system; battery swapping arms; operating system; standard wire gauge; breakers; switches [3]

(continued)

Table 1 (continued)

Parameters	AC-I	AC-II	DCFC	Inductive charging	Battery swapping
Power tariff applicable (energy/demand charges are levied based on connection type)	Required grid voltage (110 V) is not available in India	HT tariff	HT tariff	HT tariff	HT tariff
Capital expenditure on charging technology (₹)	15,000–35,000 [4]	15,200–76,000 [5]	2,100,000–3,520,000 [6]	59,000–632,500 [7]	Depends on charging point used for battery charging [8]
Expenditure on additional infrastructure (₹)	0	250,000–400,000 [9, 10]	600,000–1,250,000 [9, 10]	380,000–720,000 [9, 10]	Data not available for swapping arms and battery movement systems for e-buses (cost per swapping arm is ₹ 105,000/- for four-wheelers); for rest of the additional infrastructure: ₹ 250,000–₹ 400,000 [8–10]
Area requirement (m^2)	0.09 (wall mounted)	0.8	2	None (accommodated beneath the parking bays)	Depends on charging station capacity and chargers used; no additional area required for battery swapping arm (accommodated under parking bays)

(continued)

Table 1 (continued)

Parameters	AC-I	AC-II	DCFC	Inductive charging	Battery swapping
Maintenance cost (%)	Periodic and regular maintenance entail 10% and 2% of installation (set-up) cost, respectively[f,g]				
Requirement 11 kV ESS[h]	No	No	Yes	No	No
Cable connector length	17 m	17 m	17 m	17 m	Not applicable
Number of EVs that can charge simultaneously	2	2	2	1	1
Envisaged difficulties in drawing power from the distribution network	Required input voltage is not available in India	Moderately difficult; power can be drawn by a DT connected to an HT line	Difficult; power can be drawn only from an 11 kV substation, which is less accessible than an HT line	Moderately difficult; power can be drawn by a DT connected to an HT line	Moderately difficult; power can be drawn by a DT connected to an HT line

(continued)

Table 1 (continued)

Parameters	AC-I	AC-II	DCFC	Inductive charging	Battery swapping
Prevalence of technology for e-bus fleet charging in India (yes/no)	No (as required input voltage is not available)	Yes	Yes	No	Yes

[a] At 90% efficiency for wired charging and 75–80% efficiency for wireless [11]
[b] Effective usable energy of battery = 119 kWh, based on a detailed calculation considering several operational aspects such as depth of discharge, etc.
[c] Calculated
[d] Distribution transformer
[e] Supervisory control and automation
[f] Periodic maintenance happens once every 4 years (as per industry standards on asset management practices)
[g] Regular maintenance happens annually (as per industry standards on asset management practices)
[h] Electrical sub-station

From the practicality-assessment (summarized in Table 1), it is explicit that:

- The service (grid) voltage required for AC-I is 108–120 V, which is not available in India
- Owing to the weights of batteries for e-buses (900–2700 kg), battery swapping is not possible
- Inductive charging is not advisable due to its low efficiency (approximately 75%).

Hence, the investigation shortlists AC-II and DC Fast-Charging as the plausible charging technologies for e-buses, the latter of which can be further divided into two sub-classes depending on the EVSE (Electric Vehicle Supply Equipment) design—pantograph and cable connected.

3.3 Applying Multi-criteria Decision Analysis (MCDA)

Selecting the most suitable technology for a particular charging method (for any EV segment) is an expensive riddle to solve, especially as the efficacy and feasibility of infrastructure establishment and the use of a charging technology rely on a number of technical and economic factors. On one hand, the technology must meet the technical criteria necessary for charging the EV, and on the other hand, its deployment and operation must be cost-effective. Given that the EV market is still at an initial stage of maturity, and consequently the technologies are currently evolving rapidly, it is quite possible that from among the charging options presently available in India, none may satisfactorily meet all the requirements. Therefore, a trade-off driven approach for selecting the suitable technology(ies) as per the requirements of different charging methods becomes vital. Nevertheless, objectively deciding on such trade-offs is a composite task and hence, a logical framework is required to solve the decision-making problem. The study develops this framework in the form of Multi-Criteria Decision Matrices (MCDMs) composed of various techno-economic parameters, which are assigned weights based on their perceived degrees of importance, using the following scale (Fig. 1):

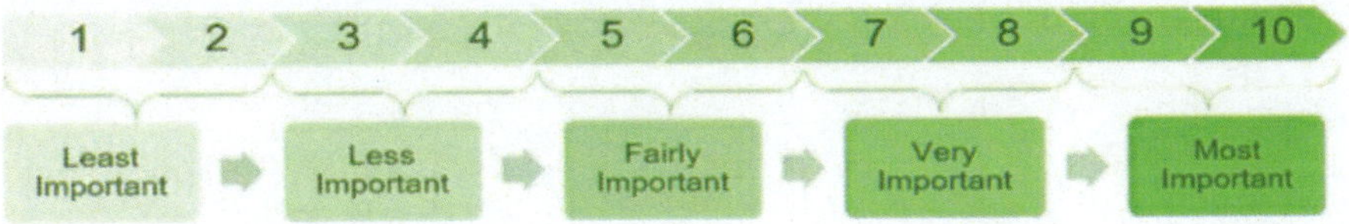

Fig. 1 Scale for assigning weight to a parameter

It must be noted here, that the economic and technical parameters that are considered in the complex MCDMs, and the weights that are assigned to them, differ with charging options/possibilities, as the criteria to evaluate the "*most suitable*" technology also vary. Further, when a weight is ascribed to a parameter, it must be treated in isolation from other competing parameters. Also, all parameters must be viewed solely in context of the concerned charging option. All e-bus charging technologies are ranked against individual parameters, wherein the technologies that best satisfy the ideal value for a particular parameter are ranked highest (e.g., out of four charging technologies, the most-suited technology against a parameter will have rank "4"). Upon ranking all technologies against all parameters, aggregate weighted ranks for all technologies are arrived at. The technology notching up the highest aggregate weighted rank is deemed as the most preferred option. Conversely, the lowest aggregate weighted rank depicts the least preferred option.

The technical and economic parameters, along with their weights, to assess the suitability of charging technologies for e-buses are shown in Tables 2 and 3 respectively.

4 Results and Discussions

Based on the listed technical and economic parameters and their ideal values, with the help of the MCDM tool, the study assesses the different EV charging technologies which are suitable for charging intra-city public e-bus fleet and presently available in Indian market.

4.1 Selection of Charging Technology for Depot Charging of e-Buses During Operating Hours

Table 4 presents the MCDM for selecting the most suitable technology for charging urban public e-bus fleets during operating hours of public transport (post completion of trips) in a depot/terminal.

Table 2 Technical parameters for selecting charging technology(ies) for e-buses through multi-criteria decision analysis

S. No.	Parameter	Ideal value	Justification of the ideal value	Depot charging during operating hours	Overnight depot charging	En-route charging
1	Charging time	30 min	Considering power output of 240 kW for a charging technology (maximum power rating of an EV charger) to charge a 200 kWh battery[a]	**Weight**: 10 **Justification**: Maximum importance is ascribed to this parameter as limiting the charging duration for e-buses will help achieve the desired service frequency and quality	**Weight**: 4 **Justification**: The practice of overnight charging would play a supportive role to charging during operating hours of the fleet. Also, during overnight charging, the demand for charged vehicles would not be immediate. Therefore, charging time here has less importance	**Weight**: 10 **Justification**: This parameter has the maximum importance, as maintaining the service headway of a route is the priority for bus operating agencies. The primary target while planning en-route charging station is to minimize the charging time
2	Potential for maintaining route frequency	Time headway of the route; represents time difference between subsequent bus arrivals	The time headway of a route may differ significantly based on features of the network as well as the route. The technology which charges quicker is more apt for keeping service headways low	**Weight**: 10 **Justification**: Since maintaining service frequencies/headways at pre-electrification levels is a pre-requisite, this parameter has the highest degree of importance as well. The importance is looked at in light of the context of depot charging post completion of trips	**Weight**: 4 **Justification**: As overnight charging of e-buses is meant to complement the charging operations during operating hours, a low weight is ascribed. This is primarily as there is no urgency for maintaining service frequencies overnight, during non-operating hours	**Weight**: 10 **Justification**: The impact on headways due to en-route charging shall be high, since the bus would halt to charge. This headway gain may also delay subsequent trips. Therefore, it needs to be minimized and has is given the highest weight possible

(continued)

Table 2 (continued)

<table>
<tr><th>S. No.</th><th>Parameter</th><th>Ideal value</th><th>Justification of the ideal value</th><th>Depot charging during operating hours</th><th>Overnight depot charging</th><th>En-route charging</th></tr>
<tr><td>3</td><td>Grid current required</td><td>16–32 A</td><td>The current range prescribed here is based on the grid (service) current in available in India's power network</td><td colspan="3">Weight: 6
Justification: In case the available service current in the grid is lower than the required level for the concerned charging technology, it may require cost-heavy replacement of service wires and could impact the accessibility to the grid also</td></tr>
<tr><td>4</td><td>Requirement of 11 kV ESS</td><td>Nil</td><td>A charging option which requires no ESS is always preferred</td><td colspan="3">Weight: 6
Justification: Setting up an ESS does not come under the ambit of the charging station developer; it is the responsibility of the DISCOM to provide. Notwithstanding, the charging station developer would require to impress upon the DISCOM and this may require substantial paperwork leading to higher administrative cost and delay in setting up the charging station</td></tr>
<tr><td>5</td><td>Areal requirement per EVSE (including area for additional infrastructure)</td><td>–</td><td>Areal requirement for EVSEs and all extra infrastructure must be as low as possible</td><td colspan="2">Weight: 4
Justification: If the areal requirement per EVSE is low, the space constraints are less and therefore, the convenience in placing an EVSE is more. Also, low areal requirements help bring down the establishment costs for a charging station. However, it would not be a major real-estate challenge in this case as depots have sufficient space to accommodate infrastructural changes and are planned for extra capacities</td><td>Weight: 6
Justification: Finding enough space for setting up a charging station could be a barrier in case of en-route charging as in the present scenario no such en-route space requirement exists</td></tr>
</table>

[a]Usable energy: 119 kWh

Table 3 Economic parameters for selecting charging technolgy(ies) for e-buses through multi-criteria decision analysis

S. No	Parameter	Ideal value	Justification of the ideal value	Depot charging during operating hours	Overnight depot charging	En-route charging
1	Capital cost per EVSE	₹ 15,000	This value is based on the minimum price of an EVSE available in India	**Weight**: 10 **Justification**: Expenditure on an EVSE may be the largest component of the capital investment on an EV charging station. The highest possible weight is ascribed to this parameter, since it impacts the financial viability of a charging station project		
2	Cost of power for charging an e-bus through an EVSE	Minimum tariff applicable	The charging technology that attracts the lowest electricity tariff is ranked highest. Energy consumption for all technologies is considered to be same	**Weight**: 10 **Justification**: The main operating expense for running a charging station is the cost of electricity used for charging. In this case, just the energy charge (a variable cost) is taken into account. The demand charge (a fixed cost) is not considered as both AC-II and DC Fast Charger will require HT connections, entailing the same demand charge		
3	Cost of additional infrastructure	Minimum cost	Cost of additional infrastructure depends on the charging technology. The technology requiring least expenditure on additional infrastructure is ranked highest	**Weight**: 5 **Justification**: Additional (or supporting) infrastructure requirements are different for different EVSEs. Certain significant cost levers are listed below: • AC-II systems require a captive/dedicated DT in order to operate at full capacity • DCFC charging systems require liquid cooled cables to operate at high power levels and temperatures, without heating up (as heat eventually could slows charging down) Costs that may not be directly incurred by the developer have not been considered		
4	Maintenance cost per EVSE	₹ 0–15,000/annum	Based on standard industry practices for such projects. Minimum cost option is ranked highest	**Weight**: 2 **Justification**: Maintenance cost is a recurring expenditure that includes repairing, servicing and inspection costs. Its impact on the total operating cost is envisaged to be low		

Table 4 MCDM for selecting charging technology for depot charging of e-buses during operating hours, post completion of trips

Parameters	Criteria	Weight (W^a)	AC-II		DCFC			
					Pantograph		Cable connected	
			$R_{AC\text{-}II}$[b]	$W * R_{AC\text{-}II}$	R_{DCP}[b]	$W * R_{DCP}$	R_{DCC}[b]	$W * R_{DCC}$
Technical	Charging time	10	1	10	3	30	3	30
	Potential for maintaining route frequency	10	1	10	3	30	2	20
	Grid current required	6	3	18	2	12	2	12
	Requirement of 11 kV ESS	6	3	18	2	12	2	12
	Areal requirement per EVSE (including area for additional infrastructure)	4	3	12	1	4	2	8
Economic	Capital cost per EVSE	10	3	30	1	10	2	20
	Cost of power for charging an e-bus through an EVSE	10	3	30	3	30	3	30
	Cost of additional infrastructure	5	1	5	3	15	3	15
	Maintenance cost per EVSE	2	3	6	1	2	2	4
	Sum of weighted ranks		139		145		151	

[a]Weight of a criterion for a particular charging requirement of a specific vehicle segment
[b]Rank of a charging technology against a particular criterion

The MCDM (in Table 4) shows that cable connected DC Fast Charger notches the highest aggregate weighted rank followed by a pantograph DC Fast Charger. This implies that cable connected DC Fast Chargers are most suitable for charging e-bus fleets at depots/terminals during operating hours, in between subsequent trips as and when required.

4.2 Selection of Charging Technology for Depot Charging of e-Buses Overnight

Table 5 showcases the MCDM for selecting the technology for charging e-bus fleets overnight at depots/terminals.

From the aggregate weighted ranks in the MCDM as shown in Table 5, one could infer that an AC-II charging system is most suitable for charging e-bus fleets overnight, at depots/terminals.

4.3 Selection of Charging Technology for En-route Charging of e-Buses

Table 6 exhibits the MCDM for selecting technology for en-route charging of e-buses.

From the aggregate weighted ranks in the MCDM as presented in Table 6, one could deduce that cable connected DCFC EVSEs are most suitable for charging e-buses en-route.

However, one should consider the outcome of the MCDMs with some caution. The EV market in India, and globally, is still nascent and the vehicular/battery-related aspects, as well as the charging technologies are still evolving, and this may affect the technical feasibility and economic viability of the concerned technologies in the near future. Therefore, it is critical to bear in mind that the given rankings of the charging options are reflective of the present scenario only and may change in future as the EV market matures. It is advisable that the stakeholders take a fresh look at the Decision Matrices and revise them after a period to make appropriate decisions.

Table 5 MCDM for selecting charging technology for depot charging of e-buses overnight

Parameters	Criteria	Weight (W)	AC-II		DCFC			
					Pantograph		Cable connected	
			$R_{AC\text{-}II}$	$W * R_{AC\text{-}II}$	R_{DCP}	$W * R_{DCP}$	R_{DCC}	$W * R_{DCC}$
Technical	Grid current required	6	3	18	2	12	2	12
	Requirement of 11 kV ESS	6	3	18	2	12	2	12
	Charging time	4	1	4	3	12	3	12
	Potential for maintaining route frequency	4	1	4	3	12	2	8
	Areal requirement per EVSE (including area for additional infrastructure)	4	3	12	1	4	2	8
Economic	Capital cost per EVSE	10	3	30	1	10	2	20
	Cost of power for charging an e-bus through an EVSE	10	3	30	3	30	3	30
	Cost of additional infrastructure	5	1	5	3	15	3	15
	Maintenance cost per EVSE	2	3	6	1	2	2	4
	Sum of weighted ranks		127		109		121	

Table 6 MCDM for selecting charging technology for en-route charging of e-buses

Parameters	Criteria	Weight (W)	AC-II		DCFC			
					Pantograph		Cable connected	
			R_{AC-II}	$W * R_{AC-II}$	R_{DCP}	$W * R_{DCP}$	R_{DCC}	$W * R_{DCC}$
Technical	Charging time	10	1	10	3	30	3	30
	Potential for maintaining route frequency	10	1	10	3	30	2	20
	Grid current required	6	3	18	2	12	2	12
	Areal requirement per EVSE (including area for additional infrastructure)	6	3	18	1	6	2	12
	Requirement of 11 kV ESS	6	3	18	2	12	2	12
Economic	Capital cost per EVSE	10	3	30	1	10	2	20
	Cost of power for charging an e-bus through an EVSE	10	3	30	3	30	3	30
	Cost of additional infrastructure	5	1	5	3	15	3	15
	Maintenance cost per EVSE	2	3	6	1	2	2	4
	Sum of weighted ranks		145		147		155	

Acknowledgements Funded by Shakti Sustainable Energy Foundation.

References

1. Buchmann I (2016) Batteries in a portable world—a handbook on rechargeable batteries for non-engineers, 4th edn., vol 1. Cadex Electronics Inc., Canada
2. Press Information Bureau (2014) PIB. Ministry of Petroleum & Natural Gas, 28 Jan 2014. [Online]. Available: https://pib.nic.in/newsite/PrintRelease.aspx?relid=102799. Accessed 26 Jan 2019
3. Mäkinen J (2016) Fundamentals of electric bus charging. ABB
4. Aerovironment (2017) EVSE-RS spec sheet. Aerovironment, Monrovia
5. Aerovironment (2018) Turbocord spec sheet. Aerovironment, Monrovia
6. Aerovironment (2014) Nissan DC fast charger spec sheet. Aerovironment, Monrovia
7. Liang Y, Zhang X, Xie J, Liu W (2017) An optimal operation model and ordered charging/discharging strategy for battery swapping stations. Sustainability 1–18
8. WiTricity Corporation (2018) The next wireless revolution: electric vehicle wireless charging. WiTricity Corporation, Massachusetts
9. SPDCTL (2018) Cost data: FY 2018–19. Southern Power Distribution Company of Telangana, Limited, Hyderabad
10. IndiaMART InterMESH Ltd. (2019) Schneider SCADA systems. [Online]. Available: https://www.indiamart.com/proddetail/schneider-scada-systems-13202900712.html. Accessed 26 Jan 2019
11. Sun Mobility (2020) Sun mobility: quick interchange station. Sun Mobility. [Online]. Available: https://www.sunmobility.co.in/technology.html. Accessed 16 Dec 2020

Utility Command and Control Center—A Platform for Utility Transformation

Sudheer Polavarapu

Abstract Beyond their intention, operations maturity in utilities is a continuous journey linked with many external factors like industry trends (shaping the journey), presence of technology (enabling the journey), and affordability and viability of the solutions by utilities and ecosystem players (implementing the journey). Currently utilities industry, at least in India, is going through rapid changes by implementing/adopting smart meters for improving their non-technical losses, billing efficiencies, customer service, forecasting and making initial steps for enabling a true smarter grid. With solar/distributed generation and EVs commercial scale implementations around the corner, utilities operational readiness to adopt these and continuously evolve with market needs a different approach/system. In this paper we will provide a point-of-view about the need for Command and Control Center within utilities and how Command Control and Communications (CCC) platform enable the setup with a single unified real-time view of utility operations (esp. last mile distribution) and assets by visualizing, analyzing and optimizing utility data (system of records from utility systems, utility assets, smart meters, IoT devices, SCADA, third-Party sources), allowing utility professionals (IT and OT) to deploy personal, timely, with relevant insights, and providing closed loop mechanism to close the events/incidents that are impacting the utility service and business.

Keywords Utility command and control center · Utility operations center · Command control and communications center for utilities · Realtime utility operations center · Smart grid operations center · Utility operations transformation

S. Polavarapu (✉)
Product Management, Fluentgrid Ltd., Level 5, 9-29-19/A, Waltair Heights, Balaji Nagar, Visakhapatnam 530003, India
e-mail: psudheer@fluentgrid.com

R. K. Pillai et al. (eds.), *ISUW 2019*, Lecture Notes in Electrical Engineering 764,
https://doi.org/10.1007/978-981-16-1299-2_22

1 Introduction—The Need for Command and Control Center Within Utilities

For any utility, either regulated or deregulated, Service, Sustainability and Smartness are important to be successful over a long period. They are interdependent, and problem in one area can impact others. E.g. Not providing reliable electricity to customers can have impact on the business. Of course, it may not be direct in regulated utilities present in India, but the impact will be shown on the revenues in the form of delayed collections, customer resistance during tariff hikes, etc.

With more and more consumers becoming discern about the quality of the service offered, utilities are continually seeking ways to provide superior customer experience and minimal distortion in service by keeping customer at the center of decision making. In this regard, utilities started adopting customer centric smart metering models allowing customers to have control over energy use. This is a great shift compared to the earlier models and systems adoption. Managing day-to-day business and technology operations in such an environment at a finer level and be on top of it require Utility Command and Control Center that enable people, process and technology operate seamlessly and with optimal performance. This has to do with the ratio of actionable information that human operator needs to deal with, compared with the amount of noise or data elements that can be ignored. Utility Command and Control Center requires to look at data coming from smart meters, sensors and utility systems that need to be correlated and synthesized. It is important for utilities to be efficient in removing noise and correlating important data to allow data driven automation (Fig. 1).

Most of the utilities, have just started down the road of modernizing technology infrastructure, are still learning about their surroundings and still responding reactively to downtime caused by operational inefficiencies or asset breakdowns. Operationally mature utilities are those that are not only thinking about today's operations, but also proactively positioned to adapt and predict potential impacts on service,

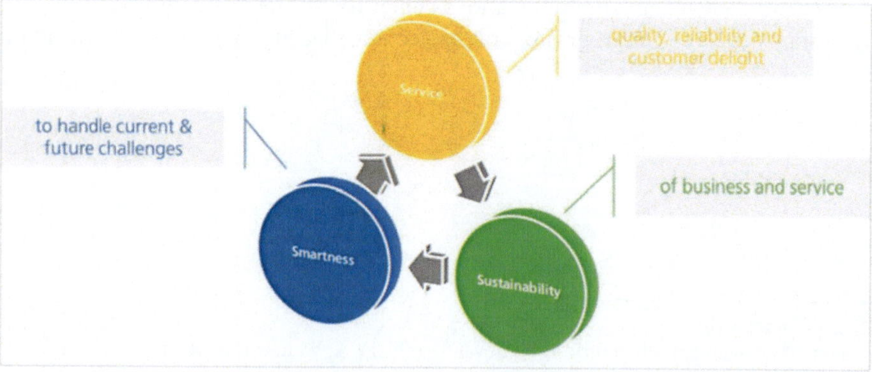

Fig. 1 Focus areas for success over a long period

reliability and performance. The glue that holds these—people, process and technology—together is when operational maturity of a utility reaches it heights. That is what an integrated Command Control and Communications (CCC) platform do, bringing people, process and technology together and enabling utility operations with actionable intelligence and a closed loop mechanism to resolve the incidents impacting the service and business.

2 Command Control and Communications (CCC) Platform for Smart and Mature Utility Operations

The foundations of the platform rely on three simple steps—Identify, Act and Measure. These steps form a closed loop mechanism for resolving the incidents and should be a recurring theme for improving. The approach of bringing these steps together pushes for more standardization across the loop (e.g. incident categorization, SOPs, industry benchmarks, etc.) and allow utility to make the progress that is clearly visible. Most often we see these steps spread across different systems, which are siloed and rarely talk to each other in real time and missing measuring part (Fig. 2).

At a high level, CCC platform for utilities

1. Provide unified view of utility operations from Billing and Customer Care, GIS, Asset Management, AMI systems, utility IoT applications and ongoing project activities
2. Identify and capture events/incidents (actionable intelligence) impacting the utility service and business, and provide an efficient means to close
3. Allow utility professionals to deploy personal, timely with actionable information to act on the incidents

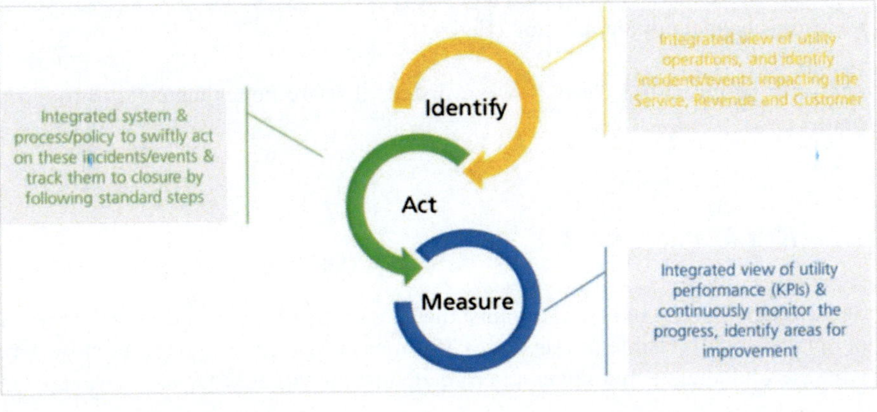

Fig. 2 Foundational steps to CCC platform

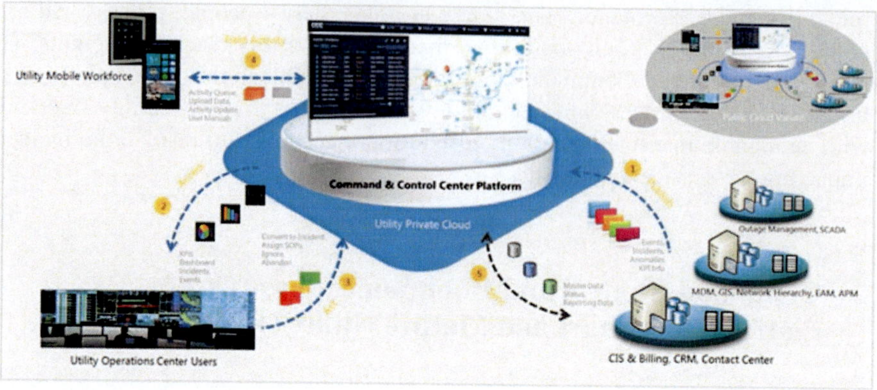

Fig. 3 Systemic view of the platform

4. Provide insights about utility performance (KPIs) and regulatory compliance
5. Surveillance/Security (Physical/Cyber) of utility assets and systems
6. Provide flexibility to quickly add emerging trends and use cases (E.g. Remote transformer monitoring)
7. Provide means (automated) to evaluate performance (Fig. 3).

At the core, CCC platform for utilities contain

- System config and administration
- Application integrations and service orchestration
- Data ingestion with VEE (real-time, schedules, on-demand)
- Dashboards
- Incident management, with Standard Operating Procedures (SOPs) and live tracking
- Analytics (descriptive, predictive and prescriptive)
- Video surveillance (with video analytics)
- Multi-channel communication (Fig. 4).

and offer potential use cases (few are listed) that are better managed through the platform.

2.1 *KPI Management*

Set key performance indicators against the benchmarks/targets, measure the progress periodically with visible trends. E.g. Customer grievance response and closure rate with SLA adherence trend over a period against the target of 98%.

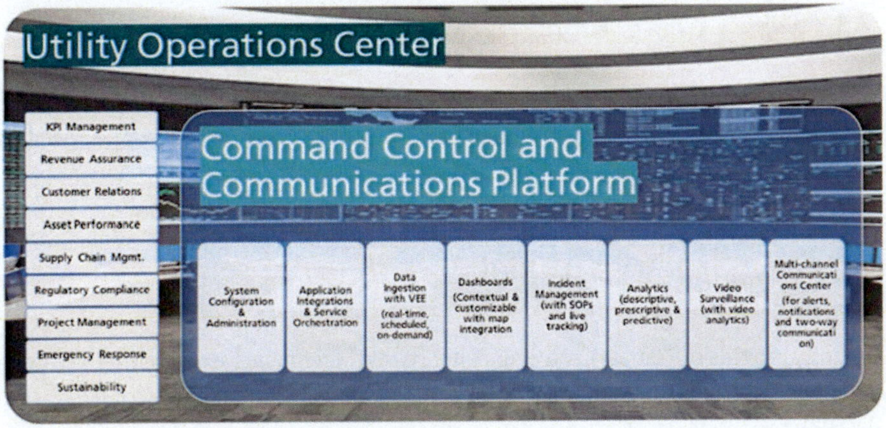

Fig. 4 The core platform and use areas

2.2 *Revenue Assurance*

Identify potential abnormally low consumption cases (through anomaly detection methods) impacting the billing/revenue, create incidents and track them to closure. Also measure the accuracy of the incident identification and improve continuously.

2.3 *Customer Relations*

Measure the overall customer satisfaction and cost to serve customer from the help desk ticket resolutions and customer service channels, identify potential areas of focus and keep improving.

2.4 *Asset Performance*

Monitor distribution transformer oil levels and condition, create incidents for potential cases that require immediate attention from field personal for a precautionary measure.

2.5 Supply Chain Management

Understand the variability in demand across various customer segments during a day/time of a day and trigger a demand response event to meet the supply. Measure the overall efficiency of such events over a period.

2.6 Regulatory Compliance

Readily provide information on energy audits, non-technical losses, service coverage, customer satisfaction, smart meter/device compliance as per CEA, SAIDI, SAIFI, CAIDI, CAIFI, etc.

2.7 Project Management

Smart meter rollout and implementation progress tracking at regular intervals (weekly/bi-weekly). Over the air firmware upgrade progress tracking.

2.8 Emergency Response

Identify a single or multiple outage events from smart meters, accurately predict outages and dispatch field personal on needed cases with actionable information so that he can quickly fix the issue and restore the service. Field person can use readily available information about the outage from his mobile while assessing the problem and finding a fix (Fig. 5).

3 Benefits

The more the benefits the more utilities cultivate the CCC platform and utility operations. The need and timing are perfect with smart meter rollouts and gird modernization initiatives in India. Utilities should look beyond transactional benefits and focus more on outcomes while adopting such a system of systems. There are plenty of game changing and sustainable benefits the platform can offer

- Bring the culture of focus (process) and provide a unified platform (across functions) to work on the issues that are critical for business and customer service

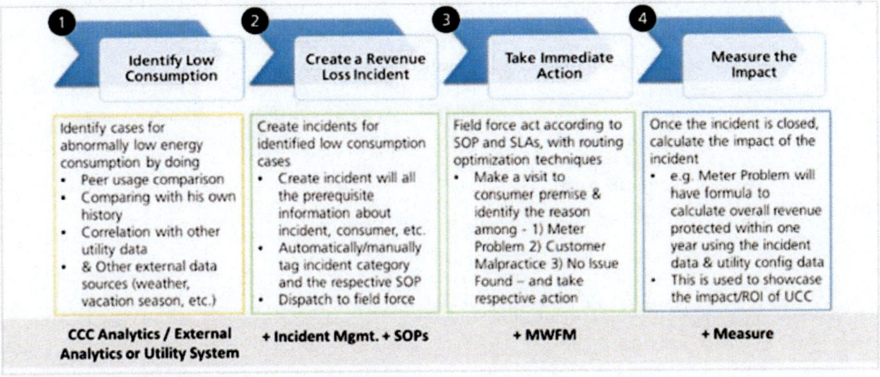

Fig. 5 Sample case breakdown

- Reduce noise around the issues, bring actionable information to staff and provide closed loop mechanism so that they can have sense of responsibility to act on the issues and close
- Square up utility operations with the required agility, and provide a platform/means to handle future challenges (future ready)
- Setup and monitor KPIs, and experience a visible transformation
- Achieve significant improvement in curtailing revenue losses, cost to serve customers, customer satisfaction, asset performance and the overall operational efficiency (Fig. 6).

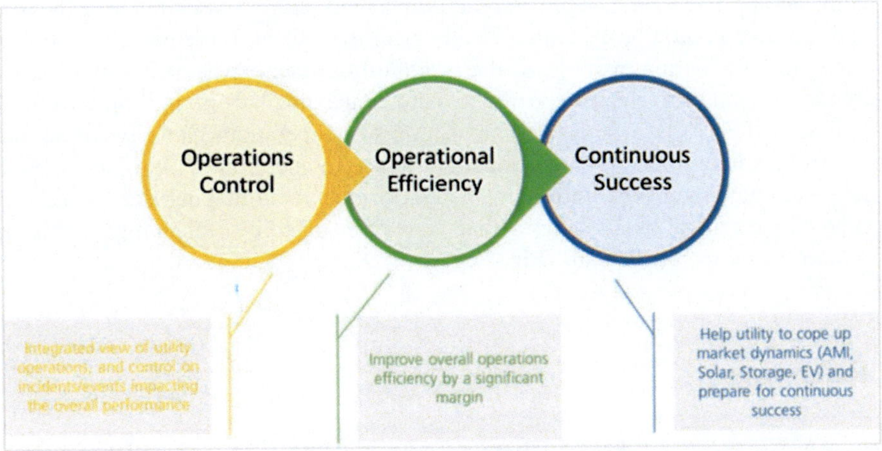

Fig. 6 Adoption path towards continuous operations maturity and success

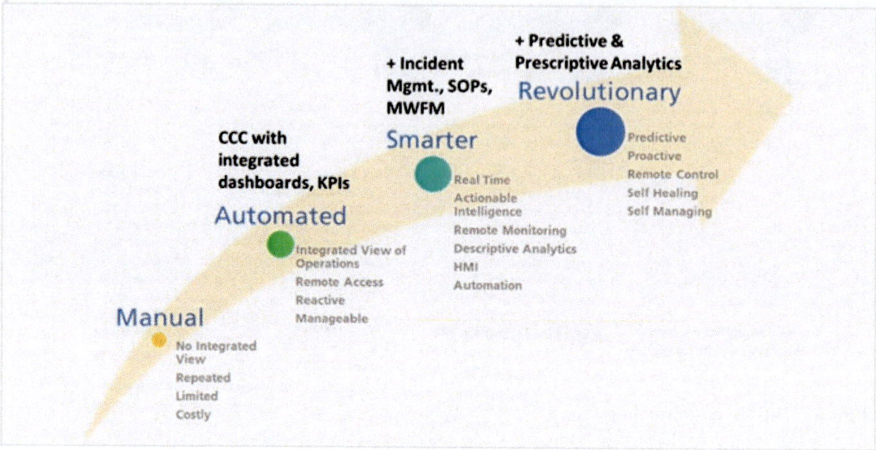

Fig. 7 Adoption path towards continuous operations maturity and success

4 Adoption Path

CCC platform for utilities works at the intersection IT (Information Technology) and OT (Operations Technology) and enable variety of use cases through utility operations center. It is quite easy to get carried away to start with many cases at a time or at the start. Remember, typical adoption requires organizational alignment across units and there exists a learning curve, especially in a way staff performs their day to day activities. So, the initiative should be a strategic one and better executed with lean approach. First, identify and solve business-critical cases, evaluate the value and improve continuously until a steady state is reached. Incrementally add more cases as the adoption evolves. At the peak, utilities can experiment with emerging trends, and innovative cases specific to them. Once the platform is adopted, utilities will be on their path towards continuous improving operations maturity and success.

Utilities may notice organizational resistance to adoption thinking the adoption may eliminate their jobs. In reality, the platform is about helping staff spend more time on the critical business tasks that cannot be solved by technology, and it may require more operations staff (Fig. 7).

References

1. https://electricenergyonline.com/energy/magazine/801/article/Operations-maturity-Smart-management-of-smart-grids.htm
2. https://www.cognizant.com/industries-resources/energy_and_utilities/smart-metering-maturity-assessment.pdf

Electric Mobility: EVs as Virtual Power Plants

Susheem Pandey

Abstract In today's energy market, consumers are becoming producers, and technology is placing more control in the hands of the many. On their own, familiar technologies—behind-the-meter energy storage, solar arrays, smart thermostats and electric vehicles (EV)—provide valuable but small-scale energy and sustainability benefits. When aggregated to become virtual power plants (VPPs), however, they become game-changers in grid management. In energy today, the watchword is "distributed." Today's massive, centralized electric power systems are being trans-formed as distributed energy resources (DER) plug into the mainstream market. As we integrate renewables, VPPs emerge as a responsive energy model that uses smart software and the so-called Internet of Things to aggregate thousands of smaller, separate power-producing sources scattered across the grid. These VPP networks, formed and managed by a growing number of companies across the globe, help utilities to match grid supply and demand by providing backup power during outages and peak-shaving. EVs hit the 2-million mark by the end of 2016 and expected to be around 11 million by 2025, some estimate. The takeaway is that when 2 million EVs (and growing) connect to a charge station, VPPs can substantially alter demands on the grid by slowing or ceasing EV charging (with permission from the vehicle owner). Likewise, with bidirectional flow capabilities, EVs will enable VPPs to deploy smart discharging, moving stored energy from the EV battery to reduce load on the grid within seconds. Robust batteries and smart technologies that support EV aggregation are reaching their prime. However, as with most nascent technology applications, there are challenges that the industry needs to overcome to maximize EV's grid value.

Keywords EV-Electric vehicles · VPP-Virtual power plants · GPS-Global positioning system

---

S. Pandey (✉)
Orion Power Limited, Dhaka, Bangladesh

R. K. Pillai et al. (eds.), *ISUW 2019*, Lecture Notes in Electrical Engineering 764,
https://doi.org/10.1007/978-981-16-1299-2_23

1 Introduction

Modern age living and advancement in science and technology has drastically improved the living and working conditions and atmosphere on one hand, while also adversely impacted the climate by global warming. Human activities have resulted in creating environmental hazards such as 1. flooding, 2. extreme, and variable precipitation, 3. extreme Hot conditions, 4. drought, 5. ocean acidification 5. Glaciers melting and 6. loss of the Arctic sea ice. Major factor in causing global warming is emission of CO_2.

One of the major contributions to CO_2 is automobile emissions. The solution lies in use of electric vehicles (EVs) in combination with renewable energy sources. However, the problem with renewable energy sources is that they are extremely intermittent: they produce electricity according to the weather—not necessarily to what is needed. For example Solar irradiation depends on clear sky, cloud, weather conditions such as rains, storm etc. It keeps changing every hour or so. Similarly wind does not flow at a constant speed all the time. It also varies with the weather and season. This places a limitation on control of electricity produced by these renewable means. This intermittency leads to differences between the production and consumption of energy which destabilize the grid, leading to blackouts. This also leads to economic and operational problems. There are other problems also with Renewable Energy sources. (A) Electricity cannot be stored as it is perishable and has to be used as it is produced in real time. Only small storages in batteries are plausible. (B) This gives rise to the need of balancing the demand and supply of electricity in the grid in order to tap and realizing the potential of volatile renewable energy sources for consumers of electricity. (C) One of the ways to balance the conditions is to use idle power plants as back-up to ensure that electricity is available when needed. (D) But this is highly inefficient and expensive as also limits the accommodation of increasing shares of renewable energy sources because the grid has to operate increasingly under variable supply from renewable energy sources. (E) This also reduces the market share of dispatchable power plants, such as coal or gas. (F) Therefore cannot guarantee back-up at all times.

An alternative solution to all the above challenges is use of EV as virtual power plants (VPP).

2 Background

The 2015 UN Climate Change Conference in Paris, resolved to mitigate the climate change and ensure that the global warming does not rise beyond 2 °C. The solution lies in increasing use of Renewable Energy sources for meeting the energy requirements. However, the biggest problem was that the power transmission and distribution grids in their present state of operations are not capable of storing and distributing the variable and intermittent renewable energy. So alternative solution is badly needed.

Researchers in the field of energy developed a software algorithm that can pool batteries in rental electric vehicles (EVs) together and turn them into 'virtual power plants' that can store electricity at peak production times and sell back when the price is right. The fleet operators can make an additional profit from running such a virtual power plant.

3 EVs as Aggregated Power Centres

Storage of Electricity is efficiently possible through battery systems. In 10 years from now all automobiles running on petroleum source are likely to be replaced with Electric Vehicles running on batteries. As EVs are more widely adopted, storage capacity for electricity becomes increasingly available. This capacity may be employed to offer balancing services to the grid. But the commercial viability of such an arrangement will hinge on the trade-off between the operating costs, such as battery wear, and potential profits. Here, we consider the electricity stored in EV batteries as inventory. The battery can be allocated to four, mutually exclusive states: 1. Increase/Add inventory (Charging of battery). 2. Decrease inventory—Discharge for renting. 3. Decrease inventory—discharge to the grid. 4. No change in inventory—left idle.

The biggest challenging task in such a model is 1. The optimal allocation of EVs over time which constitutes a multi-period inventory flow problem. 2. Uncertain energy prices at any given time. 3. Day-ahead EV allocation is much more complicated of all problems. Problem becomes highly complicated when it is not known in advance when and where EVs will be used, like for example in a car sharing fleet.

In case of a private person who owns a single EV, inventory can be substituted. The owner can decide between charging, discharging, driving, and leaving idle. But When considering fleets of EVs, the location matters, as rental demand (energy for rentals) depends on location. We make a trade-off between a class of demand where location matters (drivers want a car to be close to their departure location) and a class of demand where location does not matter (vehicles can discharge to the grid from any capable charging point in a city).

A mixed rental-trading strategy is developed to analyze the potential for EV fleet owners to gain profits from renting out cars, while at the same time using spare battery capacity and excess electricity to actively trade in the energy market. This fleet of EVs is considered as a VPP, a collection of distributed energy sources, which are centrally managed to generate power at consumption peaks and absorb excess electricity when consumption is low. There are following factors which are variables and need to be estimated with precession through the information from electricity exchange, rental car fleet companies, power distribution companies—(1) Location and numbers of car fleets on real time basis. (2) Price of electricity on real time basis. (3) Peak and lean periods of power demand on real time basis and (4) trend of car rental demand.

When EVs are charged or discharged to the grid, they are aggregated to VPPs, which act on fluctuating electricity price signals. A mixed rental-trading strategy optimizes electricity purchases of EV fleet owners in the market to charge the fleet for the purpose of driving, or discharge at a later stage at a price premium. The proposed strategy is validated with real energy market data and real data about electric vehicles, whose movements are tracked with GSM (Global System for Mobile Communications) and GPS (Global Positioning System). This technology provides real-time information about battery and location of the EVs.

There are different scenarios to be considered in this: 1. A rental car is most profitable when actually rented out. 2. Rental car that is charging its battery is a cost burden on company. So the algorithm and method of mixed rental-trade mechanism is to determine how many electric vehicles can be withdrawn from rental services at any given time to become part of the virtual power plant.

Determination of vehicles out of the entire fleets which can be taken out for VPP is successfully achieved through GPS data collection and analyzing it. This gives the answer as to the number of cars that will be rented in the span of 15 min intervals. The trade off has to be done between meeting the rental demand and using the idle cars as VPP. A remarkable balance has to be struck between the two, which is possible only if real time accuracy is ensured. It is important as the fleet owner can not disappoint the clients of rental demand and at the same time wants minimum vehicles to be standing idle. Also, once a certain number of cars has been committed to the virtual power plant, it would be breach of contract with the electricity company to rent them out at last-minute notice.

4 Profit from Peak Power

Fleet operators running a virtual power plant today can expect to see their profits increase with seven per cent. Most of that profit comes as follows: 1. At lean demand period of Discoms when the power is available in excess, the fleet vehicles can charge their batteries from absorbing excess peak power from the traditional power grid as electricity is at low prices based on Time of Use (demand and Supply situation). 2. Fleet vehicles can sell the electricity by discharging their batteries to the grid at times when the power demand on the grid is high and electricity price is also high due to peak demand again based on Time of Use (Demand and Supply conditions). They found that selling electricity back to the grid was only profitable during short time periods when demand for electricity is high and electricity prices rise.

Profits from virtual power plants are expected to grow in the future with improving battery technology thus making batteries cheaper to operate and the gross additional profit from virtual power plants may be as high as 12% in 2022. It is expected that by 2022 a substantially very large number of people will have an electric vehicle becoming a part of the virtual power plants so much so that selling electricity back to the grid and may even replace traditional power plants devoted to generating power at peak moments.

5 Virtual Power Plant

A virtual power plant concept combines a number of remotely located independent energy resources from disparate locations into a network that provides reliable power 24 h a day. This network is created based on 1. Software-based technology. 2. Smart grid. 3. Planning, scheduling, and bidding of distributed energy resources (DER) Such a VPP network provides stable and reliable power. VPP concept involves DER (Distributed Energy Resources) hence it is a departure from centralized plants.

VPP is expected to turn into most reliable and predictable power source capable of meeting the same reliability as traditional centralized power plants. And the added feature is that while in traditional plants, often coal or natural gas plants the Power flow has been in one direction; from the utility to the business or consumer in case of VPP it is bidirectional that is, in both directions depending on demand and supply situation.

The challenge in DER based on RE: In the last 12 years there has been a huge rise in Independent Power Producers who are using multiple Renewable Energy Resource to generate power. This manifold increase in Renewable Energy generating plants based on solar wind and other resources have added another feature that is bidirectional flow of power. Since all these are intermittent and independent of one another they disturb the grid stability to a large They are clean sources of power but they disturb and disrupt the grid balance and stability. Hence there arises a need for new models. The VPP is in fact a fallout of this avalanche of Renewable Energy DERs. The intermittent nature of RE DER can destabilize the grid for example an oversupply can lead to outage of the grid, and a sudden undersupply when demand is at peak calls for immediate switching of dispatch-able plants in operation which again results in destabilization of the grid. An oversupply when generation of RE sources is at surplus can also cause a challenge of reverse power flow situation on the grid. This is a technical problem. All such problems can be handled with VPP which can absorb excess power when surplus and charge its battery, discharge and provide additional power to the grid when demand is high. All these are achieved with prior planning and scheduling of loads thus maintaining grid stability. VPP is less expensive, less polluting capable of producing electricity 24 h a day and hence can be a suitable substitute to fossil fuel based generation.

A VPP concept utilizes the fleet of EV to stabilize and balance the electricity smart grid that are connected to decentralized RE sources of solar, wind and hydro etc. Since RE sources generate power based on weather conditions and not on demand basis the EVs can be used to supply and consume power based on demand basis to balance this destabilized situation. The key consideration in using VPP as balancing agent of grid is the successful car fleet management. In particular, we analyze the potential of parked EVs to absorb electricity from the grid, and provide electricity back to the grid when needed. For discharging the battery to the Grid by a vehicle the Location of the vehicle is important. Similarly for discharging the battery for rental again the location of vehicle is important. The vehicle will need to be strategically located for both the purpose so that it can immediately meet the need of (1) discharging to grid when electricity price is high as power demand is at peak, (2) when a client needs a car on rental. In case of charging it may not be located at strategic places as it can charge from anywhere nearest charging point. Actual data on location of the fleet vehicles is collected from GPS, and prices are collected from Power exchange, demand peak and lean period collected from Discom. Then a Fourier series approach is applied involving weighted objective function with asymmetric payoffs. VPP can be profitable to fleet owners, ecologically advantageous through reductions in wind power curtailment, and beneficial to consumers by reducing energy expenses.

6 Obstacles to Virtual Power Plants

There were some concerns like: 1. Such a concept of VPP involving a combination of fleet component, software, GPS, scheduling and analyzing mechanism and a networked operation is open to cyber attacks, 2. high levels of distributed energy resources can affect local voltage, unless a voltage control scheme is in place. 3. Regulatory incentives are needed to help virtual power plant operators bring their benefits to the system. These obstacles are overcome by making the VPP design insulated from any cyber attack by security software, applying voltage control devices at appropriate places to counter voltage fluctuations, and a suitably designed incentive scheme by regulatory to tap the potential. Once these are in place then the VPP can balance the imbalances created by large number of distributed RE generations and other failing fuel based plants. As an imbalance occurs, the transmission operator can immediately put VPP to increase or decrease the power so as to balance the grid. This is possible on real time basis. Since VPP can respond to demand situations on real time basis this obviates the need for load forecasting on day ahead basis.

7 Challenges Facing the EVs as VPP

Following challenges are faced by the concept of EV as VPP. 1. Need for Robust batteries and smart technologies that support EV aggregation. 2. Policies that provide a framework for market access and aggregation. 3. Regulations must be written to enable benefits. 4. Cost of participating in VPP must be reasonable and nominal. 5. Economy can be achieved by tapping the potential of Cloud computing, communication technology and volume game, by involving as many fleet in VPP as possible. 6. VPP model must be structured for maximizing its value. All stakeholders must have fair exchange of value like Fleet operators, EV drivers, Site host, Grid operator, VPP operator all should exchange values for maximizing the efficacy and penetration of VPP. 7. Policy support should expand market opportunities and with falling cost of participation more choices of compelling VPP operators will emerge. 8. RE must also support the EV drive so as to maximize the participation of EV fleet into VPP.

8 Architecture and Schematic

The philosophy of EV as VPP depends on the complementary roles of electric vehicle and Power system. They complement each other and the architecture of this model hinges mainly on the power storage and discharging capacity of battery in an EV. As a class of vehicle involving battery for power storage and running this model include Plug-in-Hybrid vehicle also. Those vehicles which cannot be plugged in cannot participate in VPP. The VPP philosophy is also driven by the need to penetrate into power distribution system for stabilization of the power grid. Climate change necessitates controlling global warming which in turn needs reduction in CO_2 emission. This calls for increasing role of renewable energy in power generation. Renewable energy is highly intermittent and destabilizes the grid. This in turn can be controlled by plugging in EV (and PHEV) for charging and discharging vehicles in and from the grid. This will reduce the spikes in power created due to intermittent DERs (Distributed Energy resources). When a vehicle connects with the grid for charging then it is called V2G (Vehicle to grid). When a vehicle is connected to the grid for discharging and supplying power back to the grid it is called G2V (Grid to Vehicle).

Though individual EV is just one small contributor to the VPP but a large number of such vehicles will be an asset to the power system. Such an asset can be intelligently managed. The VPP architecture is based on an aggregator model where the aggregated EV (PHEV) is centrally coordinated for charging and discharging in V2G and G2V process. The charging is controlled through a charge control mechanism in a Battery management system. BMS plays a crucial role by controlling the charging power through charge controller in G2V mode. When in V2G mode the batteries are discharged in a coordinated manner as per grid peak power requirement so as o smoothen the load curve. Conversely when there is low demand and power is surplus on the grid the Grid supplies power to batteries in a coordinated charging system again to smoothen the load supply curve. Entire mechanism is controlled by a central coordinator in VPP. VPP architecture includes communication and hardware interfaces and an intelligent charging and discharging system. The aggregation technology is based on the Virtual Power Plant (VPP) concept where the VPP is the EV coordinator.

The system architecture of a VPP consist of Transmission System Operator, EV, PHEV, VPP coordination system, Battery management System, Communication hardware and Software. When peak load conditions are faced by transmission grid the TSO sends signal to VPP for connecting the EV/PHEV for discharging its battery to the grid. The number of vehicles, quantum of energy to be discharged to the grids estimated by the VPP software by collecting data on position and location of EV/PHEV through GPS. Linear programming and Fourier series application is used in analyzing as to the tradeoff between vehicles for rental demands and vehicles for discharging demands. Similarly when the power demand is low on transmission grid and there is power surplus then TSO sends signal to EV through VPP for charging the batteries. In both the cases whether charging or discharging the information on electricity price on real time basis is collected from power exchange by the VPP system and decision is made by software using Fourier series and linear programming as well as transportation model. In the entire process all participating stakeholders are kept updated in respect of charging and discharging requirement which in turn is based on Peak and Lean demand periods on transmission grid.

In the VPP design architecture the vehicles considered are pure EV running on battery and PHEV which is a hybrid between Engine and battery. There are three types of PHEV considered in this architecture. 1. Series—parallel PHEV. 2. Series PHEV and 3. Parallel PHEV. A series parallel PHEV means. This classification is based on the main difference among the drive system used and the interconnection of its components, before the power is transferred to the wheels. In the series–parallel hybrid vehicle, the system is designed to operate both in a series or parallel configuration. The reconfigurable system is made possible by the use of a planetary gear, which is the mechanical coupling (MC) for the three machines. In the series hybrid vehicle, the electric traction system and Internal Combustion Engine (ICE) system operate in a series connection. In sequence, the ICE is coupled with a generator (Gen.) which generates the electric power for recharging the battery, the battery then supplies an electric motor driver to transfer power to wheels. In the parallel hybrid vehicle, the ICE and electric motor (EM) operate in parallel mode, where the ICE supports the electric traction at certain points of the driving pattern, e.g. when higher power is needed to the wheels. In the battery-powered EV class, the drive system is realized using only an electric motor and a motor driver. Therefore the only energy source is the battery pack. See Fig. 1.

9 Controllable Charging/Discharging Operation

EV/PHEV is charged and discharged in a controlled and coordinated manner managed by EV coordinator. Charging and discharging can be by home charging or public charging stations depending on classes of vehicles. Charging and discharging is done based on current rating, voltage rating, single phase and three phase supply. For a single phase supply the charging current rating is 16 A and for a three phase supply the charging current is 32 A. A simple illustration for the charging and discharging of vehicles is given in Fig. 1. The EV system architecture is planned to respond to different control signals from a centralized EV coordinator. The control signals for the vehicles can be generated by a VPP and based on different variables such as the power system frequency, the market spot price, location of EV/PHEV, segregation of rental demand or charging/discharging demand on EV/PHEV. Figure 2 depicts the complete architecture and working philosophy of a VPP. The figure shows different entities interfaced with the centralized VPP system. Different RE sources are also interfaced with, EV is one of them. Other entities like solar, wind, hydro etc. are also interfaced. But the study is restricted to EV only. A generic VPP can aggregate and control various distributed energy resources (DERs), like solar, wind, biomass, hydro etc. including EV. TSO gives the activation commands for accepted regulating power reserves contracts. Distribution System Operator (DSO) gives the grid status

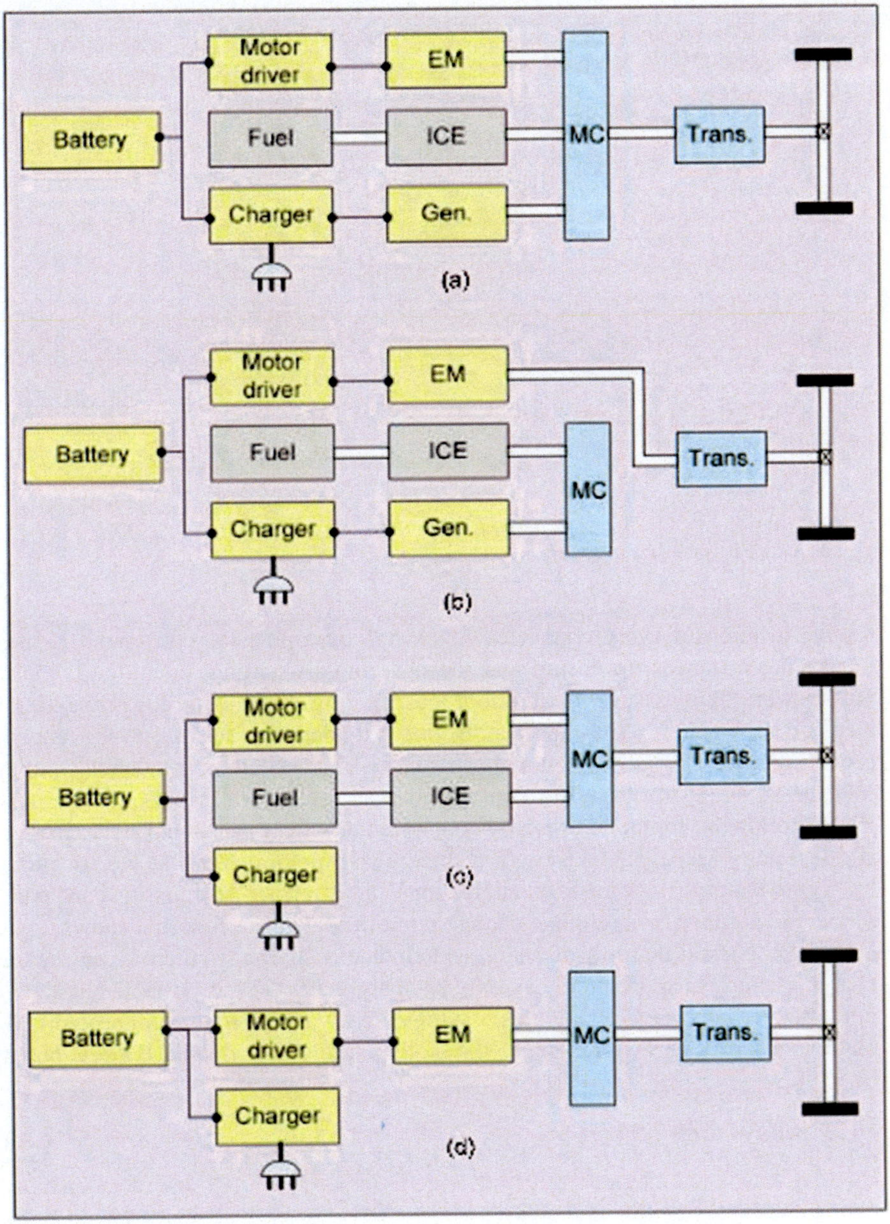

Fig. 1 Series Paralle Hybrid Drive system

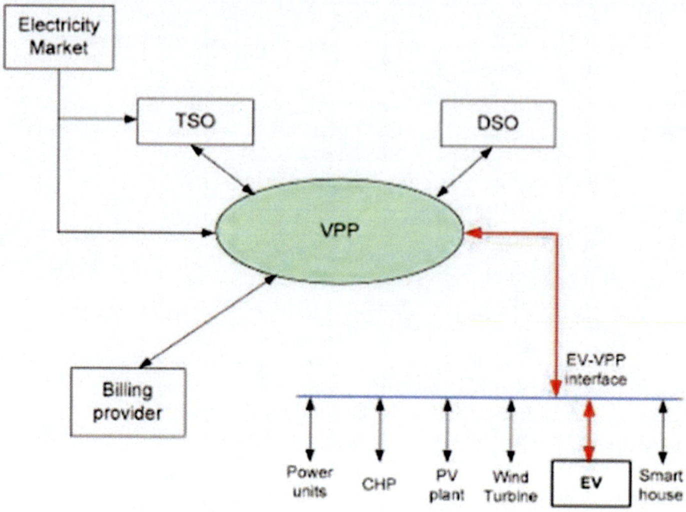

Fig. 2 EV-VPP system Architecture

for the location of every connected EV. Grid constraints are considered at these interfacing instances, to ensure that the charging/discharging operation complies with power quality issues. In addition, the metering information for accounting is also collected via the DSO interface. Billing provider provides bills to the resource providing services. The control requests for the EV test bed are generated from the VPP, based on the grid needs of regulating power reserves of the TSO. In a battery-EV the following components can be considered: a battery pack—a battery charger a BMS (Battery Management System) a three-phase motor driver—an electric motor.

A pure battery EV is the ideal choice for VPP as it has zero emissions and hence serves the purpose from climate change point of view also. Figure 3 shows a pure battery EV backed charging discharging architecture. In Fig. 1d there is only battery charging which means battery can only be charged i.e. G2V is possible but V2G is not possible. In order to enable this V2G feature a V2G inverter is connected to the battery system (See Fig. 3). Hence the architecture will include a Battery, battery

Fig. 3 Battery Charge controlling system

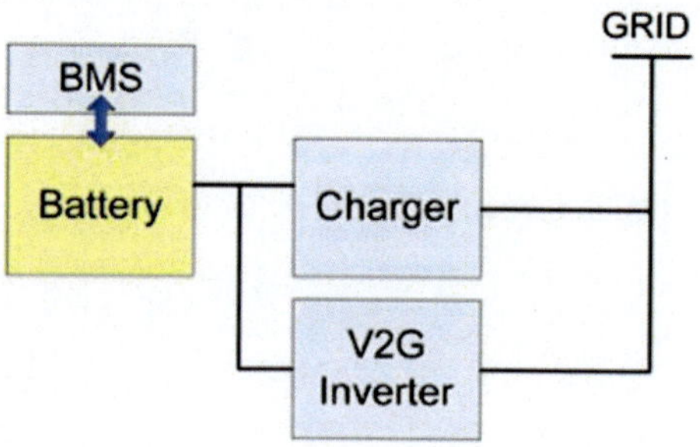

charger, a V2G Inverter and a Battery management System (software controls) to control and monitor the process of charging and discharging while interacting with the Grid and VPP coordination software.

The battery pack is interfaced to a battery management system (BMS) which monitors its status. The battery charger is an AC to DC converter and V2G interface is a single phase DC to AC inverter. The AC to DC converter is directly connected to main grid, by means of a three phase cable connection; The design of the battery pack takes into account the 1. V2G operation requirement and 2.The choice of a battery technology.

Choice of battery technology depends on two vital aspects A. Greater energy-to-weight ratio, B. Greater power levels and C. Low self-discharge when not in use. On all these criterion the Li-Ion battery fits the best. Lots of research is being done on battery types. The latest technology is Li-sulphur battery with higher Power, rapid charging, deep density of discharge etc.

10 Social Perspective of EV as VPP

Using EV as VPP is overall beneficial for the society in following ways. 1. Consumer gets power at a cheaper price. 2. Grid stability is maximized with EV discharging to and charging from the grid as per requirement of grid balancing. 3. Pure battery EV is a zero emission source of power hence suits the climate change reduction targets also. 4. Power plants' operations are optimized. There is a flip side to it for EV fleet owners because of the increased competition and depreciation cost of battery. However, if the participation of EV increases to more than 50% then this concept is a win win situation for both fleet owners as well as power consumers and overall power trading market also. As more number of fleet owners connect with the grid for discharging then the power to the grid increases while demand remains the same. This reduces the clearing price of the electricity in the trading exchange. Whereas when the same fleet connect with the grid for charging then then the demand for power increases while the supply remains same. This increases the electricity price. In order to make this concept a commercially viable and socially acceptable, the government has to play a vital role through its regulatory policies by keeping tariff incentives suitable for EV fleet owners so that they partake in the VPP and make profits. Incentivising the tariff will be the game changer. If the tariff for battery charging is lower then EVs shall charge at lower tariff, and when the power demand goes high at peak hours and so does the tariff, they discharge to the grid. The difference is the profit. However, they make most of the profit while charging at lower than average market price, because they cannot just forgo their car rental demands for the sake of discharging to the grid at the option of VPP coordinator, because VPP may not call EV drivers at their availability whereas when VPP do call them for discharging the EV may be having rental clients to serve. The secret of success lies in proper coordination between rental demands and Charging and discharging demands. This is accomplished by the entities like VPP coordination, TSO, DSO, Power exchange, Cloud analytical system, Communication hardware and software etc. The social acceptance of EV as VPP depends on each player maximizing and sharing his information with the fellow participant. The win win situation for society will happen when Fleet owners, Grid operators, Power exchange, and Power consumers, as well as climate change, all take benefit out of the VPP concept. Technology, Management, Legislation and Finances, all have to play a big role if EV as VPP is to become socially acceptable.

11 Conclusion

Distributed energy resources pose both a challenge as well as an advantage for the power system. Advantage because they provide Renewable Energy for which the government is already committed to increase its share in total energy mix. But the challenge is that they are intermittent and weather based. This negative feature of RE can be compensated by using EV as VPP. EV can stabilize the power grid which are destabilized due to intermittent nature of RE. This is possible because EVs can respond to the grid demand on real time basis. Thus also obviating the need for Load/Demand forecasting (day ahead). This benefit is however not without cost. The cost of EV fleet participating in VPP, cost of batteries, depreciation on batteries, tax structure of government. There is an opportunity cost also to fleet owners, that is the rental income. If they make a wrong choice as to whether to take up a rental consumer or to take up a discharging opportunity at the grid, it may be a loss making proposition for them. That is why a very close coordination and information exchange is needed at the level of VPP coordinator. This is achieved by a the model of mixed rental trading scheme. In this scheme the proper tradeoff between rental clients and power trading opportunities is effected through EV VPP coordination. The VPP architecture uses many components in order to ensure successful mixed rental trading scheme. The VPP coordination sits in the center handling interfacing of other components like TSO, DSO, RE generators including EV/PHEV, BSO, Cloud computing, Communication hardware and software, GPS system. The EV driver gets the status of tariff and power status from the DSOP. Power exchange gives the spot market price of electricity. TSO gives the status of demand and supply on the grid and gives signal to EV through VPP network. EV responds to it by charging/discharging to the grid. This stabilizes the grid which was otherwise destabilized due to intermittent RE sources connected to the grid. All this happens on a real time basis that is why the need for day ahead forecasting is also not there. This paper presents a strategy, which is both profitable for electric vehicle fleet owners and sustainable for society and planet. The proposed mixed rental- trading strategy allows fleet owners to charge their electric vehicles more cheaply, use their storage capacity for arbitrage trading, and rent out these vehicles as usual. The mixed rental-trading strategy recommends the optimal states for all electric vehicles in the fleet across charging (adding inventory), discharging for driving (decreasing inventory), discharging to the grid (decreasing inventory), or being idle (no change in inventory). Fleet owners make a trade-off between a class of demand where location matters (drivers want a car to be close

Table 1 Charging power rates

AC current (A)	AC voltage (V)	Grid connection	Power (kW)
10	230	Single phase	2.3
16	230	Single phase	3.7
32	230	Single phase	7.4
16	400	Three phase	11
32	400	Three phase	22

to their place of departure) and a class of demand where location does not matter (vehicles can discharge to the grid from any capable charging point). Lastly, Battery is a key component in this concept. Its huge storage and discharging capacity plays a critical role in balancing the grid. As more and more advancements are being made in this field we have batteries that can not only provide higher power and charge but also get charged very rapidly as also discharge speedily without compromising battery efficiency and life. Today the most successful storage battery is Li-ion which has high energy to weight ratio, higher power levels and low self discharging when not in use. Next technology is Li-sulphur battery. As the development promotes more battery and EV into power market the price of battery will also fall while efficiency will go up. This will result in greater market share of EV fleet owners in VPP which will increase the viability of this concept of EV as VPP (Table 1).

Smart Street Lighting

Vijay Barve

Abstract The type of light sources have evolved over a period of time and there has been an improvement in the luminous efficacy of the light sources as they evolved. However the more important aspect would be how efficiently the lighting system is controlled during its lifetime to achieve energy efficiency during operations giving maximum benefit to the customer. The smartness in a lighting system relates more to the control part of the lighting system. These control technologies are also evolving rapidly. The system architecture consists of a Main server, local controller and the street lighting fixtures. There are different modes of communication like GSM, RF or the power cabling network itself for data/signal transfer between different components of the street lighting system. It is possible to provide different types of control like individual control, group control or a combination of both. The control would involve either ON/OFF control or dimming. The smart street lighting pole has multi-purpose uses like for mounting of various sensors, video camera, wi-fi system, information display etc. The smart street lighting provides a solution which is energy efficient, having high up time wherein fault location is instantaneous and remedial actions need bare minimum time, provides a quick and easy user interface for system data, Visual graphical environment to track failures, check system health etc.

Keywords Server · Controller · Sensor · Smart · Signal

1 Introduction

As the light sources have evolved over a period of time there has been an improvement in luminous efficacy. The focus is now on how efficiently the lighting system can be controlled. Smart street lighting is an automated solution for remote operation and monitoring such that it functions in an energy efficient manner resulting in monetary benefits to the Customer.

V. Barve (✉)
TATA Consulting Engineers Limited, Pune, India
e-mail: vvbarve@tce.co.in

R. K. Pillai et al. (eds.), *ISUW 2019*, Lecture Notes in Electrical Engineering 764,
https://doi.org/10.1007/978-981-16-1299-2_24

Power consumption of street lighting is estimated to be around 40% of the energy budget of a city [1]. In India the contribution of street lighting is @ 2% of the total energy consumption. As per estimates street lighting contributes to 6% of global emission of green house gases [1].

2 Smart Street Lighting System

2.1 *Salient Features*

Salient features of Smart street lighting are as follows:

(1) Web/App based Contol and monitoring facility
(2) Ease of fault detection, reduction in manpower and logistics requirement
(3) Emergency override facility for both local and remote modes to a predfined account holder
(4) On/Off/Dimming functionality
(5) User friendly software system with cyber security and access control, generation of statistical data, generation of alert messages

2.2 *Modes of Communication for Control*

The modes of communication are as follows:

(1) Driver control based on Power line communication system [2, 3]

Power cable acts as a mode of communication for data transfer between local controller and individual fixture. Group control is possible (Fig. 1).

(2) Driver control based on RF communication [3]

Communication happens through Radio Frequency (RF) mode for data transfer between local controller and individual fixture. Group or individual control is possible. Possibility of weakening of signal due to obstructions in line of sight (Fig. 2).

(3) Driver control based on GSM communication [3]

Communication happens between the Main server and the street light fixture through GSM. Group or individual control is possible.

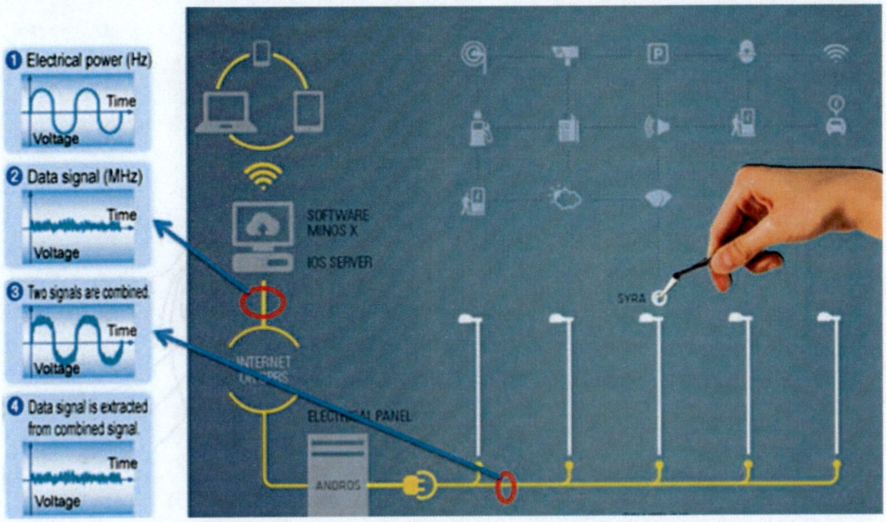

Fig. 1 Data transfer via powerline

Fig. 2 Dimming by RF communication

2.3 Components (Fig. 3)

It comprises of the following:

(1) Central server

Located at the Customers premises or is cloud based. Communicates with local controller in group control and directly with fixture in individual control.

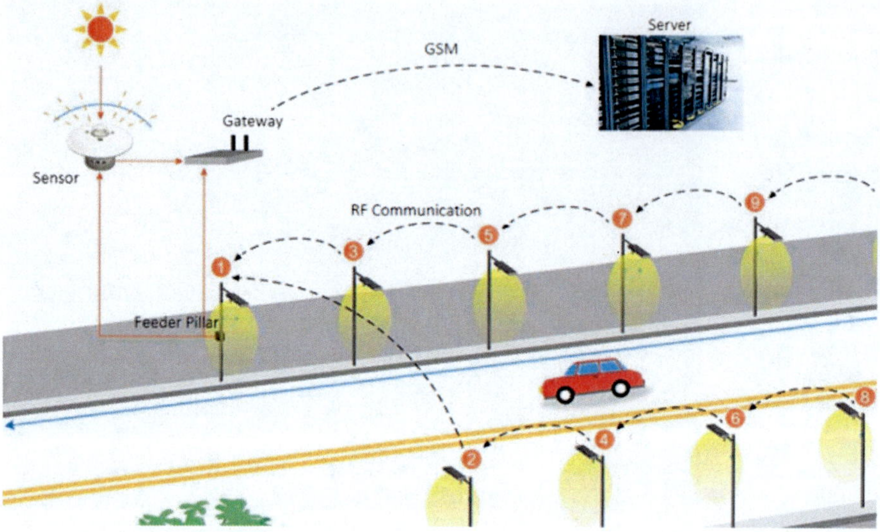

Fig. 3 Components of system

(2) Local control module (feeder pillar + controller)

It will have timers/sensors for control of fixtures. It acts as a gateway to Main server for all fixtures.

(3) Communication system (RF/GSM/Powerline)
(4) LED Street light Luminaries

Will have dimmable/non-dimmable driver and communication module. It will also have sensors giving input to the communication module.

(5) Sensors/Timers
(6) Web based software for lighting management.

2.4 *Methodologies for Control*

These can be classified based on type of control (ON/OFF or Dimming), based on source of command (Timer or sensor control) e.g. Photocell, motion sensor, occupancy sensor, based on extent of control (group or individual).

(1) Control by timer

The ON and OFF timing of the lights is set in advance (Fig. 4).

18:00Hrs to 22:00Hrs all street lights ON

22:00Hrs to 1:00Hrs 66% street lights ON

1:00Hrs to 6:00Hrs 33% street lights ON

Fig. 4 Control by timer

(2) Control by Photoelectric cell

The light intensity is set in advance for switch ON or OFF of the lights (Fig. 5).

(3) Control by Dimmable driver

Light output of individual street light fixture can be controlled with the use of a dimmable driver. It is possible to have a fine control over the light output (Fig. 6).

(4) Control by movement sensor

The sensor senses the approaching vehicle, the light is switched ON and the brightness is controlled based on the nearness of the vehicle to the pole. The fixture is dimmed

All street lights switched ON after sunset and during overcast conditions

All street lights switched OFF after sunrise

Fig. 5 Control by photoelectric cell

18:00Hrs to 22:00Hrs all street lights ON with Zero dimming

22:00Hrs to 1:00Hrs all street lights dimmed by 33%

1:00Hrs to 6:00Hrs all street lights dimmed by 66%

Fig. 6 Control by dimmable driver

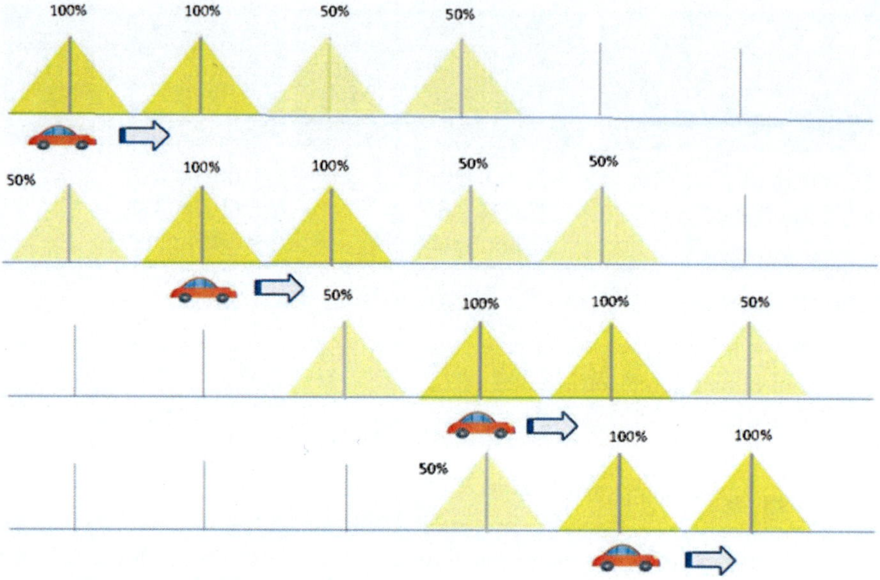

Fig. 7 Control by movement sensor

as the vehicle passes by and is finally switched OFF. The process repeats for all the lights along the road as the vehicle moves past (Fig. 7).

(5) Group/individual control for an area

For this type of control every fixture will have a driver with dimming features. Web based monitoring and control can be given (Figs. 8, 9 and 10).

Fig. 8 Group control of lighting fixtures for an area

Fig. 9 Group control of lighting fixtures for a road

Fig. 10 Individual control of lighting fixtures for an area

2.5 *Holistic Approach to Smart Lighting*

It is possible to consider add on features while implementing smart street lighting system. The smart poles and the communication network spread across the city can be useful to create a platform wherein it is possible to connect IoT enabled devices to monitor and control other non-lighting applications of a smart city infrastructure. The features of smart pole are provisions for Sensors like noise and air pollution, Video camera monitoring, display boards, battery charging point for electric vehicles, Wi-Fi network, Call points for emergencies.

3 Pros of Smart Lighting

The pros of smart street lighting are as follows:

(1) Energy saving
(2) System ON time is more and identification of fault location is easier
(3) Flexibility in load control
(4) Customer friendly interfaces for fast accessibility of data
(5) Collection, storage and organization of data on system working and visual display of the same
(6) Web based alarm generation platform for faults for distribution via SMS or e mail
(7) Automatic generation of statistical data for analysis
(8) Integration of non lighting applications of a smart city infrastructure on the same network.

4 Conclusion

Considering the pros mentioned above like energy optimisation, flexibility and simplification of Operation and Maintenance, auto generation and retrieval of data etc., a smart lighting option can be evaluated for a Smart City project. The cost analysis shall be worked out for a given case based on the options available and depending upon Client's specific requirements if any. A holistic approach with integration of non lighting systems could give added benefits.

References

1. Cacciator G, Fiandrino C, Kliozovich D, Granelli F, Bouvry P. Cost analysis of smart lighting solutions in smart cities
2. Pradeep Kumar (2013, Mar 21) Municipal street lighting energy efficiency scenario in India. Available : https://www.superefficient.org
3. Product information from Vendors

Intelligent Street Light Controller with Security System

N. VinothKumar, S. Shyam Sunder, T. Viswanthan, and M. Mathan Kumar

Abstract Two distinct issues in India are reduction of the power consumption and security of humans. Lots of initiatives are taken from Indian government to ensure the safety, security for Indian citizen, especially for women and children. This paper aims at efficient energy saving method for solar based autonomous street lights and also to provide the security for human during emergency situations. The street lights are automated with brightness control based on the illumination intensity, tracking of vehicle or human movement. This paper includes the implementation of theft indicator for solar panel and battery of the street lights. The human security is ensured by continuous monitoring of security camera which is kept at street light pole. It supports to call emergency services such as ambulance, fire and police during the emergency situations. The fast alertness can be achieved due to sharing the location using Global Positioning System (GPS). To realize the concept, the hardware is implemented and the results are verified. Thus, the street light energy is saved about 35% in comparison with existing system. Also, the overall security for the streetlights as well as the humans can be achieved intelligently.

Keywords Energy conservation · Emergency alert · Street-light · PWM · Fault detection · Security

N. VinothKumar (✉) · T. Viswanthan · M. Mathan Kumar
EEE Department, Kumaraguru College of Technology, Coimbatore, India

T. Viswanthan
e-mail: Viswanathan.t.eee@kct.ac.in

M. Mathan Kumar
e-mail: Mathankumar.m.eee@kct.ac.in

S. Shyam Sunder
Distribution System Division Central Power Research Institute, Bangalore, India
e-mail: shyamsundar@cpri.in

R. K. Pillai et al. (eds.), *ISUW 2019*, Lecture Notes in Electrical Engineering 764,
https://doi.org/10.1007/978-981-16-1299-2_25

1 Introduction

The street light plays a major role in the society for safety purpose during night hours. It ensures the provides illumination during night, ensures the safety of human and also helps in reduction of crime activities [1, 2]. Street lights indirectly assisting for the reducing of accidents during the night hours. Most commonly used street lights are high pressure sodium vapor (HPSV), Low Pressure Sodium Vapor (LPSV) and Metal Halide (MH) lamps. The consumption range is about 50–400 W depending on the lumen's requirement. The street lights consume energy about 18–40% in total energy. More than 2 lakhs street lights are installed in major cities of India. The total consumption is about 19 MW in major cities and the equivalent cost depends upon the tariff of the location which is much higher [3]. It is noted in many places, the lights are even ON during day time which cost additional power consumption. Hence, the reduction of cost and energy can be saved in two ways. One is to reduce the power consumption using latest lighting technologies such as using Light Emitting Diode (LED). The LED lights provides best result than the conventional lights. It has many advantages such as energy efficient, Long lifetime, rugged, ecofriendly and no warm up time, etc. [4, 5].

The second solution is usage automation and Internet of Things (IoT) [6]. Automation is preferred over manual operation everywhere because of its advantage like remote ON/OFF, power reduction, increase in efficiency and efficient operations. The reduction of energy not only results in energy cost but also reduction in carbon di oxide emission [7]. According to the CNCF the amount of carbon-dioxide produced for the production of 1KWH of power is about 0.94 kg. Hence the global warming can also be reduced. The power consumption of the street lights can be reduced largely by replacing existing HPSV, MH lamp with the LED lamps. Thus, the energy consumed by the lamp can be reduced around 50–75%. Another method is to use the energy saving circuits which shall be provided 20–30% of reduction in power consumption [8].

Initially, the automation is to be considered by extracting the information of sensors to turn ON/OFF the light during operation hours. Further, the utilization of smart sensors in street light, the efficient operation of street light has been improved [9, 10]. The researchers have proposed different techniques and technologies to improve the intelligence, reliability and controllability to reduce the operational and maintenance cost [11, 12].

The major concern of any country is the security for the citizens. Most of the criminal activities, such as robbery, murder, chain snatching etc. takes place in public places like streets, major roads and bazaars. Most of the places, a women walking alone is unsafe to them. Another issue is accident where the maximum numbers of accidents are occurring at streets or road. The last issue is that fire accidents which cost major causalities and wealth lost. Such emergency conditions, timely actions shall reduce the human and wealth loss. Hence, it is required to coordinate to the concern places with Global Positioning System (GPS) to reaching the place to rescue the victim from the emergency condition.

2 Methodology

The methodology of the proposed work can be explained with the Fig. 1. There are two types of system is presented here as follows.

1. Solar Street Lights (SSL)
2. Utility Street lights (USL).

In solar street lights the solar power can be used to glow the street light. The solar power can be saved in battery during day which can be utilised in the night. Whereas the Utility street lights are used with existing AC power from utility grid ie. Nothing but normal street light which is followed traditionally. Here, the absence of solar panels and battery.

The working methodology is same for both the systems and the only difference is that the input power supply where it come from.

The major part of the proposed work can be classified into two.

1. Power Saving Unit (PSU)
2. Security system Unit (SSU).

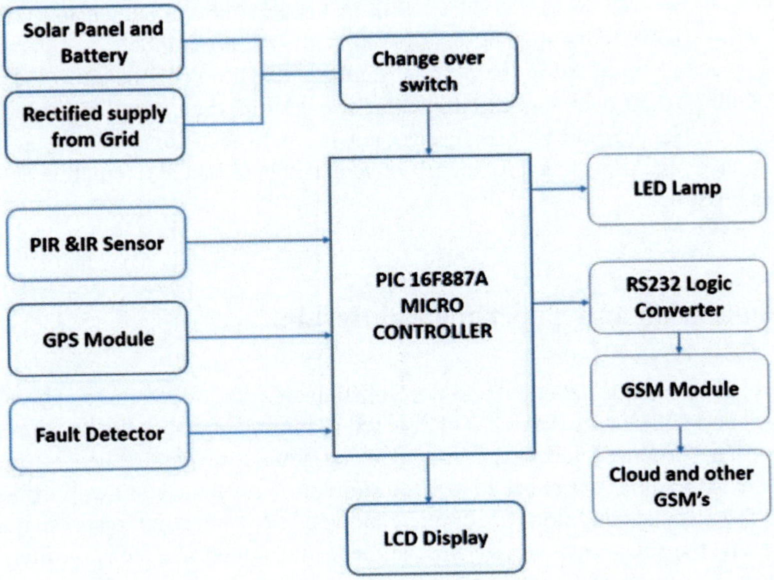

Fig. 1 Proposed methodology

2.1 *Power Saving Unit*

Passive infrared sensor and infrared sensors are used to monitor human and vehicles movements. This intern to provide the signal to control the intercity of the light. Since the intercity of the light is proposal to voltage which can be varied from minimum to maximum. Pulse Width Modulation (PWM) technique is used to control the voltage.

Conditions for PIR and IR sensors as mentioned below.

1. If PIR only detects, the intensity shall be 100%
2. If IR only detect, the intensity will be 50%
3. If both PIR and IR detects, intensity will be 80%
4. If none is detected the intensity will be 40%.

2.2 *Security System Unit*

SSU is fixed on the street light pole and the emergency service such as police, ambulance and fire service can be called during emergency situation. There are three switches for each operation. If the switch is selected, then emergency message will be sent to nearby emergency services with its accurate coordinates using GPS and GSM. Every street pole shall be addressed with unique number. To operate this service, user has to mention the mobile number. It will be connected to the cloud. If any fault or theft occurs in the particular street lamp, then it will indicate on the cloud server. Hence, if the lamp is fused or any problem occurs in the street lamp then maintenance operator can easily identify and can able to take the remedial action as soon as possible.

3 Simulation and Experimental Results

The proposed work is designed and developed using proteus version 8. The sensors are designed with logics 0 and 1 which is fed to micro controller for its appropriate actions. The proposed idea was verified using simulation before development of hardware. The logics are checked and the simulation results are shown in Figs. 2, 3 and 4. The various conditions are applied and verified as shown in respective figure. For alerting the emergency service a predefined value is stored in the controller, when emergency button was pressed, its location is displayed using the virtual terminal. The program is done with the help of MPLab IDE. The predefined values stored for the emergency service are

ambulance—108.
police—100.
fire service—101.

Fig. 6 Fault occurring in the lamp will be indicating in the cloud

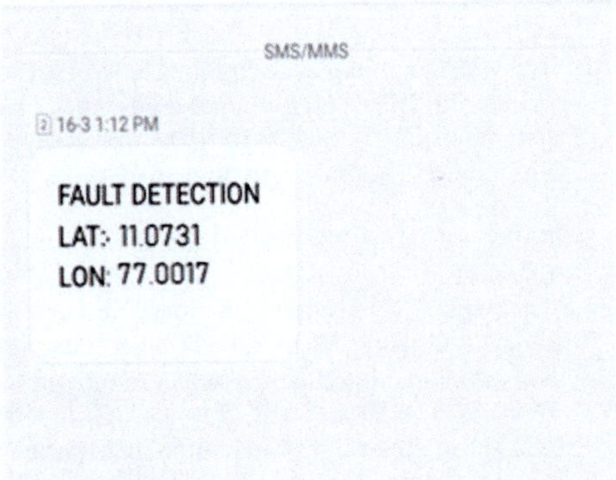

Fig. 7 SMS sent to emergency service with its coordinates

4 Conclusion

The proposed work addresses the solution for energy saving solutions for the street light. This method saves the energy around 40% in comparison with the existing LED lighting system and the almost about 70% with HPSV. Also, the security such as theft, accident and fire issues are addressed. The message sent to corresponding services with accurate coordinates which help the victim to rescue without technical delay. Battery theft has been solved in proposed street lighting system.

Fig. 10 Prototype acknowledgment

References

1. Yue W, Changhong S, Xianghong Z, Wei Y (2010) Design of new intelligent street light control system. In: IEEE international conference on control automation
2. Jettanasen C, Yoomak S (2010) Comparative study of lighting quality and power quality for LED and HPS luminaires in a roadway lighting system. In: IEEE region 10 conference, pp 2173–2178
3. Kaleem Z, Yoon TM (2010) Design and fabrication of automatic street light control system. Eng e-Trans 5(1):27–34
4. Chakrapani DNB, Naveen Kumar V, Padma Sai Y, Archana Bhange K (2016) Street light data acquisition using VLSI based object recognition. In: 2016 IEEE international conference on computational intelligence and computing research (ICCIC), pp 1–4
5. Devi DA, Kumar A (2012, Nov) Design and implementation of CPLD based solar power saving system for street lights and automatic traffic controller. Int J Sci Res Publ 2(11)
6. Chun C, Hung C, Kun L, Kuan C, Chun Y (2014) Energy efficient street light supervising and control system. In: Proceeding of FIT
7. Elejoste et al (2013) An easy to deploy street light control system based on wireless communication and LED technology. Sensors (Basel) 13(5):6492–6523
8. Chushan L, Jiande W, Xiangning H (2010) Realization of a general LED lighting system based on a novel Power Line Communication technology. In: Proceeding of IEEE applied power electronics conference and exposition, pp 2300–2304
9. Ozcelebi C, Lukkien J. Exploring user-centered intelligent road lighting design: a road map and future research directions. IEEE Trans Consum Electron 57:788–793
10. Siddiqui AW, Ahmad HK, Yang, Lee C (2012) ZigBee based energy efficient outdoor lighting control system. In: Proceeding of ICACT, pp 916–919
11. Wu H, Tang M, Huang G (2010) Design of multifunctional street light control system based on AT89S52 single-chip microcomputer. In: IEEE 2nd international conferences on industrial mechatronics and automation (ICIMA)
12. Mohamaddoust R, Haghighat AT, Motahari Sharif MJ, Capanni N (2011) A novel design of an automatic lighting control system for a wireless sensor network with increased sensor lifetime and reduced sensor numbers. Sensors 11(9):8933–8989

Cyber Physical Security of the Critical Information Infrastructure

Using Behaviour Analysis

Dharmendra Kumar, Himanshu Nayyar, Darshana Pandey, and Aamir Hussian Khan

Abstract Due to thrust in adoption of smart grid technologies and practices, the normally isolated grid is now being turned into a more dynamic, digital and real time system requiring integration of various Information Technology (IT) and Operation Technology (OT) technologies. Although integration shall ensure improvement in operational efficiency but at the same time shall also bring a unique challenge in the form of cyber security. The cyber physical components of IT and OT networks are now being exposed to public network which poses a major risk to critical infrastructure. Any failure to address cyber vulnerability in any of the component of critical infrastructure may create havoc in the entire network resulting in unavailability of system and services. Due to its large scale and varied nature of deployment, it is therefore imperative for the utility to develop practices to address such vulnerabilities in an innovative, cost effective and proactive manner. In order to develop a holistic approach of safeguarding the different components of critical infrastructure, **a study is required to be conducted on various known and possible security vulnerabilities, sources of infection, attack vectors, ports exploited and frequency to understand the behavior pattern**. An in-depth analysis of these behavioral patterns shall give utility a great deal of insight on what approach is required to be adopted to select solutions and practices that are best fit to protect the critical infrastructure from any potential and unknown cyber-attacks.

Keywords CII · SCADA · OT · IT · Physical security

D. Kumar · H. Nayyar (✉) · D. Pandey · A. Hussian Khan
Cyber Security and Quality, Tata Power-DDL, New Delhi, India
e-mail: himanshu.nayyar@tatapower-ddl.com

D. Kumar
e-mail: dharmendra.kumar@tatapower-ddl.com

D. Pandey
e-mail: darshana.pandey@tatapower-ddl.com

A. Hussian Khan
e-mail: aamir.hussain@tatapower-ddl.com

R. K. Pillai et al. (eds.), *ISUW 2019*, Lecture Notes in Electrical Engineering 764,
https://doi.org/10.1007/978-981-16-1299-2_26

Abbreviations

NCIIPC	National Critical Information Infrastructure Protection Centre
ADMS	Advance Distribution Management system
RTU/FRTU	Feeder/Remote Terminal Unit
PLC	Programmable Logic Controller
SCADA	Supervisory Control And Data Acquisition
CII	Critical Information Infrastructure

1 Introduction

CPS of critical infrastructure comprise of secure digitized and analog components engineered for functioning in an integrated manner (eg. ADMS, Smart Meters, SCADA). These solutions are required to be exposed to networks or environments which have normally functioned in an isolated environment with minimal cyber security controls in place. The security issue demands for a risk assessment of CPS and development of a framework to address the vulnerabilities and threats on the CII.

Cyber Physical Systems must often meet strict timing requirements during normal operation as well as during recovery. Cyber Physical security is crucial to maintain stable and reliable operation during the contingency situation in-case of failure of any critical system component.

Cyber attacks on CII may lead to operating failures and synchronization issues, damaging critical physical system components which may interrupt the services and make the system unstable resulting in debilitating impact on national economy, public health and safety. To ensure CII operates in a safe, secure and reliable manner, cyber security is required to be implemented at each layer of cyber physical systems and hence it demands for in depth analysis of data transmission and storage procedures.

2 Challenges

Cyber Security industry has developed various cyber security solutions for Information Technology network which are based on the latest threat model and security techniques. The solutions are now being developed keeping Artificial Intelligence and machine learning based concepts in mind which shall help in anomaly detection in a proactive manner.

The development of security solutions on the ICS/OT systems has been slow. Hence exposing the network to advance cyber attacks. Subsequently, CII face some of the below mentioned critical challenges in order to protect its infrastructure.

- Limited skilled man power
- Lack of awareness on cyber security
- Limited information on applicable threats to Critical Infrastructure
- Technological solutions are still evolving
- Lack of funds

3 Methodology to Address Challenges

The proactive mechanism to defend critical infrastructure is to do an analysis of various known vulnerability, advisories, and exploits related to the Organisations threat landscape.

A. *Port Based Analysis (Total—3359)* (Fig. 1)
B. *Vulnerability Type Analysis (Total—4959)* (Fig. 2)
C. *Domain Based Analysis (Total—3321)* (Fig. 3)
D. *Application Based Analysis (Total—50)* (Fig. 4)
E. *Location Based Analysis (Total—112)* (Fig. 5)

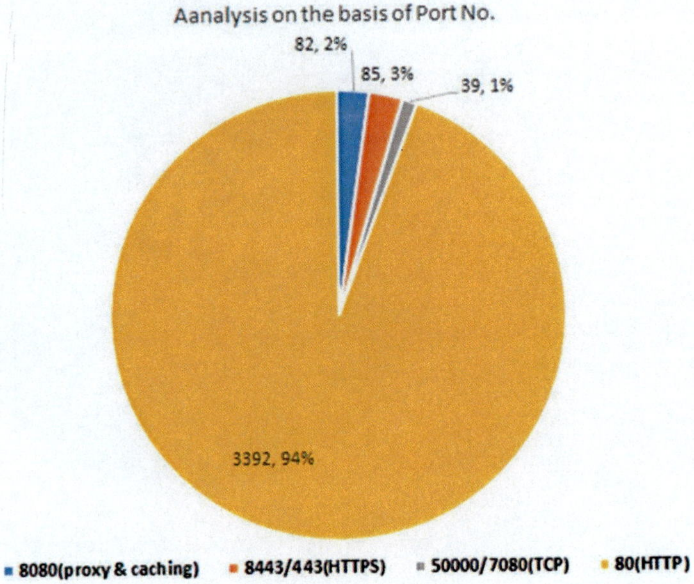

Fig. 1 Analysis of most exploited port numbers

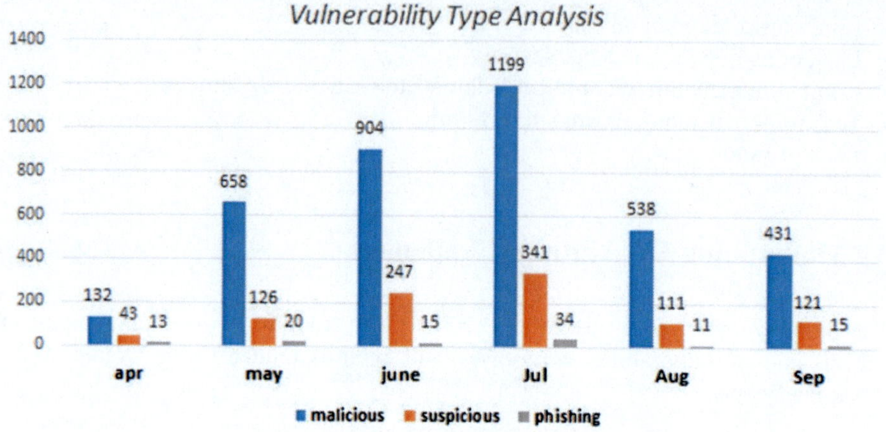

Fig. 2 Analysis of type of vulnerabilities

Fig. 3 Analysis of most targeted domain

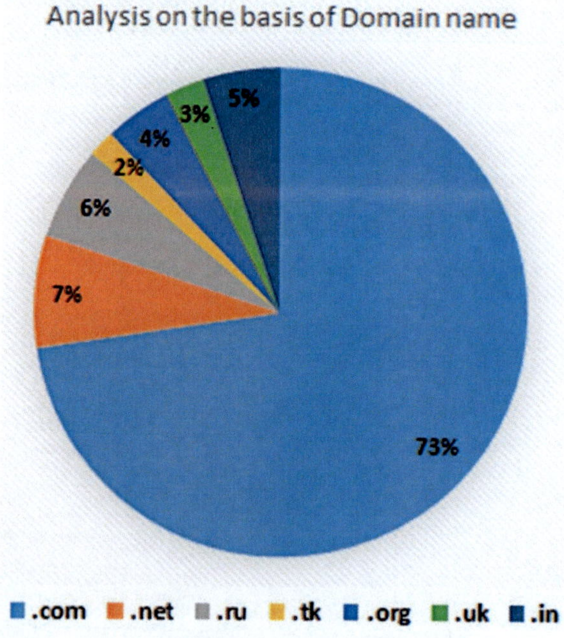

4 Conclusion

Post carrying out an extensive analysis of the vulnerabilities based on various factors, it is suggested that following approach may be adopted by an organization to develop a safe and secure network for cyber physical system of critical infrastructure.

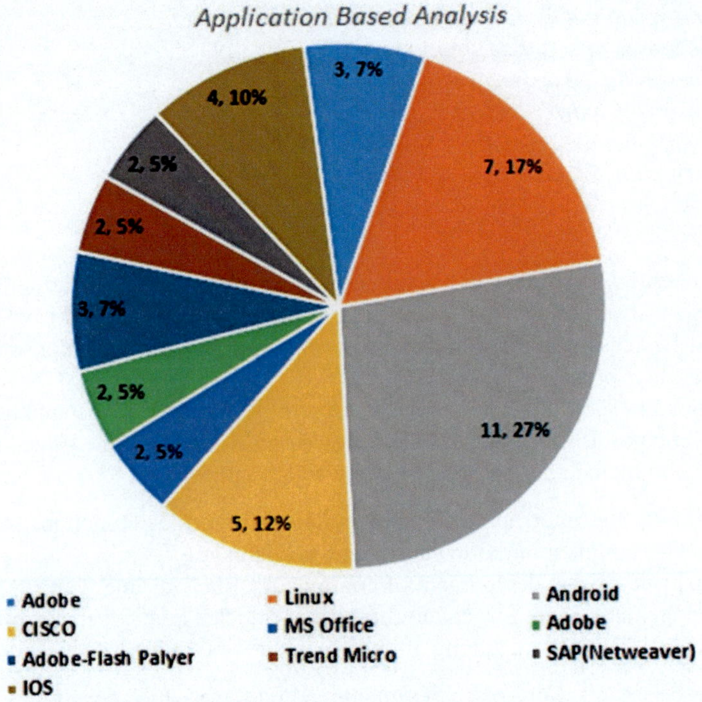

Fig. 4 Analysis of most targeted applications

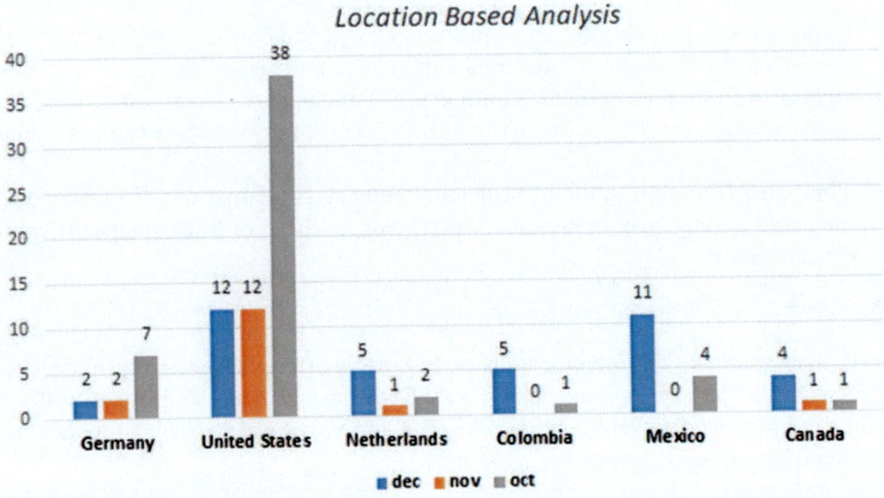

Fig. 5 Analysis of location of attack trigger

A. *Identification of Critical Information Infrastructure*
B. *Risk Assessment and Mitigation*
C. *Vulnerability Assessment and Penetration Testing*
D. *Industrial Control Firewalls*
E. *Network Behavior Analysis Solution*
F. *Information Security Incident Management Process*

A. *Identification of Critical Information Infrastructure*

CII are the critical resource, and destruction of which have a debilitating impact on national security, public safety. The logical approach to identify CII is on the basis of critical business processes and the information infrastructure supporting these critical business processes.

Stakeholders should ensure to protect critical infrastructure by reducing the risk of IT/OT outages as much as possible followed by recovery of the same. The other purposes may include but not limited to the following:

(a) To lower the interruption or manipulations of critical functions/services in critical systems in order to cause minimal damage.
(b) Department wise deployment of contingency plans in line with Cyber Crisis Management Plan for countering cyber-attacks and cyber terrorism, and ensuring compliance among all the organizational units

As part of organizations risk assessment methodology which constitutes a holistic view of all risks to national/organizational security, CII identification is a part of the same.

- **Functionality** includes the set of procedures and capabilities associated. It may be viewed at two levels—Functional Uniqueness and Dependency.
- **Degree** of Complementarities is unique in CII and Failure of one system can lead to potentially shut down other Critical Information Infrastructure in a cascading effect.
- **Time and Duration** in the identification and categorization of CII is crucial as the same system may or may not be critical at different times/under different circumstances.

B. *Risk Assessment and Mitigation*

Risk assessment is the identification of hazards that could negatively impact an organization's ability to conduct business. These assessments help identify these inherent business risks and provide measures, processes and controls to reduce the impact of these risks to business operations.

Companies can develop a risk assessment framework to prioritize and share the details of the assessment, including any risks to their critical infrastructure.

Risk assessment steps:

How a risk assessment is conducted varies widely depending on the risks unique to the type of business, the industry that business is in and the compliance rules applied to that given business or industry. However, there are five general steps that utilities can follow.

Step 1 Risk assessment is to identify any potential hazards that would negatively influence the organization's ability to conduct business.

Step 2 Determine the business assets that would negatively influenced if the risk came to function. IT systems, business operations and company reputation should be taken into consideration.

Step 3 Evaluate the risks and develop control measures.

Step 4 The risk assessment findings should be recorded.

Step 5 It is important for companies to update their risk assessments mechanism regularly to adapt the changes with potential hazards, risks and controls.

Quantitative versus qualitative.

Risk assessments can be quantitative or qualitative. In a quantitative risk assessment, numerical values is assigned to the probability an event occurrence and using the result we calculate the risk factor.

Qualitative risk assessments does not involve numerical probabilities and is based on severity level wise risks categorization.

The goal of risk assessments.

Similar to risk assessment steps, the specific goals of risk assessments will likely vary based on industry, business type and relevant compliance rules. An information security risk assessment, for example, should identify gaps in the organization's IT security architecture, as well as review compliance with infosec-specific laws, mandates and regulations.

Some common goals and objectives for conducting risk assessments across industries and business types include the following:

- Developing a risk profile that provides a quantitative analysis of the types of threats the organization faces.
- Developing an accurate inventory of IT assets and data assets.
- Justifying the cost of security countermeasures to mitigate risks and vulnerabilities.
- Understanding the return on investment, if funds are invested in infrastructure or other business assets to offset potential risk.

C. *Vulnerability Assessment and Penetration Testing*

It is recommended that organization may opt for pan independent and third party based Vulnerability Assessment and Penetration Testing activity which shall give organization a clear view of current state of affairs in terms of security.

Vulnerability assessment process that is intended to identify threats and the risks they pose typically involves the use of automated testing tools, such as network security scanners, whose results are listed in a vulnerability assessment report.

It is atmost necessary for enterprises to identify and remediate weaknesses before they can be exploited. A comprehensive vulnerability assessment along with a management program can help companies improve the security of their systems.

Penetration testing involves identifying vulnerabilities in a network, and it attempts to exploit them to attack the system. The primary aim is to check whether a vulnerability really exists and to prove that exploiting it can damage the application or network.

D. *Industrial Control Firewalls*

Firewalls blocks threats and control the flow of communications by filtering the packets which contain malicious traffic.

ICS/OT technology include switches, routers, and firewalls along with features like Access Control Lists, Stateful Inspection and Deep Packet Inspection, each of which filters traffic differently.

1. The internet protocol address of the device that send the messages.
2. The IP address of the device receiving the message
3. Priority information

E. *Network Behavior Analysis Solution*

NBA is required to ensure service uptime, data integrity, regulatory compliance and public safety for the CII.

1. Network Segmentation

The most prevalent weakness on ICS networks is the weak or missing boundaries between ICS and enterprise networks. We will provide Boundary protection between the networks and even micro segmentation among the micro-grids.

2. Automated Asset Discovery and Management

- We can Discovers level 2 control devices: operator stations, engineering workstations, and servers (Windows/Linux-based),
- Discovers level 1 control devices: PLCs, RTUs, DCS controllers,
- Discovers level 0 devices (I/Os),
- Discovers non-communicating assets,

Provides detailed information on asset type, specific models, OS and firmware versions, and more (for level 1 and level 2),

Provides interactive asset map displaying assets, communication patterns, protocols used, and conversations.

3. Continuous Network Activity Monitoring, Anomaly, and Threat Detection

- Detects threats and anomalies by monitoring device communications and protocols (both external and internal)
- Out of the box security policies for threat and anomaly detection

- User-friendly granular policy customization engine for threat and anomaly detection,
- OT data-plane and control-plane engineering protocols coverage.

4. Controller Integrity Validation

- Identify changes to controllers made over the network, including configuration changes, code changes, and firmware downloads.
- Identify changes made to controllers by physically connecting to the devices (via serial cable or USB device)

5. Vulnerability Assessment and Risk Management

- Risk score by device
- Vulnerability assessment for all control devices
- Real-time alerts on suspicious activities and threats detected in ICS networks
- Full audit trail of ICS activities
- Historical controller information to support backup and recovery

All the above mentioned when integrated together with the Industrial Control Firewalls data flowing through the micro-grids will be reaching network behavior analysis solution & any threat vector can be identified.

F. *Information Security Incident Management Process*

The purpose of this document is to establish and maintain a documented procedure for mitigating the, risks dealing with information security incidents. It describes the various security incidents that can occur, reporting mechanism for security incidents, handling and analysis of these incidents.

Incident management is a process to identify, manage, record and analyze threats to landscape in real-time and gives a comprehensive view of security issues within the infrastructure. Incident is defined as an active threat to attempt intrusion, violation of policy and unauthorized access to data.

The 4 step process for incident management system are:

1. Identification of security incidents through 24X7 logging and monitoring.
2. Detect identified incidents and take appropriate steps for mitigation.
3. Respond to the incident by containing it and through proper investigation
4. Resolve: Learn key takeaways from every incident

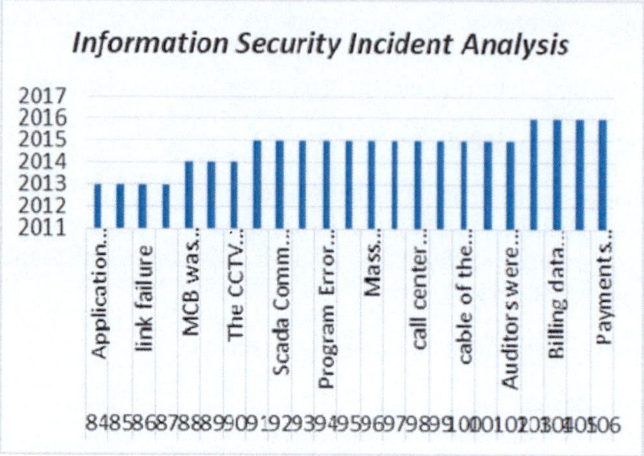

The incidents will be categorized on basis of severity:

1. **High Severity**—An incident would be categorized as high if it has a high impact on the organization's business impacting few critical systems. For example, **malicious code attacks, denial of service attacks, unauthorized access, password cracking attempts, misuse of systems would be high severity incidents**.
2. **Medium Severity**—If any business process (financial or operational) affected and have considerable **impact on availability of one/two systems**.
3. **Low Severity**—An incident would be categorized as low if it has a **less impact on the organization's business or services**.
4. **Very Low Severity**—An incident would be categorized as low if it has a nearly **no impact on the organization's business**.

A strong security incident management process is imperative for reducing recovery costs, potential liabilities, and damage to the victim organization.

5 Way Forward

The future work shall require carrying out demonstration of most common attack types on Critical Infrastructure systems and studying the various cyber-attack techniques carried out on cyber physical systems so far. The outcome of the demonstration shall help in developing the system hardening best practices. The outcome of the demonstration shall also provide great insights on how hackers can develop ways to exploit the critical infrastructure of the organization. In addition to this studying the various attack techniques, shall provide critical infrastructure organization to focus on their vulnerable areas and align their cyber security strategy towards closure of the same.

References

https://www.nciipc.gov.in/
https://www.cert-in.org.in/
https://www.us-cert.gov/

Roll-Out Strategy for Smooth Transition of Traditional Meters to Smart Meters

Dharmendra Kumar and Aamir Hussain Khan

Abstract Power Industry is going through a paradigm shift of replacing its traditional electric meters with a real time and two-way communicating smart meters. Smart Meter roll-out is the first milestone in traversing the journey of Smart Grid. This transition shall lead to major behavioral change of consumers and utility professionals because smart meters will be more than a mere electricity monitoring device. Smart meters will empower consumers by monitoring their consumption pattern, carrying out on-demand read and enable them to use power resources efficiently. To carry out smooth transition, utilities are expected to prepare themselves and streamline their routine practices by strengthening processes to manage highly diversified nature of smart metering eco system.

Keywords Smart meter · Roll-out strategy · Change management · Smart grid

1 Introduction

As a pre-requisite, every organization is expected to develop processes and practices or mature its already existing processes and practices to bring in required operational efficiency to manage the highly demanding nature of smart metering ecosystem. The success of smart meter roll out shall depend upon the availability of required processes with necessary efficiency and proper controls in place.

One of the best ways to develop a roll-out strategy is to deploy a process based maturity model for the key processes which shall not only improve current processes but also bring in the required maturity for achieving the operational efficiency. As a pre-roll out strategy, this model shall evaluate the current maturity status and ensure

D. Kumar (✉)
IT Infra DC, Data Comm. & AD, Tata Power Delhi Distribution Limited, Delhi, India
e-mail: dharmedra.kumar@tatapower-ddl.com

A. H. Khan (✉)
IT Infra DC-Cyber Security & Quality, Tata Power Delhi Distribution Limited, Delhi, India
e-mail: aamir.hussain@tatapower-ddl.com

R. K. Pillai et al. (eds.), *ISUW 2019*, Lecture Notes in Electrical Engineering 764,
https://doi.org/10.1007/978-981-16-1299-2_27

synergy amongst the identified stakeholders. Some of the key points to develop a resilient roll-out strategy are:

- Good planning is the key to a successful roll-out. Planning includes the requisite resources, expected process performance, responsibilities and necessary authorizations explicitly recited in the roll-out strategy to avoid the potential bottlenecks.
- Imparting appropriate trainings to the identified stakeholders, enhance their skills and enable them to work efficiently with the implied set of qualities.

2 Objective

The smart meter implementation will bring about extensive changes in to energy industry. In order to ensure successful implementation within the stipulated schedule with the defined quality aspects, a holistic roll-out strategy should be prepared.

The prime objective to prepare a roll-out strategy is to decide the course of actions in line with the regulatory framework and envisage challenges to manage them at the initial level. A roll-out strategy acts as a baseline to evaluate the progress of the project. The strategy also ascertains the authorities and responsibilities of identified stakeholders because stakeholders play a vital role in the implementation.

3 Challenges

- **Supplier Management**: The supplier management is one of the key areas that addresses the acquisition of products, services, and components which can finally be delivered to the customer or included in a product or service system. It also addresses the arrangements of processes and policies to ensure on time availability of necessary products and services for a smooth roll out of millions of smart meters. This area's practices can also be used for other purposes that benefit the project. Typically, the products to be acquired by the project are determined during the early stages of planning and development which is the reason why the department managing the suppliers should align its processes and policies in accordance with the project timelines.
- **Change Management**: Major aspect of managing change is to manage the expectations of the stakeholders. Resistance from consumers is another biggest challenge for the DISCOMs. Although companies are coming up with various incentive schemes for the consumers but managing consumer behavior on a long term basis is a real challenge which DISCOMs has to gone through.
  In addition to the consumers, resistance from employees towards the change in the working approach has to be managed.
- **Return on Investment (ROI)**: Financial model is a challenge because technology is expensive in its early stage and cost is a scarce resource for any organization.

Electricity distribution is a regulated business and to choose between the lowest cost implementer and best quality implementer could be a challenge for the organization. After the selection the OEM/ implementer, receiving the expected ROI within the expected time period is a real challenge.

- **Selection of Wrong Solution**: A poorly planned project may result in the worst nightmare for a utility to come out as true which is a solution not able to meet the business requirement. Therefore to address this issue, requirement analysis stage plays a pivotal role in capturing all the necessary requirements. Other factors which may also result in selection of wrong solution could include: poor testing, no stakeholder feedback and lack of business owner involvement.
- **Late Delivery**: Another challenge that a utility faces today is late delivery of final product, solution etc. as per the defined timelines. This issue may have rippling effect on the entire Project Management life cycle and hence can be the deciding factor in success or failure of the project. Factors that influence late delivery may include the following—poor/no estimating or planning, poor progress tracking, lack of critical resources, excessive overtime, too much rework, constantly changing requirements and unexpected external incidents or other disruptions.
- **Stretched Resources**: Resource Management is one of basis steps of project management and an experienced project manager is expected to devote significant amount of time in ensuring proper resource management. Some of the reasons why resource management may become a challenge include the following reasons- poor/no estimating or planning, excessive overtime, wrong resources and acceptance of too many changes without understanding impact (Fig. 1).
- **Task Prioritization**: One of the other important issues that a project might face is poor prioritization of tasks. Lack of focus on what is important to the business or performance and channeling all the energy on all the tasks can also create lot of issues, finally giving an impression that things are not moving in the right direction.
- **Never Ending**: When involved in large scale projects, few of the tasks become never ending which may lead to poor morale of team members. Reasons attributing to this issue are incorrect scoping, incorrect estimation methods, failure to revise plans and schedules based on changing customer demands, inadequate resources/incorrect resource estimation or planning and lack of obtaining commitments from relevant stakeholders.

4 Methodology

One of the recommended methodology to develop a roll-out strategy is to deploy a process based maturity model for the key processes which shall not only improve current processes but also bring in the required maturity for achieving the operational efficiency. As a pre-roll out strategy, this model shall evaluate the current maturity

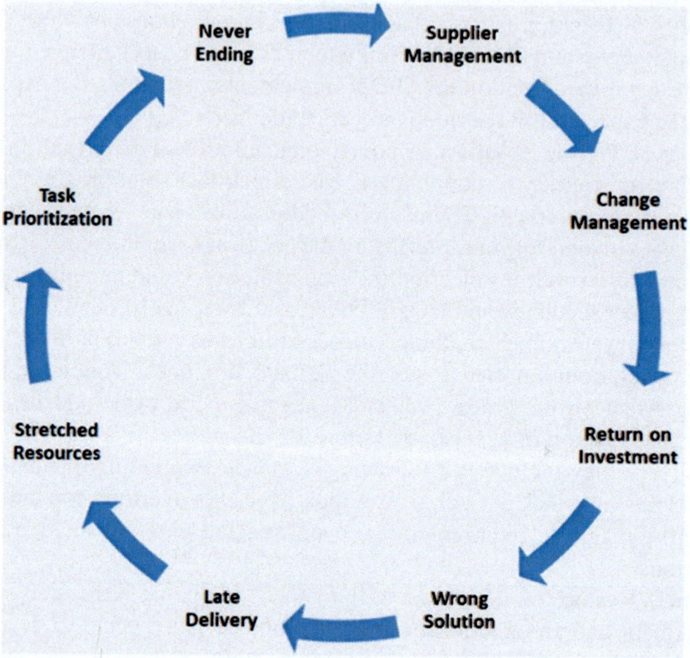

Fig. 1 Challenges in smart meter roll out

status and ensure synergy amongst the identified stakeholders. Some of the key points to develop a resilient roll-out strategy are:

- **Identification of Processes**: Identification of Processes which have direct impact on project deliverables is key to ensuring a successful roll out. All such processes which will be part of ecosystem should be analyzed thoroughly to ensure success parameters like input, output, ownership, checkpoints etc. are in place.
- **Gap Analysis**: A survey on gap analysis is required to be carried out on the identified processes to ensure whether processes are designed to meet the expected outcomes and also to create new processes if some are not available. This analysis does more than just identifying gaps in the processes being used. It also involves determining if the processes are utilized, persistent, and habitual. A well-crafted business process is of little value if it is not used.
- **Planning**: Planning is a continuous process. Good planning is the key to a successful roll-out. Planning includes the requisite resources, expected process performance, responsibilities and necessary authorizations explicitly recited in the roll-out strategy to avoid the potential bottlenecks.
- **Training**: Imparting appropriate trainings to the identified stakeholders, enhance their skills and enable them to work efficiently with the implied set of qualities.
- **Risk Management**: The first step to mitigate any risk is to identify it. The identified risks can be:

- transferred to insurance companies
- treated by taking appropriate actions to reduce either its impact or the likelihood
- tolerated by taking no action if either the mitigation action is not cost effective or the impact of risk is very low
- terminated by ignoring the activity involved substantial risk value

Risks can be identified based on the context or the assets involved in the implementation (Fig. 2).

- **Monitor and control**: Monitoring is continual checking, supervising, critically observing or determining the status in order to identify change from the performance level required or expected. Monitoring and controlling of the activities against the plan to ensure successful implementation is a crucial task for the successful and timely completion of the project. Activities and process can be monitored by regular reviews. All the processes are expected to establish required controls to ensure the process is effective and efficient. Periodic audit is one of the best way to ensure the compliance of the established processes and practices. In case of deviation, adequate corrective actions are taken.
- **Decision Analysis**: Various decisions have to be made during the implementation of such a colossal project such as vender evaluation, technology evaluation,

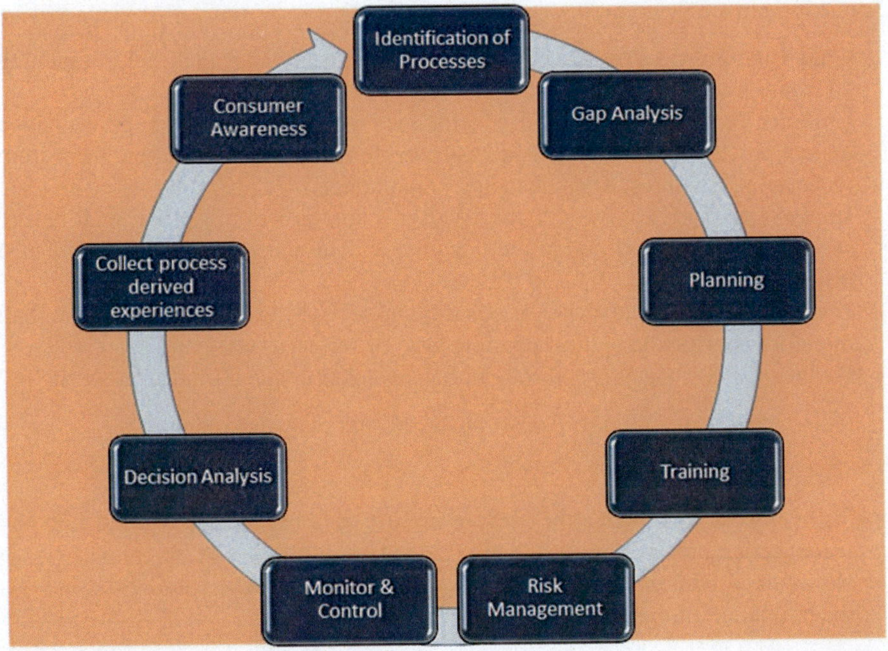

Fig. 2 Methodology to address challenges

etc. A criteria should be established and maintained for evaluating the available alternatives. Evaluation of effectiveness of the selected alternative must be ensured.

- **Collect process derived experiences**: While implementing the high maturity processes, the implementer must collect the best practices, challenges and lessons learnt for future analysis and improve the processes. Process derived experiences played a significant role in the continual improvement of the established procedures and practices.
- **Consumer awareness**: Smart meter rollout cannot be successful without the empowerment of consumers and awareness is the only way to empowerment. An aware consumer is the empowered consumer. Awareness shall also help to address the challenge of resistance from consumers. The avenues of consumer awareness are: benefits can be printed on bills, published on websites, discussed in RWA meetings, etc. Hence, plan for consumer awareness strategy holds an important place in roll-out strategy.

5 Expected Benefits

Successful smart meter roll out leads to the immense benefits to all the stakeholders. Some of them are listed below.

Gain Customer Trust: Exceeding customer expectations shall help in gaining customer trust.
Enhance Brand Reputation: Reducing the impact of service disruptions with a mature process to identify and address potential incidents and prevent them from reoccurring, shall result in enhancing brand image.
Improve Service Delivery Time: With a robust process in place shall ensure services are delivered quickly, efficiently, and in accordance with established timelines.
Increase Quality: Due to the various steps involved in delivering a service, it shall therefore provide the highest possible level of service quality and availability.
Reduce Cost: Lower costs through proactive planning and limited re-work.

Other benefits of successful smart meter roll-out may also include:

Utility:

- Smart meters will reduce the manpower cost of bill readers, disconnection and reconnection orders.
- No. of complaint calls will get drastically reduced because DISCOMs shall get the intimation of meter failure proactively.

Customers:

- Consumer can monitor their own consumption pattern so that they can change their electricity consumption to reduce electricity bills.
- Consumers can check their real-time bill by using ABD (Anytime Bill Display) option in smart meters.
- Moreover, the industrial consumers can analyze their productivity pattern based on the pattern of consumption.

6 Conclusion

Smart meter rollout is a major transformation, DISCOMs are going through. DISCOMs have to prepare a strategy not just for the smart meters implementation but also for its maintenance and sustenance. Managing post implementation challenges of smart meters should be the major focus of the implementation strategy.

Having a process based maturity model in place shall ensure consistently delivering services more closely aligned with the organizational requirements or business objectives (Fig. 3).

It shall help build confidence among workforce, synergizes efforts, and brings organization to a level of minimal defects. In addition to this, it also helps in transforming internal work culture from operating in silos to functioning as a unified, holistic arm aligned to a single common vision.

Fig. 3 Adoption of a model to achieve business goals

Appendix

ABD	Anytime Bill Display
DISCOM	Distribution Company
OEM	Original Equipment Manufacturer
RWA	Resident & Welfare Association
ROI	Return on Investment

Reference

https://www.cmmiinstitute.com

Pan Asia Grid Interconnection Project (PAGIP)

S. K. Ray and Shuvam Sarkar Roy

Abstract Globally, the experiences have proven that interconnection of smaller power systems to form a large power pool or regional grid is beneficial to all participants in terms of efficiency, economy, reliability and resilience. In this paper, an attempt was made to list the numerous benefits that exist for taking in interconnecting regions, in general, and in particular the GCC, SAARC/BIMSTEC and the ASEAN region. Further, an attempt has been made to identify all activities and tasks to be completed to ascertain the feasibility of interconnecting the regions. It does appear from the data and facts analysed that such an interconnection between these three regions will be very beneficial to all the Nations in the Pan Asia Grid. The narration has been kept simple for easy understanding by all.

Keywords Interconnection of grids · Regional grids · Cross border electricity trading · Interconnection framework · Grid integration

1 Background

Post-independence, India developed state-wide power systems in 1950s and 1960s. Later, India started interconnecting the state grids with the adjacent states for sharing surplus power or importing to meet shortages. These interconnections were initially operated in radial mode which were not very efficient. Following the model of power pools in Europe and America, India also setup five regional grids in the 1980s. Later, these regional grids were interconnected to form a national grid. Today, India operates one of the largest power systems in the world covering 3 million km^2 with connected generation resources aggregating over 350 GW and about 270 million electricity customers serving a billion plus population. After the interconnection of all the regions and creation of a national grid, there have been no grid failures in the past

S. K. Ray · S. S. Roy (✉)
ISGF, New Delhi, India
e-mail: shuvam@indiasmartgrid.org

S. K. Ray
e-mail: skray@indiasmartgrid.org

R. K. Pillai et al. (eds.), *ISUW 2019*, Lecture Notes in Electrical Engineering 764,
https://doi.org/10.1007/978-981-16-1299-2_28

6 years. This experience underscores the rationale for larger interconnected power systems. India also has operating power exchanges, and the uniform market price is discovered for the entire country on the power exchanges. This can be interpreted to mean that the national resources are being utilized optimally and demand is met in an equitable manner because of the national grid.

In general, owing to peak and non-peak hours in every power system, the additional capacity to be reserved for peak hour demand can be reduced considerably in an interconnected system as that capacity can be drawn from another area where peak maybe experienced at different hours. If a power plant or transmission line in a power system fails, the capacity can be provided instantly from the adjacent power systems if they are interconnected. Similarly, in case of disasters, both natural and man-made, the chances of grid failure are much less in a large power pool.

2 Regional Grids in South Asia, Southeast Asia and Middle East

Two decades ago, the United States Agency for International Development (USAID) mooted the idea of interconnection of electricity grids in South Asia and Southeast Asia and supported the feasibility studies for interconnections and prepared master-plans for SAARC Grid and ASEAN Grid. In South Asia, the Indian grid is already connected to the grids of Nepal, Bhutan and Bangladesh. Interconnection with Sri Lanka is also proposed through undersea cables, which is likely to happen in the near future. These interconnections have benefitted all the nations in the region. Bhutan can sell its clean hydropower; Bangladesh has been able to import a part of their electricity demand from India. Similarly, there is a master plan for interconnection of the grids of ten ASEAN countries in the Southeast Asia and many interconnection lines have already been built. These regional interconnections have led to economic development and contributed to improving access to electricity for millions of people in these regions.

Almost during the same time, the countries in the Arabian Gulf region made the master plan for a common grid for the Gulf Cooperation Countries (GCC). An interconnector was built connecting the Kuwait, Saudi Arabia, Bahrain, Qatar, UAE and Oman. This interconnector has been in operation for the last ten years. Gulf Cooperation Countries Interconnection Authority (GCCIA) is operating and maintaining these interconnector systems and also running a power market. Before establishing the interconnector, the cost–benefit analysis of the interconnector was very thoroughly analysed. It was found that the interconnector will provide enough benefits in terms of sharing of generation capacity and operating reserves to justify its cost. In addition to this, the countries have also benefited through the power trading taking place through these interconnections. Thus, one can see that the technical benefits of the Gulf interconnector is enough to justify the investments. One important feature of the GCC region is that the Saudi grid operates at 60 Hz and the rest of the nations, operate at 50 Hz. Interconnecting these regions hence required an HVDC station. The HVDC station helps in frequency and voltage control by automatically responding to changes in these parameters on either side of this region. Thus, one can infer that difference in the operating frequency need not be a hurdle for building interconnections. The Bahrain grid is connected to the interconnector by a sub-sea cable. This has performed satisfactorily after initial teething problems.

ASEAN established the electricity interconnecting arrangements within the region through the ASEAN Power Grid (APG) under the ASEAN Vision 2020 adopted in the Second ASEAN Informal Summit in Kuala Lumpur on 15 December 1997. HAPUA (Heads of ASEAN Power Utilities/Authorities), as SEB (Specialised Energy Body), is tasked to ensure regional energy security by promoting the efficient utilisation and sharing of resources. The construction of the APG is first done on cross-border bilateral terms, then expanded to a sub-regional basis and finally to a total integrated regional system. It is expected to enhance electricity trade across borders which would provide benefits to meet the rising electricity demand and improve access to energy services in the region.

3 Rationale for Interconnection of Regional Grids in ASEAN, SAARC and GCC Nations

The relevance of interconnection of regional grids has gained attention in the era of increasing share of variable renewable energy (VRE) resources on the grids. Integration of intermittent VRE, especially solar and wind, is efficiently handled in a larger balancing area that offers better forecasting of VRE generation and opportunities for intra-day and hour-ahead mitigation measures for managing the variability of VRE resources. Larger power systems could also offer large quantities of demand flexibility and dispatchable generation resources at lower cost. Interconnected grids could also offer the opportunity to replace their own costly generation by a relatively cheaper imported power. Bangladesh can be quoted as an example for such a benefit. Nepal, which has so far been a net-energy importer from India, can use the same transmission lines to export its surplus power when new plants are commissioned in Nepal. Even in power deficit and surplus scenarios, due to demand profile diversity, opportunities of mutually beneficial energy trade exist. These variations are not only due to difference in time zones, but also due to seasonal differences and also difference in weekends and festivals in each interconnected country.

The Gulf region did not have much fuel diversity at the time of establishment of the interconnector but now there is fuel diversity in GCC nations as Saudi Arabia, Oman and UAE plans to build GW scale solar plants.

Between the eastern parts of ASEAN Grid to the western end of the GCC Grid, there is a 5-h time zone difference which can be leveraged efficiently in the interconnected operations. As solar generation diminishes and evening peak starts in ASEAN region, solar generation will be at its peak in ASEAN and GCC regions. Later, when evening peak load increases in western parts of India and GCC region, the base load plants in ASEAN grid could support it. Also, the morning peak in ASEAN can be supported by based load plants in the SAARC/BIMSTEC region. This manner of interconnected operations helps not only in integration of renewables, but also efficient operation of base load plants in all the regions. The hydro generation in SAARC and ASEAN could also improve the power quality in GCC grid when there will be tens of gigawatts of solar generation in GCC nations. The interconnection could also have the potential to reduce fossil fuel consumption in these regions through sourcing the clean solar, wind and hydel power.

Figure 1 shows the likely places where the interconnectors needs to be built to interconnect the three regions. There are two options as far as achieving interconnection is concerned. A separate interconnector can be built or the national grid of a nation can be used for regional power transfer. To decide on the suitable choice, detailed analysis is required and it is felt that a mix of the two option may be the optimal choice.

For connecting GCC region to the SAARC/BIMSTEC region, there are two options as shown in Fig. 1. We can have a line from SAARC region say from Afghanistan to Kuwait connecting the intervening nations also. However, it will require connecting, Afghanistan and Pakistan to India and through India to other SAARC/BIMSTEC nations. On the other side of Afghanistan and Pakistan, if GCC region is to be connected, the interconnector will pass through Iran, Iraq and other nations before reaching Kuwait.

On the other hand, a subsea cable connecting the Gujarat coast in India in SAARC/BIMSTEC region to Oman can be considered. This has got its own challenges as the distance is about 1300 km and the depth is at least 3000 m. At the same time there is a HVDC undersea cable, 1600 m below sea level in Italy, though it is a 420 km long cable. The longest electrical subsea cable is 580 km long and connects Norway with Netherlands. The depth is much less but it is in service since 2008. The viability of

Fig. 1 Interconnection of regional grids in ASEAN, SAARC and GCC nations

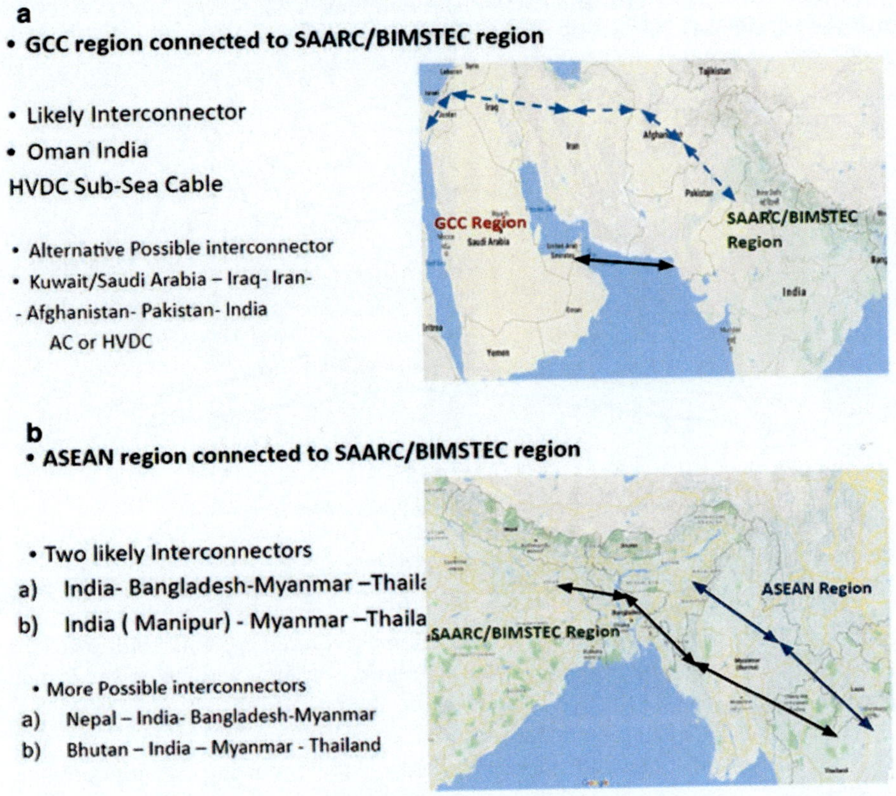

Image 1 Regional interconnections: ASEAN-GCC-SAARC

the India–Oman subsea cable can be looked into as perhaps the technology is now available.

Similarly, for connecting the ASEAN region with the SAARC/BIMSTEC region, some of the options can easily be envisaged. One option is connecting Manipur in India to Thailand via Myanmar. Another option is connecting Indian Eastern and/or North eastern grid to Thailand and beyond through Bangladesh and/or Myanmar. Image 1a and b shows the likely points of interconnection.

4 Benefits of Interconnecting GCC, SAARC/BIMSTEC and ASEAN Region

Graph 1 shows the demand curves of three regions and the combined demand curve of the interconnected region. As it can be seen, the combined Peak Demand of the interconnected region is less than the same of the individual demand curves. This

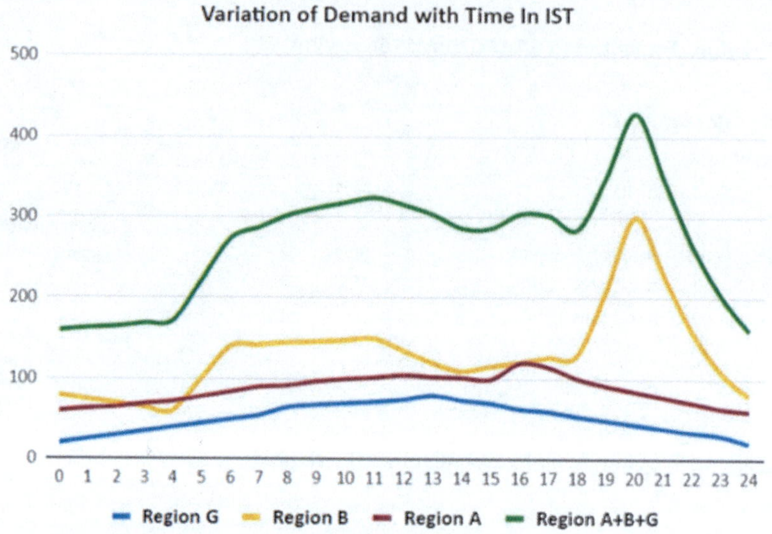

Graph 1 Individual and combined regional demand curves

reduction in the peak demand leads to reduction of required capacity addition as the installed capacity can be utilized across the regions through various power market mechanisms.

The same can also be seen in Table 1.

In this paper, the demand curve is based on the following assumptions and adjustments.

(1) The figures are assumed and do not actually represent the actual demand curve of any region or nation.
(2) The demand curves of Region A and C have been shifted to align them with a common standard time of region C. Thus, the times mentioned are not that of the individual regions but of the Region B.
(3) It is assumed that there is no grid congestion and the adequate transfer capacity, if available, for flow of electricity throughout the three regions.
(4) All system losses are assumed to be negligible.
(5) All units are in MW or GW as the case may be.

Remarks

(1) Regional figures are assumed figures and do not represent any actual region.
(2) * Reduction assumed to be proportional to average demand.

5 Benefits to the Nations of the Interconnected Regions

Certain conclusions can be drawn from the above table and the graph.

Table 1 Regional peak demand capacity

Region	Peak demand equal to required capacity (GW)	Minimum demand (GW)	Approximate average demand (GW)	Approximate reduction in installed capacity requirement due to interconnection* percentage reduction
G	80	20	50	16 = 80 – [(400/500) * 80)] 20%
B	300	60	160	60 20%
A	120	60	90	24 20%
Interconnected region G + A + B	400	160	300	100 20%
Algebraic sum of three regions	500	140	300	100 20%

[a] All figures in GW

A. The required installed capacity of the interconnected region is less than the algebraic sum of the required capacity of the individual regions without interconnection. If grid congestion is present, this benefit will be reduced but will definitely be there in any case.
B. The percentage reduction in installed capacity requirements is same for all the three regions with no transmission constraints, and losses assumed as zero. This figure may vary if some capacity addition at particular nodes is required to maintain the voltage profile or for any other system security requirement.
C. The reduction of capacity in GW terms is proportional to the capacity required in that region assuming that the capacity requirement is equal to the peak demand.
D. It can be seen that the load curve has been flattened. The peak demand of the combined system is less than and algebraic sum of the individual regions. Similarly, the Off-Peak demand of the combined system is more than the algebraic sum of the individual regions. It can be mathematically proven that the deviations of the demand are reduced in case of the combined system.

6 Economic Benefits

A. The cost of Interconnecting the regions is much less than the cost of addition of the required capacity, both yielding the same ultimate result. The money can then be utilized for other developmental activities.

B. In the figures and the table across, one can see that the generation deficient condition existing in region A has been removed by this interconnection. Under deficit generation situations, load shedding is resorted to which adversely impacts the economy. This allow the industries to grow and also other sectors like agricultural, commercial and domestic are impacted in a positive manner. All this leads to increase in GDP, which benefits all the inhabitants of that region.
C. The surplus generation in the Regions G and B can be sold to the Region A, which was power deficit. Thus, the idle generation capacity is utilized and it also leads to economic gain to the seller region, with a positive impact on their economy due to additional earning from power export.
D. A vibrant power market can be established in the interconnected region. A properly designed power market, can lead to the maximization of the benefits of Interconnecting the three regions.
E. One aspect of economic gain which is not immediately apparent from the above, is the gain due to replacement of costlier power by cheaper power. Due to variations inherent in the load curves, the generation capacity on bar and the generation level keeps on changing. The difference in the load curves gives an opportunity of replacing costlier generation by imported cheaper generation, leading to economic benefits to both the regions. This leads to the generation of producer and consumer surpluses which is an indicator of economic benefit.
F. Conservation of natural resources like coal, oil and gas as the generation from these sources at partly replaced by generation from renewable sources.
G. The interconnected grid, if utilized with proper planning and coordination can also reduce the frequent shut down and restart of generating stations. This will improve the heat rate, auxiliary consumption and also extend the life of the generating units.

7 Technical Benefits

A. The impact of tripping of a large generating unit on the power system is minimised. For example, if there is a 500 MW unit connected to a 5000 MW grid, it's outage will lead to a loss of 10% of the generating capacity. However, the same is a part of larger system say 20,000 MW, its outage will lead to a loss of only 2.5% of the generating capacity. Thus, the impact of the power system will accordingly be less.
B. The above logic is also applicable in case of loss of demand due to the load throw off. The larger is the system, less is the adverse impact as in the case of generator tripping as mentioned above.
C. Sharing of operating reserves, also leads to a benefit. Generation deficient regions are compelled to operate their grids with little or no operating reserves due to various other reasons. Interconnecting the regions leads to sharing of operating reserves and lead to benefits in a manner very similar to the impact

on capacity requirement. Thus, there can be an adequate operating reserve in a generation deficit region if it is interconnected to a generation surplus region. The primary, secondary and tertiary reserve, that has to be maintained, has an economic impact also. Interconnecting regions will reduce the cost of maintaining these reserves.

D. The system operator has many more options for maintaining system parameters in a large interconnected region with large number of demand and injection points spread throughout the combined region.

8 Integration of Renewable Sources of Power

A. The large interconnected system will promote integration of renewable sources in all the regions. The inherent variability of certain renewable sources will be catered to as mentioned in case of generator tripping of load throw off. This will be more frequent, almost continuous, but less severe.

B. Due to the difference in time zones, the variation of solar generation in the three regions will lead to the availability of solar power in the individual regions for a longer period of time. This will also lead to mitigation of the adverse impact of solar generation due to its variation in the very short term (within a minute) which is hard to predict and quantify.

C. Solar and wind generation prediction methodologies have improved significantly in the past few years. Still the prediction is not quite compared to capacity of a fossil fuel plant to adhere to it's given schedule. For the reasons explained earlier, these variations will be better addressed in a large interconnected grid. This is also true to a lesser extent in case of a run of river hydro.

D. The large interconnected system will lead to better utilization of hydro sources with reservoirs and pumped storage pants. With proper planning, these resources can be utilized when the generation deficit exists in the interconnected region or to replace the costlier generation.

E. In a small system, the variation in climatic condition and weather is limited. However, in a very large, geographically spread out system, these may be significant. This can be utilized to utilize the renewable sources in a much more efficient manner. For example, the run of river hydro generation in one part can be utilized to operate a pumped storage plant in another part. Thus, the cost of generation from the pumped storage plant is minimised when the run of river hydro power is utilized even when low demand which exists in the run of river hydro region.

9 Environmental Benefits

The combined interconnected regions will lead to better integration of renewable energy sources. They will also replace the costly generation as explained earlier. This will lead to many environmental impacts.

A. Reduction in release of greenhouse gases. This will reduce global warming.
B. Reduction of pollution due to cutting down on usage of fossil fuel.
C. Reduction in carbon footprints.

The benefits of interconnection these regions will be the same as that enumerated above. However, the benefits have to be quantified or at least estimated. This will then have to be compared to the cost of the building the interconnections. For this, the various alternative locations for building interconnectors have to be considered. One needs to decide the likely points of connection, kind of connection (AC or DC), capacity of the interconnections etc. The whole process may require a period of eight to ten years before the interconnection is commissioned. Hence, the one needs to plan based on the likely scenario from 2028 onwards.

10 Activities and Tasks to be Completed

A. Garner support and disseminate the concept by interacting with all key stakeholders in the three regions. This will include the ministry, power sector utilities, regulators, system planner, etc. and also key experts.
B. To study power sectors, the three regions and their constituent nations for analysing and forecasting key parameters like demand, generation, nation grid augmentation plan, energy export and or import plan, generation potential and other relevant. The cost of generation and the strength of the grid is also to be ascertained in consultation with the key stakeholders of the region and the constituent nations.
C. The savings in capacity addition and the operation reserve sharing to be ascertained for, based on the demand, generation forecasts, fuel mix etc., for a period of 20 years starting 2030.
D. To study and forecast the interregional flows (Or Trades) expected on these interconnectors for a period of 20 years starting 2030.
E. The reduction in carbon emissions, integration of renewable sources and saving of natural resources to be quantified.
F. To work out the desired capacity of various interconnections required based on the above information.
G. The alternate routes for interconnection may be studied. considering all factors like cost, system condition and other social, economic and political aspects.
H. To study and estimate the costs of building these interconnectors and the payback period based on the outcome in (a) and (b) above.

I. Recommend suitable incorporation of smart grid technology in the interconnected region Grid.
J. To analyse and recommend a feasible mode of establishing a vibrant power market between these regions. Power exchange is one option, banking is another. Other options like long, short and medium term PPA etc. also need to be consider and suitable recommended.
K. To suggest the required regulatory, policy and legal requirements including a dispute resolution mechanism, unscheduled interchange settlement mechanism and all other commercial terms, and conditions including tariff, charges and penalties.
L. A draft of a recommended agreement to be signed by all the parties is to be prepared.

11 Next Steps and Way Forward

It is suggested that the project should be executed and structured as below:

a. A suitable entity should run the project secretariat with a strong team of technical people. They should also have presence in all the three regions of GCC, BIMSTEC/SAARC and ASEAN. They should have a proven track record of executing large transmission related project of this or similar nature.
b. A high-level Project advisory body will be formed including representatives from GCCIA, BIMSTEC, ASEAN secretariats the Project Secretariat. This high-level body will supervise the project and also resolve any issue that may arise.
c. The Project Secretariat will be responsible for proper execution of the project.
d. It is necessary to have adequate funds to execute the project. Developmental agencies in the Region may be approached to secure adequate funding.
e. A Technical committee with two to three members each from all the three regions will execute the technical tasks of the project. Secretary of the committee will be from the Project Secretariat.
f. A Finance committee with two members each from all the three regions will execute the financial tasks of the project. Secretary of the committee will be from the Project Secretariat.
g. A Regulatory and Commercial committee to be formed, sometime after the technical and finance committee is formed, to execute the relevant tasks of the project. Secretary of the committee will be from the Project Secretariat.
h. Other than the head office of the Project Secretariat, their regional office to be established in the three regions. This will help in overcoming language and other such barriers. A national representative may also be established at some of the nations if the requirement is felt for the same. This will also help the project to more inclusive of the various nations and the Regions.
i. The project secretariat and the members of the committee will ensure stakeholder consultation in all the Regions as a continuous process.

j. Each of the committee will come out with a comprehensive report after completions of all tasks entrusted to them.
k. The project secretariat will come out with the Final Report combining the individual committee reports, feedback from key stakeholders etc.

Initial deliberations between different stakeholders have been initiated on the feasibility of potential interconnections between India and Myanmar as well as India and Oman. However, no feasibility studies have been taken conducted yet. If the buy-in of all stakeholders on this idea of an interconnected grid between these three regions can be achieved, then the concept can be further taken ahead by conducting first a Pre-Feasibility Study, and based on the same, a detailed feasibility study for the potential interconnections needs to be conducted. Once the feasibility study is over and a Detailed Project Report (DPR) is prepared, the actual erection of the interconnectors can start.

References

1. HartatiRS, Linawati et al Hourly load forecasting of electricity in Bali, Indonesia using adaptive neuro fuzzy inference system. Electrical Engineering Department, Udayana University, Bali, Indonesia
2. Gaur K, Rathour HK et al Analysing the electricity demand pattern. Power System Operation Corporation Limited, New Delhi, India
3. Ray SK, Jain G Roadmap for South Asian Regional Power Exchange (SARPEX). https://sari-energy.org/wp-content/uploads/2017/05/Roadmap-for-South-Asian-Regional-Power-Exchange.pdf
4. ASEAN POWER GRID powering the region. https://www.unescap.org/sites/default/files/Session%204.1.%20Bambang%20Hermawanto_ASEAN.pdf
5. Ray SK, Jain G, Tyagi N (2017) Development of regional electricity markets. The Institute for Policy, Advocacy, and Governance (IPAG)
6. SAARC Framework Agreement on Energy Cooperation (Electricity). https://powermin.nic.in/en/content/saarc-framework-agreement-energy-cooperation-electricity
7. Guidelines for import/export (cross border) of electricity-2018. https://powermin.nic.in/sites/default/files/uploads/Guidelines_for_ImportExport_Cross%20Border_of_Electricity_2018.pdf
8. Discussion paper on market based economic dispatch of electricity. https://www.cercind.gov.in/2018/draft_reg/DP31.pdf